STREAM, LAKE, ESTUARY, AND OCEAN POLLUTION

STREAM, LAKE, ESTUARY, AND OCEAN POLLUTION

Nelson Leonard Nemerow, Ph.D.

Professor and Environmental Engineer

VNR VAN NOSTRAND REINHOLD COMPANY

New York

Library of Congress Catalog Card Number: 84-13191
ISBN: 0-442-26720-7

Manufactured in the United States of America

Published by Van Nostrand Reinhold Company Inc.
135 West 50th Street
New York, New York 10020

Van Nostrand Reinhold Company Limited
Molly Millars Lane
Wokingham, Berkshire RG11 2PY, England

Van Nostrand Reinhold
480 Latrobe Street
Melbourne, Victoria 3000, Australia

Macmillan of Canada
Division of Gage Publishing Limited
164 Commander Boulevard
Agincourt, Ontario M1S 3C7, Canada

15 14 13 12 11 10 9 8 7 6 5 4 3 2 1

Library of Congress Cataloging in Publication Data

Nemerow, Nelson Leonard.
Stream, lake, estuary, and ocean pollution.

Includes bibliographical references and index.
1. Water–Pollution. 2. Water–Analysis. I. Title.
TD425.N36 1985 628.1'61 84-13191
ISBN 0-442-26720-7

There is a phenomenal resiliency in the mechanisms of the earth. A river or lake is almost never dead. If you give it the slightest chance by stopping pollutants from going into it, then nature usually comes back. When we deal gently with the earth—even when we have thoroughly damaged it—we can repair our friendship with it.

Dr. Renè Dubos
(*U.S. News and World Report*, Feb. 23, 1981)

PREFACE

It is with deep humility that I give you a textbook on this vast and comprehensive subject. I do not profess to be an expert on all aspects of receiving water evaluation. But, on the other hand, few if any persons now living can claim in all honesty to be such an expert. At the same time I recognize that my students, as well as workers worldwide, urgently need to be aware of every major facet of receiving water analysis.

The first five chapters represent the foundation for all modern stream analysis. One could consider each chapter an ample subject for an entire book. Therefore, each of these chapters contains selective rather than comprehensive coverage of all the material available. If the reader masters the subsection theories presented in each chapter, he/she should be in a position to comprehend most other subjects of significance to the major theme of the chapter.

In the past, and even at the present, too many scientists and engineers treat the subject of receiving water analysis as a single discipline, to be solved either with a biological survey or a chemical series of analyses or simply by a dissolved oxygen survey. Workers in today's analyses of streams programs might also presume, erroneously, the single discipline approach. Therefore, this text was written with a view toward overcoming this false presumption. The writer has observed more than once that "highly polluted" streams often show no oxygen sag characteristic at the time of analysis. Only a biological analysis would confirm the fact of pollution. On the other hand, a biological analysis seldom, if ever, provides the analyst with design criteria with which to overcome the polluted state of the stream. In the first three chapters I have endeavored to give equal emphasis to the biological, hydrological, and biochemical aspects of stream analysis. I recommend that the reader complete all three chapters before deciding how to undertake a comprehensive survey of a particular receiving water. In Chapters 4 and 5, I have presented the mathematics of computing first—the deoxygenation rate and then the reaeration rate—commonly referred to as stream reaction rates. Next, they are considered collectively in Chapter 6 in order to plot the oxygen concentration profile for a stream receiving some organic wastewater. Although many procedures are given for computing k_1 (the deoxygenation rate), k_2 (the aeration rate), and the oxygen sag curve plots, a summary of the author's recommended procedures on receiving water quality objectives is also given Chapter 7. Since many contaminants are being discharged near or in estuarine waters, and since more and more multipurpose uses are being made of these special waters, I describe in Chapter 8 two generally used techniques for analyzing estuarine waters. No pollution abatement program is successful without the proper administration of the stream resources. Since our first major federal legislation in 1948, some evidence of this fact can be observed in the United States. Administration of these resources from many different viewpoints and by many political entities is described in Chapter 9. I have even taken an author's liberty of submitting my proposed administrative system of marketing stream-pollution-capacity resources in this chapter. In case the reader believes that carrying out the doctrines described in the first six chapters is an easy one solved merely by routine sampling and analysis, I suggest the analyst follow the practical illustration given in Chapter 10. Here the frustrations of a stream analyst are vented for public display and consideration. The mistakes of an analyst are not easily forgotten by the reader, but successful undertakings, on the other hand, are likely to be taken as normal practice, and easily overlooked. The purpose of this chapter is not to discourage the stream analyst, but to encourage the use of knowledge and scientific tools in analyzing and evaluating the condition of a receiving water in the most equitable and proper manner possible.

More and more of land-based contaminants pass through lakes of various sizes and shapes, causing many different types of pollution. Today, a major concern is that of hazardous wastes, which is presented in Section 4 of Chapter 7. Lake pollutants are discussed and analyzed in Chapter 11. Finally, land contaminants reach the coastal area causing unusual and dramatically visible forms of pollution. These are the subjects in Chapter 12. As the reader can note from studying Chapter 10, mountainous quantities of data are collected in assessing the effect of contaminants on watercourses. It is becoming necessary to derive programs and to utilize computers to predict the result of changes in quantities and qualities of contaminants on receiving waters. An introduction to model preparation for predicting river and lake watercourse reaction to changes in land contributions is given in Chapter 13.

I would like to thank all researchers who have taken the time and made the effort to write, either in paper or text form, their ideas and experiences in stream analysis. From them I have gained much foresight and information. Equally vital, however, have been the last 35 years spent sampling, analyzing, and assessing the pollution capacity of many of our streams, rivers, and lakes. Experiences in these endeavors, although not always entirely satisfactory, still make for excellent teachers. Many of my graduate students have labored over semifinished material from these chapters distributed during their classes. Their suggestions for improving clairty and content are deeply appreciated. They can take pride in knowing that their contributions will help future generations of students.

Once again, I have prevailed upon the patience of my wife, Joan, who persevered while I spent long hours writing. She typed and edited this entire book, and served as an inspiration to me in completing the text in its present form.

NELSON LEONARD NEMEROW

CONTENTS

STREAM, LAKE, ESTUARY, AND OCEAN POLLUTION

1
BIOLOGICAL ASPECTS OF STREAM POLLUTION

The biological consequences of stream pollution often persist long after the evidence of physical or chemical contaminants has disappeared. For this single reason, streamwater quality evaluations must include a detailed biological analysis. Therefore, the stream analyst familiarizes himself with the basic biological parameters and phenomena which are involved. The intent of this chapter is to provide the reader with a description of the role of biology in stream pollution evaluation.

1-1. BACTERIAL GROWTH PHASES IN STREAMS

Water pollution control investigators have understood for more than two decades that bacterial growth proceeds in continuous phases. One popular system describes this growth as occurring in seven distinctly separate and significant phases. It is shown graphically in Fig. 1-1. The sigmoid growth curve is very useful, since by analyzing a stream microbiologically we can obtain a good indication of our location on a relative pollution scale. However, this curve is many times more "ideal" than "real."

Butterfield (1929*b*, 1931) found that when he grew bacteria in a series of flasks in which the food concentration was progressively varied, the numbers of bacteria always rose rapidly to a fixed ceiling, after which they remained stationary or slowly declined. Reducing the bacterial population by filtration, chlorination, or heating restored the high rate of biochemical activity, and the population again rose toward the ceiling value. The ceiling value was definitely related to the food concentration, although the buildup of metabolic waste products may also influence the ceiling value of bacterial numbers. He also found that it was the growth rate of the bacteria, rather than their mere numbers, that determined the rate of oxidation. When the population became static

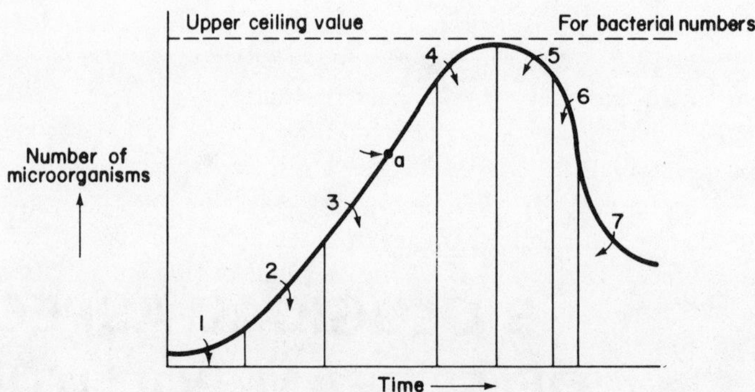

FIG. 1-1. Bacterial growth phases. a = Maximum rate of increase; 1 = Stationary phase; 2 = Accelerated growth phase; 3 = Logarithmic growth phase; 4 = Decreasing growth phase; 5 = Negative growth phase; 6 = Accelerated death phase; 7 = Logarithmic death phase.

at exceedingly high numbers, the rate of oxidation became nearly zero, showing that oxidation results only from bacterial growth, and vice versa.

In its early stages, with a constant rate of multiplication per unit of population, the process is autocatalytic in that the rate of addition of new individuals is proportional to the total population, which is progressively increasing. This is the well-known exponential or geometric rate of increase in population [see Fig. 1-1(3)].

$$N(t) = Ae^{kt}$$

where

A = a constant
$N(t)$ = No. of bacteria at time t
$k = \log a$

Ultimately, overcrowding occurs—a situation which can best be described as one exceeding the carrying capacity of the stream due to lack of food supply and accumulation of toxic metabolic products. The population changes from an increasing rate of growth to a decreasing one, approaching, finally, an upper ceiling value.

The important fact to remember is that at the midpoint in the logarithmic growth curve (see Fig. 1-1, pt. a) there is a maximum rate of increase. The rate of increase increases for the duration of exponential growth. When the bacterial population is kept somewhere near this midpoint where mortality equals reproduction, rate of multiplication remains proportional to the actual food concentration (organic matter). Underlying this relationship is the theory of the Biochemical Oxygen Demand (BOD) reaction, in which oxygen is used by microorganisms in direct proportion to the organic matter remaining.

Predators such as protozoa help to keep the bacterial numbers at the logarithmic growth phase. If the food concentration were constantly replenished, an equilibrium would exist between the bacterial numbers and rate of oxidation. This is why we continuously add soluble food to a sewage oxidation pond. However, in a laboratory BOD bottle and in a stream contaminated with organic matter at one point and undergoing self-purification, the food concentration is being continuously reduced downstream from the point of contamination. *Thus there is a continual readjustment of the bacterial population to the steadily decreasing food supply, in which the rate of bacterial reproduction is automatically maintained at a maximum level, and approximately in proportion to the concentration of available food.* This phenomenon is shown in Fig. 1-2.

We have just shown how bacteria live, grow, multiply, and die under conditions where a good deal of food is usually present. However, there is one barrier to this oxidation of organic matter which the bacteria carry out: the death rate of bacteria themselves. It is generally accepted today that bacteria of all types, and especially those of intestinal origin, tend to die out, even under conditions of gross pollution. It is common knowledge that storage, whether it be in a flowing river or an impounded reservoir, eliminates organisms of sewage origin and other bacteria as well. The death rate appears to be a function of time modified by a marked temperature coefficient.

In the ideal situation when no toxic elements are present, bacteria die in a logarithmic phase. Therefore, they die at a rate according to the numbers of

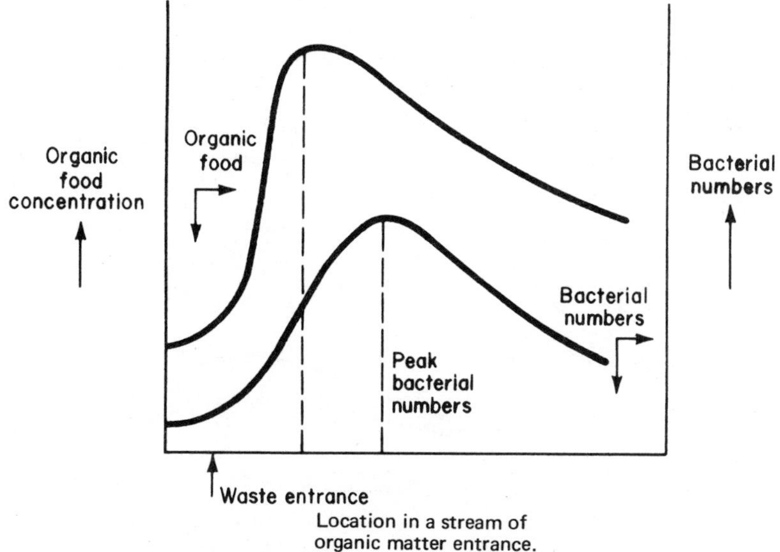

FIG. 1-2. Bacterial self-purification as related to watercourses.

organisms remaining. They don't quite match the logarithmic portion of the death curve, because some of the organisms remaining are more resistant than the ones that died off in earlier stages (see Fig. 1-1). The rate of decrease is more rapid at higher temperatures, and the initial rate of increase is greater (see Fig. 1-3), since bacterial growth is enhanced as temperature rises to about 40°C. Generally, in summer the flow in a stream is low and sedimentation occurs, whereas in winter sedimentation is lower in the areas of increased velocity of flow. Thus, more bacteria will settle to the bottom in summer than in winter, and the BOD bottle in the laboratory will be a better representation of natural conditions in the winter.

Two general types of bacteria are described by Heukelekian (1953) as present in a stream: the *native water population* and the *pollutional forms*, such as coliform microorganisms. In the biochemical oxidation of the pollution, the coliform organisms play only a minor role. The slight increase in bacteria below a sewer outlet is said to be due most probably to possible multiplication or dispersion of bacterial aggregates. The decrease in numbers thereafter parallels the decrease of organic material by oxidation.

The forces that affect the decrease of bacterial numbers in a stream are:

1. *Sedimentation.* Bacteria slowly settle and attach themselves to other aggregates which settle faster. This results in an apparent decrease in bacterial numbers in the flowing water.
2. *Protozoa.* Ciliated protozoa ingest bacteria.
3. *Food supply.* Not as abundant as in culture medium. Coliform organisms inoculated into sterile sewage will not multiply or transform organic materials, which shows the relatively poor food value of sewage for coliform bacteria only. Food is always decreasing due to oxidation.
4. *Stream temperature.* Below optimum for growth of pollutional bacteria even in summer. Higher temperature stimulates bacterial growth in presence of adequate food and favorable environmental conditions (see Fig. 1-4). A 10°C

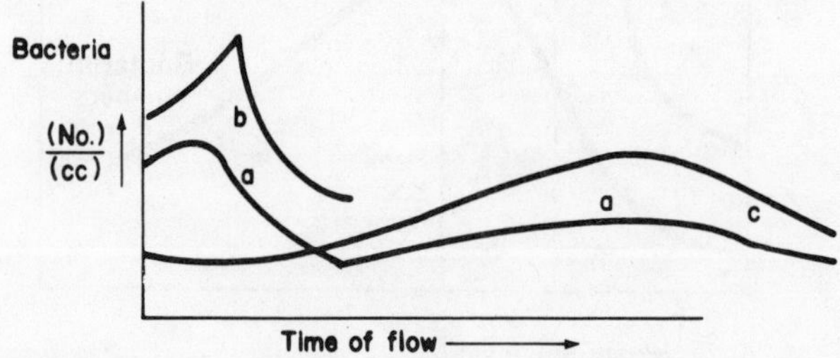

FIG. 1-3. Comparative growth of microorganisms under winter and summer conditions. (a) = stream; (b) = summer conditions (higher temperature) BOD bottles; (c) = winter conditions (lower temperatures) BOD bottles.

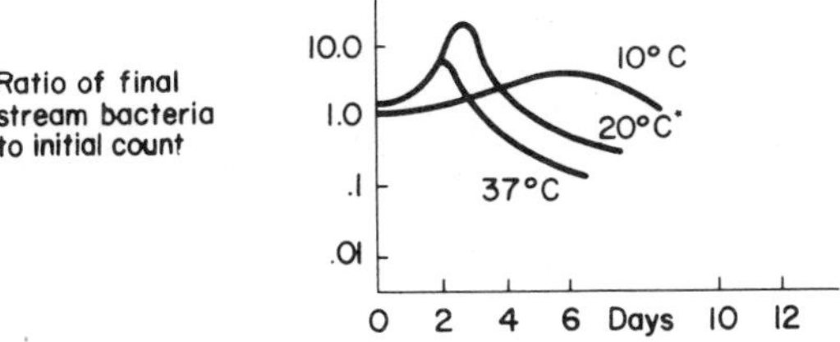

FIG. 1-4. Effect of temperature on bacterial growth. *20°C appears to be near an optimum for growth of stream bacteria.

rate in temperature results in an approximate doubling of the biological reaction rate.

5. *Sunlight.* Sunlight has bactericidal properties but is probably insignificant because of the poor penetration of ultraviolet rays, especially in turbid waters.

6. *Bacteriophage.* Destruction of bacteria by the lytic action of virus-like bacteriophage, which is parasitic on bacteria. However, this lysis occurs only when rapid multiplication of bacteria exists. (Therefore, it does not exist for intestinal groups of bacteria in receiving waters.)

7. *Industrial waste.* Seldom contributes many bacteria and mostly results in an immediate and sharp decrease in numbers. Some exceptions exist, such as in potato and other food product wastes.

8. *Dilution* of both food and bacteria by streamwater.

Butterfield's Contribution to Basic Bacteriological Stream Phenomena

Until his retirement in 1951, C. T. Butterfield was quite active in the field of sanitary microbiology. His main work in this field deals with the aspects of water microbiology and the interrelationship of bacteria and protozoa. His work with activated sludge and trickling filtration will not be considered here. Some of his major findings and areas of research are:

1. He advanced the theory that the chief function of certain zooplankton in the biochemical oxidation process is to keep the bacterial population reduced below the saturation point and thus to provide conditions suitable for continuous bacterial multiplication, which in turn results in more complex oxidation.

2. Both the rate and extent of oxidation are increased if bacteria are kept dispersed throughout the medium, such as in a completely mixed activated sludge treatment system.

3. Bacterial processes are the same regardless of whether the water is moderately or grossly polluted. However, the species of bacteria acting and their methods of operation vary. In moderately polluted waters the bacteria are of

the type which grow uniformly dispersed. In grossly polluted waters under maintained aerobic conditions, the bacteria are of the type that grow in masses or colonies. This latter type is capable of increased purification efficiency due to its mode of growth and adsorptive extracellular capsular substance.

Since his work is basic to much of our biological consideration on stream pollution, the major references containing his work are listed at the end of this chapter. The reader—especially one who would advance the biological concept of stream assessment—is urged to examine Butterfield's work in more detail.

1-2. ZONES OF STREAM REACTION TO POLLUTION

One effect of bacterial growth in streams has been the creation of definitive and distinguishable stretches below a source of organic pollution. One of our earlier stream analysts, F. J. Brinley (1942), classifies streams receiving wastes into five separate zones. These are described in summarized form in three zones as follows:

Zone I—Zone of Active Bacterial Decomposition

This zone is immediately below the source of pollution and is usually characterized by a low dissolved oxygen (0-3 ppm), especially during critical low flows, a high biochemical oxygen demand, a high bacterial count, the presence of bacterial-eating protozoa such as *Paramecium, Colpidium, Vorticella,* and *Carchesium,* and a few flagellates. The total number of plankton forms is low. *Tubifex* and *Liminodrilus* worms are found in the bottom deposits. Gas evolution often brings sludge to the surface. Sewage fungus is present. Only a few coarse fish such as carp and buffalo are found at the mouth of the sewer.

Zone II—Zone of Intermediate Bacterial Decomposition

The rate of biochemical respiration decreases in this zone, and as a result of the dissolved oxygen goes up to 3-5 ppm. The plankton volume is higher, but the plankton is still made up of pollutional forms requiring a rich medium of bacteria and solid particles for feeding. There is an increase in green and blue-green algae forms. In addition to carp and buffalo, shiners, minnows, suckers and sunfish are found.

Zone III—Recovery

The water gradually becomes clearer, green plants reappear, small animals serve as food for fish. Oxygen increases and fish return.

One of the difficulties with this classification is that the first two zones are rather arbitrary and subjective. It is often impossible to distinguish between Zone I and II as defined above. However, it is apparent that a determination of the appearance can reveal considerable information about the relative pollution of a stream.

The length of any biological zone of a stream is also affected by the physical characteristics such as stream gradient. These are more fully described in Chapter 2.

Patrick's Biological Measures of Stream Conditions

It is a well known fact that in a normally healthy stream there is a biodynamic cycle which results in a balance of plant and animal life. The effect of sanitary and industrial pollution is to alter this cycle. Patrick believes that because the physical, chemical, and environmental characteristics in a stream are so variable, tests involving these would not accurately predict the effects of an effluent on a given stream (1950). She has developed a system of observing organisms in streams and assuming the degree of pollution from an analysis of the groups and relative numbers present.

Patrick created seven different taxonomic groups of organisms to be used as biological measures of stream conditions:

1. The blue-green algae, some green algae, some rotifers
2. Oligochaetes, leeches, snails
3. Protozoa
4. Diatoms, red algae, and most of green algae
5. All rotifers not in (1) plus clams, worms, and some snails
6. All insects and crustacea
7. All fish

From observations and enumerations of the seven groups she has arrived at five classifications of a river:

1. *A Healthy Stream.* Balance of organisms: the algae are mainly diatoms and green algae; the insects and fish are represented by a variety of species. Numbers 4, 6, 7 (above) are all above the 50% level, based upon levels found in a natural station upstream.

2. *A Semi-Healthy Stream.* Balance somewhat disrupted. Pattern is an irregular one; a given species will be overrepresented. The following possibilities may appear: either or both 6 and 7 (above) below 50% and 1 and 2 under 100%. Either 6 or 7 below 50% and 1, 2, and 4 100% or above; or 4 is double-width column (one which has many more species than normal), resulting in the dominance of a single organism.

3. *Polluted.* Stream in a condition in which the balance of life found in a healthy station is upset. However, conditions are favorable for some groups of organisms, such as 1 and 2. The following may be observed: species of either or both 6 and 7 are absent, and 1 is 50% or better. Species of 6 and 7 are both present but below 50%, so that columns 1 and 2 must be 100% or more.

4. *Very Polluted.* Stream in a condition that is definitely toxic to plant and animal life. Often many groups may be absent. This state occurs if 6 and 7 are absent and 4 is below 50%, or if 6 and 7 are present but 1 or 2 is less than 50%.

5. *Atypical.* This group cannot be compared, either due to general ecological conditions or to methods of collection, to the healthy stations used as a basis; i.e., healthy stations may be in eutrophic conditions, while cool oligotrophic water could hardly be rated on the same basis. Likewise, a stream with a steep, newly eroded bank would not be comparable with a normal stream with shallow water near edges. The extent of sampling and the competence of the laboratory technicians will undoubtedly influence the final decision of the classification of the stream.

Patrick concludes that the general effect of pollution seems to be a reduction in species number, with the most tolerant surviving. It is also interesting to note that groups 3 and 5 are not used specifically in defining any of the above five classifications.

In relation to stream sanitation, Dr. Patrick did propose a method for measuring stream conditions by the presence or absence of species of all major groups that play a role in the biodynamic cycle of the stream, instead of the classical examination of the physical and chemical characteristics of the streamwaters. In Dr. Patrick's method the bacteria and fungi present are considered only as to total count and as indicator organisms, respectively.

This method was based on a survey conducted on the Conestoga basin in Lancaster County, Pennsylvania, in 1949. The area comprises about 475 square miles with a bedrock of limestone and shale and a "rolling" topography. It is intensively farmed and has many streams, the margins of which are bordered by cultivated fields. There are, however, some indistrial and residential areas in the basin, so that while some streams receive only farmhouse wastes, others receive industrial and sanitary wastes. The larger streams are usually turbid, owing to the erosion of clay soil, and the waters are of eutrophic type. At each stream two stations were selected: one station was to determine general conditions, and a riffle, slack water, and a pool were included; the second station was used to determine the effect of an effluent on the stream by studying and comparing the ecological conditions just below the confluence with those above it. Samples of water, mud, and of the various organisms were taken at each of the 77 stations, and were identified with the assistance of 22 scientific consultants. The organisms so found were grouped into seven columns as listed on page 7.

At each of the stations the results of chemical analysis for different times and a species list of bacteria were obtained. Histograms were prepared for each of the stations, and, based on them, five general classifications of streams were prepared (also listed on page 7).

The results show that toxic agents, and not depletion of dissolved oxygen, cause the most damage to plant and animal life in the streams. Although none of the stations had a limiting amount of dissolved oxygen, those subjected to toxic effects showed an alteration in their biodynamic cycle ranging from a reduction in taxonomic groups to a complete lack of animal and plant life.

All indications point to the fact that in a healthy stream a great many species should be present, but no one species should be predominant. The effect of pollution seems to be a reduction in the number of species, with an increasing abundance of individual species among those surviving. Therefore, the percentage present or the absence of taxonomic groups can be used as an index of stream pollution, since healthy streams require balanced physiological activity. Moreover, this method reflects the conditions existing over a period of time and not at the moment of sampling, as may occur in the case of physical and chemical analyses. However, in some cases, the use of bioassays to evaluate streams may be preferable to a lengthy stream biological analysis. The latter is often time-consuming and quite costly, without providing all the answers required by the stream analyst.

Since her work deserves to be read by all advocates of the biological system of stream classification, a list of many of Patrick's most valuable references will be found at the end of this chapter.

MacKenthun (1969) gives an excellent graphic picture (see Figure 1-5) of the kinds and number of animal life found in typical streams following the introduction of organic, toxic, and inert forms of pollution.

Palmer (1962) gives an enumeration of both clean water forms of algae and algae associated with organically enriched watercourses.

Clean Water Algae	*Pollution Algae–Algae Common in Organically Enriched Areas*
Group and Algae	*Group and Algae*
Blue-Green Algae *(Myxophyceae):*	Blue-Green Algae *(Myxophyceae):*
Agmenellum quadriduplicatum, glauca type	*Agmenellum quadriduplicatum, tenuissima type*
Calothrix parietina	*Anabaena constricta*
Coccochloris stagnina	*Anacystis montana*
Entophysalis lemaniae	*Artrospira jenneri*
Microcoleus subtorulosus	*Lyngbya digueti*
Phormidium inundatum	*Oscillatoria chalybea*

(Continued)

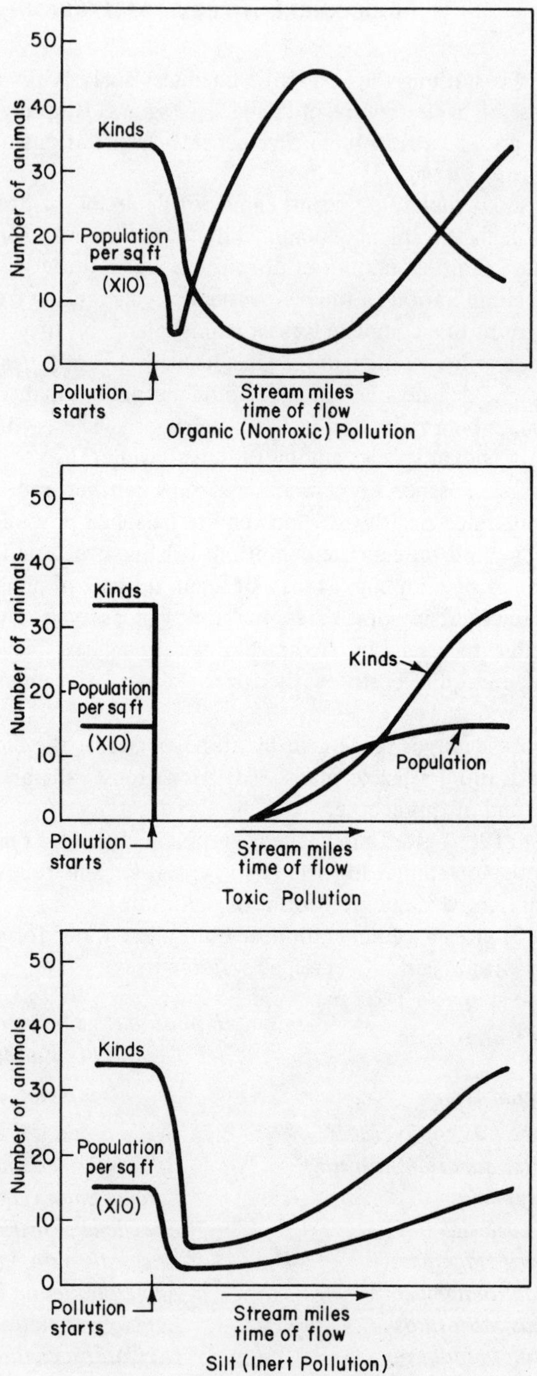

FIG. 1-5. Pollutional effects on animals.

Clean Water Algae

Group and Algae

Green Algae *(nonmotile Chlorophyceae):*
- *Ankistrodesmus falcatus, var. acicularis*
- *Bulbochaete mirabilis*
- *Chaetopeltis megalocystis*
- *Cladophora glomerata*
- *Draparnaldia plumosa*
- *Euastrum oblogum*
- *Gloeococcus schroeteri*
- *Micrasterias truncata*
- *Rhizoclonium hieroglyphicum*
- *Staurastrum punctulatum*
- *Ulothrix aequalis*
- *Vaucheria geminata*

Red Algae *(Rhodophyceae):*
- *Batrachospermun vagum*
- *Hildenbrandia rivularis*
- *Lemanea annulata*

Diatoms *(Bacillariophyceae):*
- *Amphora ovalis*
- *Cocconeis placentula*
- *Cyclotella bodanica*
- *Cymbella cesati*
- *Meridion circulare*
- *Navicula exigua var. capitata*
- *Navicula gracilis*
- *Nitzschia linearis*
- *Pinnularia nobilis*
- *Pinnularia subcapitata*
- *Surirella splendida*
- *Synedra acus var. angustissima*

Flagellates *(Chrysophyceae, Cryptophyceae, Euglenophyceae and Volvocales of Chlorophyceae):*
- *Chromulina rosanoffi*
- *Chroomonas nordstetii*
- *Chroomonas setoniensis*
- *Chrysococcus major*
- *Chrysococcus ovalis*

Pollution Algae–Algae Common in Organically Enriched Areas

Group and Algae

Blue-Green Algae (*Cont'd.*)
- *Oscillatoria chlorina*
- *Oscillatoria formosa*
- *Oscillatoria lauterbornii*
- *Oscillatoria limosa*
- *Oscillatoria princeps*
- *Oscillatoria putrida*
- *Oscillatoria tenuis*
- *Phormidium autumnale*
- *Phormidium uncinatum*

Green Algae *(nonmotile Chlorophyceae):*
- *Chlorella pyrenoidosa*
- *Chlorella vulgaris*
- *Chlorococcum numicola*
- *Scenedesmus quadriculla*
- *Spirogyra communis*
- *Stichococcus bacillaris*
- *Stigeoclonium tenue*
- *Tetraedron muticum*

Diatoms *(Bacillariophyceae):*
- *Gomphonema parvulum*
- *Hantzschia amphioxys*
- *Melosire varians*
- *Navicula cryptocephala*
- *Nitzschia acicularis*
- *Nitzschia palea*
- *Surirella ovata*

Flagellates *(Euglenophyceae, Volvocales of Chlorophyceae):*
- *Carteria multifilis*
- *Chlamydomonas reinhardi*
- *Chlorogonium euchlorum*
- *Cryptoglena pigra*
- *Euglena agilis*
- *Euglena deses*
- *Euglena gracilis*
- *Euglena oxyuris*
- *Euglena polymorpha*
- *Euglena viridis*

Clean Water Algae	*Pollution Algae–Algae Common in Organically Enriched Areas*
Group and Algae	*Group and Algae*
Flagellates (*Cont'd.*)	Flagellates (*Cont'd.*)
Chrysococcus rufescens	*Lepocinclis ovum*
Dinobryon stipitatum	*Lepocinclis texta*
Euglena ehrenbergii	*Pandorina morum*
Euglena spirogyra	*Phacus pyrum*
Mallomonas caudata	*Pyrobotrys gracilis*
Phacotus lenticularis	*Pyrobotrys stellata*
Phacus longicauda	*Spondylomorum quaternarium*
Rhodomonas lacustris	

Mackenthun (1969) also presents a table comparing clean water animal forms with those commonly associated with polluted waters (see below).

Organism Associations (MacKenthun, 1969)

Clean water association		Polluted water association	
Algae	*Cladophora* (green)	Iron bacteria	*Sphaerotilus*
	Ulothrix (green)	Fungi	*Leptomitus*
	Navicula (diatom)	Algae	*Chlorella* (green)
Protozoa	*Trachelomonas*		*Chlamydomonas* (green)
Insects	Plecoptera (stoneflies in general)		*Oscillatoria* (blue-green)
	Negaloptera (hellgrammites, alderflies, and fishflies in general		*Phormidium* (blue-green)
			Stigeoclonium (green)
		Protozoa	*Carchesium* (stalked colonial ciliate)
	Trichoptera (caddisflies in general)		*Colpidium* (non-colonial ciliate)
	Ephemeroptera (mayflies in general)	Segmented Worms	*Tubifex* (sludgeworms)
			Limnodrilus (sludgeworms)
	Eimidae (riffle beetles in general)	Leeches	*Helobdella stagnalis*
		Insects	*Culex pipiens* (mosquito)
Clams	Unionidae (pearl button)		*Chironomus* (-Tendpipes) *plumosus* (bloodworms)
Fish*	*Etheostoma* (darter)		*Tubifera (Eristalis tenax)* (rat-tailed maggot)
	Notropis (shiner)		
	Chrosomus (dace)		
		Snail	*Physa integra*
		Clam	*Sphaerium* (fingernail)
		Fish*	*Cyprinus carpio* (carp)

* Names from: American Fisheries Society Special Publication No. 2, "A List of Common and Scientific Names of Fishes from the United States and Canada" (Second Edition) Ann Arbor, Mich. (1960), 102 pp.

We have already seen that stream biological systems are highly influenced by and even dependent upon the type and extent of organic matter present. In the next section we will see how the degree of organic matter (saprobicity) influences specific species of biological life.

1-3. SAPROBICITY SYSTEM[1]

Kolkwitz and Marsson (1908) developed the original *Saprobiensystem* for assessment of organic pollution. They postulated that when a river received a heavy load of organic matter, the normal processes of self-purification would result in a series of zones of decreasingly severe conditions succeeding one another downstream, and each containing characteristic animals and plants:

Zone 1. Polysaprobic. The zone of gross pollution with high molecular weight organic matter; very little or no dissolved oxygen; formation of sulfides; abundant bacteria and other organisms; few species of animals living on decaying organic matter or bacteria. The reader may note the similarity of this classification to Heukelekian's (1953) zone of degradation.

Zone 2. Mesosaprobic. This zone contains simpler organic compounds; steadily increasing oxygen; the upper portion contains many bacteria and fungi with more types of animals, but few algae; the lower portion possesses more mineralization (the conversion of organic to inorganic matter) suitable for algae and tolerant animals and some rooted plants. Although this class is divided into an upper and lower, it resembles Heukelekian's decomposition zone. Some later workers have found an increase in algae in the lower section of this zone—the resulting high daytime oxygen and low nighttime oxygen contents sometimes leading to fishkills. For this reason it was referred to as the zone of partial recovery.

Zone 3. Oligosaprobic. The zone of recovery where mineralization is complete and oxygen is back to normal. It contains a wide range of plants and animals.

Liebmann (1951) began to apply the ideas of Kolkwitz and Marsson in his handbook, published in German. Since then he has endeavored to refine his biological classification system. Much objection to both the original work and to Liebmann's later work has centered about the placing of certain organisms in the particular zones, and in disagreements with the identification of some of the fauna. Some workers argue that the mere presence of certain species does not signify the true ecological condition, and that the entire population is more significant. Furthermore, many biologists claim that changes in numbers of dom-

[1] Saprobicity is a measure of biodegradable organic matter.

inant species are more important than just changes in given species. Liebmann's system takes some of these objections into account.

Hynes (1960) points out that the reactions of different creatures to different aspects of organic pollution vary considerably. Some react to deoxygenated water, some to products of decomposition such as ammonia or sulfides. He concludes that to adhere to a rigid system of biological zone classification as it relates to water quality is to "bemuse oneself with the idea that complex ecological changes can be subjected to simple classifications." Further, he believes that "the Saprobiensystem is applicable only to the particular conditions produced by heavy sewage pollution in a slow and evenly flowing river. If the effluent is not sewage, or if the river is turbulent, it breaks down."

Hynes sums up the case for biological indices of stream pollution as follows:

1. Biological analyses take less time than chemical "because a single series of samples reveals the state of the animal and plant communities, which themselves represent the results of a summation of the prevailing conditions."

2. Biological analysis reveals the effects of intermittent or previous pollution, whereas chemical analysis must be made at the precise moment pollution occurs and then must be averaged over many samples at many locations.

However, Hynes realizes that biological analyses require expertise both in analyzing and evaluating. He maintains this to be true also for the chemist. Since 1960 many laboratory improvements have been made to streamline chemical analysis, such as atomic absorption, autoanalyses, and rapid forms of spectrographic and other analyses. At the same time, biological methods seem, to this author, to have increased in complexity. The major drawback to biological analysis, in Hynes' own words, is that

> biological study can reveal only the general type of the pollution; it does not indicate the exact substances involved. It can distinguish between organic and poisonous pollution, but it does not identify, except in certain cases, the particular poison which is causing damage. Biological analysis also has the great disadvantage that it does not deal directly with concentrations.

The designing engineer must know what contaminants to limit in wastes and the percentage reduction of these contaminants for which he must strive. Although Hynes does not suggest this, it may be possible to determine quantitative removals required by bioassays. He does suggest that the 20/30 Royal Commission standard (20 ppm BOD and 30 ppm suspended solids) does not offer a valuable goal for prevention of stream pollution.

Sladecek (1965) continued and refined the previous work done on the saprobicity system. He found that although this system is only an empirical one and shows many insufficiencies, it can be considered valid. He arrived at 10 major findings to improve the workability of the system:

1. "As long as no better system or method is placed at our disposal it is necessary to develop, to correct and to improve the existing system of Kolkwitz and Marsson. There is only one such system, but there are several modifications and interpretations of it. The unification of all contributions, the selection of good ideas as well as the rejection of all wrong statements have to be solved urgently." (p. 629) Sladecek attempted to do this by constructing the circular diagram shown in Fig. 1-6. Sladecek's own description of the diagram is included in the figure.

Sladecek presents his view on the saprobicity system in tabular form (Table 1-1). He has attempted to provide substance and specific criteria for each separate classification as far as bacteria, dissolved oxygen, hydrogen sulfide, and BOD are concerned. These are shown in Table 1-2. He further presents clearer definition of the biological features of the eusaprobity and transsaprobity areas (Table 1-3).

2. "One of the very promising ways is the recognition of the whole community of producers, consumers and decomposers inhabiting a certain habitat.

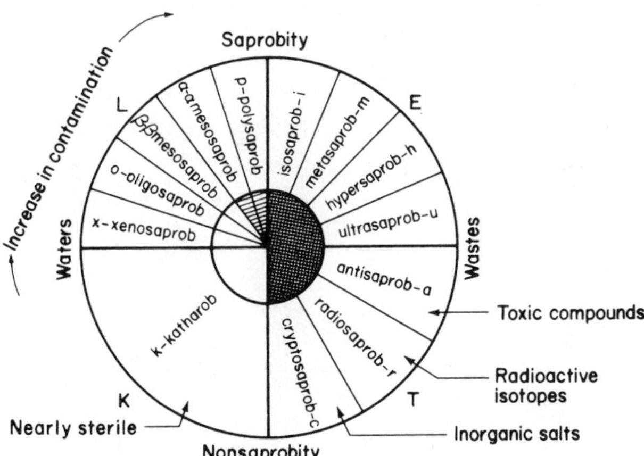

FIG. 1-6. The system of saprobity. All waters can be included in the circle. They are divided into four main groups: K = Katharobity, L = Limnosaprobity, E = Eusaprobity, T = Transsaprobity. The circle is divided by two main lines into four quadrants. The vertical line divides it into the left half including "waters" and into the right one including "wastes." The horizontal line divides the circle into the upper half including saprobic conditions and into the lower one including nonsaprobic (asaprobic) conditions. The value of BOD is increasing from x to u. The self-purification proceeds from u to the left. In the small circle, white = clean water, hatched = polluted water, black = wastes. Original. (*After Sladecek, 1965.*)

TABLE 1-1. A Survey of the Saprobity System. [*After Sladecek, 1965. By permission of publishers.*]

Main group	Abbrev.	Degree (zone)	Abbrev.	Evaluation		Trivial names
Katharobity (most pure waters)	K	0. katharobity	k	Drinking water		Purity
Limnosaprobity (polluted surface and ground waters)	L	1. xenosaprobity 2. oligosaprobity 3. β-mesosaprobity	x o β	Positive zones * P/R > 1	According to the system of saprobic organisms of Kolkwitz & Marsson	Purity
		4. α-mesosaprobity 5. polysaprobity	α p	Negative zones * P/R < 1		Pol-lu-tion
Eusaprobity (sewage and industrial wastes undergoing bacterial decomposition	E	6. isosaprobity 7. metasaprobity 8. hypersaprobity 9. ultrasaprobity	i m h u	degree of ciliates degree of colorless flagellates degree of bacteria and mycophyta azoic degree (but non-toxic)		Wastes
Transsaprobity (wastes inaccessible to bacterial decomposition)	T	10. antisaprobity 11. radiosaprobity 12. cryptosaprobity	a r c	toxic degree radioactive wastes wastes containing inorganic non-toxic substances, special cases		Wastes

* P/R = Producers/Reducers

TABLE 1-2. Approximate Data on Correlation of Biological, Bacteriological, and Chemical Values.
[*After Sladecek, 1965. By permission of publishers.*]

Item	Degree of saprobity	Abbrev.	Total psychrophilic bacteria per 1 ml, less than	Coliforms per liter, less than	DO ppm more than	DO satur. more than	H_2S ppm (less than)	BOD_5 ppm (less than)	Specific substances
0.	katharobity	k	500	20	various	various	0	0	(residual chlorine)
1.	xenosaprobity	x	1,000	10,000	8	60	0	1 (2)*	
2.	oligosaprobity	o	10,000	50,000	6	50	0	2,5 (4)	
3.	β-mesosaprobity	β	50,000	100,000	4	40	0	4 (6)	
4.	α-mesosaprobity	α	250,000	1,000,000	2	20	0	7 (9)	
5.	polysaprobity	p	2,000,000	20,000,000	0,5	10	traces	40 (80)	E_h = +200 m V and more
6.	isosaprobity	i	10,000,000	3,000,000,000	traces	0	1	40–400 (600)	E_h = +50–200 m V
7.	metasaprobity	m	20,000,000	10,000,000,000	0	0	1–100	200–700	E_h = less than + 50 m V
8.	hypersaprobity	h	50,000,000	1,000,000	0	0	10	500–1,500 (2,000)	ptomains present
9.	ultrasaprobity	u	10	0	0	0	0	1,000–60,000	
10.	antisaprobity	a	0	0	various	various	0	0	toxic compounds
11.	radiosaprobity	r	various	various	various	various	various	various	radioactive isotopes
12.	cryptosaprobity	c	various (0)	various (0)	various	various	various (0)	various (0)	inorganic substances

*Note: The conditions within limnosaprobity (items 1–5) show different values in stagnant and running waters.

TABLE 1-3. Biological Features of Waste Waters (Eusaprobity and Transsaprobity) in Relation to Technological and Hygienic Aspects. [*After Sladecek, 1965, By permission of publishers.*]

Item	Degree	Abbrev.	Quantity of microscopical organisms per 1 ml	Examples	Technological standpoint; Treatment	Hygienic standpoint
6.	isosaprobity	i	Ciliata 10–50,000 Flagellata 1,000–20,000 (Amoebina 0–1,000) Bacteria in mass (Fungi in mass)	Raw sewage	Biological oxidation treatment (biofilters, activated sludge, irrigations, oxidations ponds) applicable with or without mechanical pretreatment	Great danger of infection by pathogenic germs
7.	metasaprobity	m	Flagellata 5,000–300,000 Ciliata 0–5 Bacteria in mass	Septic sewage; waters containing much H_2S	Before application of biological oxidation processes, the wastes have to be aerated (hydrogen sulphide)	Great danger of infection by pathogenic germs; also toxic compounds present
8.	hypersaprobity	h	Bacteria in mass Fungi in mass Flagellata 0–5	Concentrated industrial wastes; digestion of sludge	Anaerobic treatment, lagooning; before application of oxidation processes, chemical treatment inevitable	Danger of infection by pathogenic germs and of poisoning by ptomaines in some cases
9.	ultrasaprobity	u	Bacteria 0–10 (Fungi 0–10) Abiotic	Industrial liquids; sulphite liquor; beet-sugar process wastes	Anaerobic treatment; chemical treatment and/or dilution are suppositions for further aerobic biological treatment	Spores of pathogenic germs can be present
10.	antisaprobity	a	Abiotic. Only spores, cysts and other resting stages can survive	Toxic wastes	Chemical treatment of dilution must eliminate the toxic effects	Poisons present; pathogenic germs mostly destroyed
11.	radiosaprobity	r	Various	Radioactive wastes or contaminated waters	Special treatment including biological methods	Radioactive isotopes present; danger invisible
12.	cryptosaprobity	c	Various, mostly abiotic	Wastes containing inorganic compounds or coal	Special treatment, mostly mechanical	Various conditions

As far as possible all links of the community must be determined into species and they must be evaluated quantitatively in numbers of individuals per unit volume or area, or in standard units. This way is very troublesome and fatiguing, but if we wish to consider our branch as a scientific one, we have to work hard."

3. "The saprobiological (i.e., in reality ecological) valency of each water organism must be studied in nature, i.e., as a constituent of certain communities and in the dependence on mechanical, other physical, chemical and biological properties of its environment." Sladecek is evidently encouraging the study of the organisms only in their natural interdependent state in a watercourse.

4. The laboratory studies have to be directed to experiments dealing with the life history, physiology, and ecology of the main representatives of water organisms. Not only the tolerance limits in respect to the toxic effects of certain chemical compounds have to be studied, but also their behavior under natural conditions, to experience the influence of such factors as water temperature, pH, velocity of current, overpopulation, etc. It is necessary to distinguish the factors indicated as "pollution" from nonpollutional interfering factors. The duration of individual life of aquatic organisms and also the duration of different larval, etc., stages must be ascertained to enable the answering of different questions.

5. A special system for the evaluation of the toxicity must be developed. This system may be used in conjunction with and to complement the saprobity system.

6. A special system for the evaluation of the radioactivity of the water environment must also be proposed.

7. It is necessary also to evaluate special cases of detrimental effects of inorganic matter which are not toxic and not saprobic, e.g., suspensions of coal, of small-grained mineral powders, oil films, etc.

8. After the unity of all systems in the limnology, (shown in Table 1-4) it is necessary to endeavor to classify the lakes and streams in accordance with the trophic state and in the same accord with the saprobic organisms which must be unified in a general scheme. "Not only the descriptive aspect has to be regarded, but also the dynamic aspect regarding the general metabolism of a water body." He suggests that the ratio of "producers" to "reducers" such as described by previous biologists is not the only criterion to be applied.

9. "The biological productivity studies in lakes, fishponds, artificial reservoirs, etc., can and must be enlarged to also include polluted waters, wastes and, especially, to treatment plants."

10. The main direction of applied hydrobiology must be seen in the prognosis of the future state of the water body in question and the preventive control of its quality. Not only the pollution and self-purification, but also other phenomena occurring in water, such as vegetative colorings, water blooms, excessive growths of sessile organisms, aquatic macrophytes, etc., are to be studied and understood.

TABLE 1-4. A Modification of the Table of Kolkwitz (1935, p. 249) to Show the Relation of the Saprobic and Trophic Degrees.
[*After Sladecek, 1965. By permission of publishers.*]

Example		Saprobic degree	Trophic degree
Drinking water		katharobic	—
Pure mountain waters, water from thawing of snow	Limnosaprobic	xenosaprobic	oligotrophic
Lake Geneva	Limnosaprobic	oligosaprobic	oligotrophic
Lakes situated on the Havel River near Potsdam	Limnosaprobic	β-mesosaprobic	eutrophic
Polluted fishponds near irrigation fields, several pools with decaying macrophytes	Limnosaprobic	α-mesosaprobic	eutrophic
Heavy polluted streams with *Sphaerotilus* growth	Limnosaprobic	polysaprobic	polytrophic
Sewage and industrial wastes undergoing biological decomposition		eusaprobic (4 special degrees)	hypertrophic
Industrial wastes inaccessible to biological decomposition		transsaprobic (3 special degrees)	atrophic (antitrophic)

Sladecek extended the 1955 work of Pantle and Buck to set a Saprobity Index (S) for all waters which fall in the limnosaprobic and eusaprobic portions of the circular classification system (Fig. 1-6). The lowest value of S is 0.5 for xenosaprobic waters, and the highest value is 8.5 for ultrasaprobic waters. A more detailed presentation of S values for each of the waters which contains some contaminants from organic wastes is shown in Table 1-5.

Rao et al. (1978) utilized the European Saprobity Index originally derived by Kolkwitz and Marsson to verify Stations I and II as polysaprobic. The first was dominated by euglenoid flagellates and contained very low dissolved oxygen; the second; Station II, was also dominated by Euglenae and other blue-green algae forms as well as diatoms, and falls between polysaprobic and mesosaprobic. Chlorophyceae and diatoms predominated at Station III, while Euglena was absent, probably due to the high dissolved oxygen level (8.5). This station was classed between β mesosaprobic and oligosaprobic zones. Station IV was classed as obligosaprobic because of Desmidaceae and Chlorophyceae being the dominant algal forms. Throughout the entire reach the biological saprobity index was supported by the chemical water and the benthic community numbers, types and quality at each station.

TABLE 1-5. Comparative Data Limiting the Individual Saprobic Degree. [*After Sladecek, 1965. By permission of publishers.*]

Degree of Saprobity	S	BOD_5 mg/l	Coliforms per liter	Psychrophilic heterotrophic bacteria (agar plate method) per milliliter
1. xeno-	0.5	1.0	10,000	1,000
2. oligo-	1.5	2.5	50,000	10,000
3. beta-meso-	2.5	5.	100,000	50,000
4. alpha-meso-	3.5	10.	1,000,000	250,000
5. poly-	4.5	50.	30,000,000	2,000,000
6. iso-	5.5	400.	3,000,000,000	10,000,000
7. meta-	6.5	700.	10,000,000,000	100,000,000
8. hyper-	7.5	2,000.	1,000,000	1,000,000,000
9. ultra-	8.5	120,000.	0	10

Note: S = saprobic index (Pantle and Buck, 1955, extended). BOD_5 values in limnosaprobical *stagnant* water bodies (items 1-5) can be up to twice higher than indicated in this table, if vegetative water coloring or water-bloom occur.

Howmiller and Scott (1977) proposed a system for assessment of water quality based on the species composition of the oligochaete taxocene, a group they found dominant in the benthic fauna of the Great Lakes and in inland lakes and rivers. They proposed using a formula developed by Brinkhurst et al. (1967) to provide an index known as "Trophic Condition" as given below

$$\text{Trophic Condition} = \frac{\Sigma\eta_1 + 2\Sigma\eta_2}{\Sigma\eta_0 + \Sigma\eta_1 + \Sigma\eta_2}$$

where η_0, η_1, and η_2 are the total number of oligochaete worms which are intolerant to eutrophic conditions, characteristic of slightly enriched areas, and characteristic of species tolerant to gross organic pollution, respectively. They give a classification of three groups of oligochaetes in Table 1-6.

Based on the Green Bay studies they conclude that an aquatic index of environmental quality (in order to be both sensitive and sensible) must consider both the composition of the fauna and the ecological attributes of the constituent species. Your author suggests that the same potential danger in using this index system exists as with Patrick's system. For although certain groups may exhibit ecological affinities, species typical of the group may develop tolerance to the pollutant, and may not truly represent a unique group. One might not detect pollution occurrence because of this evolution. Also, certain indicator species may not survive this evolution, and it will be unnoticed. When using a specie variety index we may be able to detect such an evolution.

TABLE 1-6. A Classification of Oligochaete Species According to the Degree of Enrichment of the Environments in Which They Are Characteristically Found.
[*After Howmiller and Scott, 1977*]

Group 0

Species largely restricted to oligotrophic situations:

Stylodrilus heringianus
Peloscolex variegatus
P. superiorensis
Limnodrilus profundicola
Tubifex kessleri
Rhyacodrilus coccineus
R. montana

Group 1

Species characteristic of areas which are mestrophic or only slightly enriched:

Peloscolex ferox
P. freyi
Hyodrilus templetoni
Potamothrix moldaviensis
P. vejdovskyi
Aulodrilus spp.
Arcteonais lomondi
Dero digitata
Nais elinguis
Slavina appendiculata
Uncinais uncinata

Group 2

Species tolerating extreme enrichment or organic pollution:

Limnodrilus angustipenis
L. cervix
L. claparedeianus
L. hoffmeisteri
L. manumeensis
L. udekemianus
Peloscolex multisetosus
Tubifex tubifex

The authors claim that since similar values of trophic indices (or specific oligochaete types and numbers) are not found under very different environmental conditions, the Trophic Index eliminates the possibility of misleading conclusions that might otherwise be reached on the basis of species diversity analysis. They used the environmentally different waters of three sections of

TABLE 1-7. Average Values of Five Indices of Pollution Compared for Three Areas of Green Bay.
[*After Howmiller and Scott, 1977.*]

	Area		
System or organism used	I	II	III
State of pollution	Grossly polluted	Less polluted	Higher water quality than others
Species diversity	1.00	1.62	1.66
Oligochaete worms/m^2	1085	1672	1152
Oligochaete worms, %	63	53	53
L. hoffmeisteri, %	73	50	42
Trophic index	1.92	1.84	1.53

Green Bay, Lake Michigan to validate their system. These results are shown in Table 1-7.

Cairns (1978) proposed a Recovery Index derived from the following equation:

$$\text{R. I.} = a \times b \times c \times d \times e \times f$$

where a, b, c, d, e, and f are the recovery factors which Cairns considers most important and described as follows:

a. Existence of nearby epicenters (e.g., for rivers, these might be tributaries) for providing "seed" organisms to reinvade a damaged system. Rating system: 1 = poor; 2 = moderate; and 3 = good.

b. Transportability or mobility for movement dissemules (dissemules might be spores, eggs, larvae, flying adults that lay eggs, or any other stage in the life history of an organism which permits it to move to a new area, either voluntarily or involuntarily. Rating system: 1 = poor; 2 = moderate; and 3 = good.

c. Condition of the habitat following pollutional stress. Rating system: 1 = poor; 2 = moderate; and 3 = good.

d. Presence of residual toxicants following pollutional stress. Rating system: 1 = large amounts; 2 = moderate amounts; 3 = none. The presence of residual toxicants should result in a decrease in the recoverability of a displaced system.

e. Chemical/physical environmental quality following pollutional stress. Rating system: 1 = in severe disequilibrium; 2 = partially restored; 3 = normal. The return to an approximation of the chemical/physical conditions originally present is an important prerequisite for the reestablishment of a community characteristic of that particular locale.

f. Management or organizational capabilities for immediate and direct control of the damaged area. Rating system: 1 = none; 2 = some; 3 = thriving, with strong enforcement prerogatives. When the degree of displacement from normal is known, the necessary corrective steps by good management may be clear and the available resources can be efficiently directed toward the desired goals.

Cairns indicates that with values of 400 and over, the chances of rapid recovery are excellent; 55-399 the chances of rapid recovery are fair to good; less than 55 the chances of rapid recovery are poor. In his view, a relatively rapid recovery would mean 40 to 60 percent of the species might become reestablished, under optimal conditions, in the first year following a severe stress; between 60 and 80 percent in the following year; and perhaps as high as 95 percent of the species by the third year.

1.4. BENTHAL DECOMPOSITION IN STREAMS

Fair and Moore (1941), in extensive research efforts, showed that sludge deposits in a stream decompose slowly, partially by anaerobic and partially by aerobic means. This "facultative" decomposition process is known as *benthal* stabilization. The result of the action of these organisms decomposing the sludge affects the stream oxygen balance, and hence, the entire stream ecology. Sludge bottom decomposition is depicted very simply in Fig. 1-7.

There are various forces or agents which bring about intimate reactions between the surface of the sludge deposits and the bottom of these beds.

1. Compaction or consolidation
2. Gas production, which channels (opens holes) in the sludge, effecting more definite mixing
3. Invertebrates, which also make channels to cause better mixing

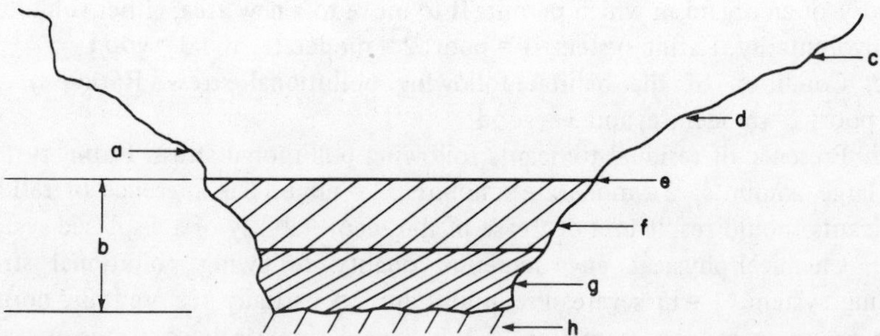

FIG. 1-7. Sludge deposits. (a) = side bank of stream; (b) = bottom sludge deposit in which benthal decomposition is occurring; (c) = flowing water surface; (d) = side bank of stream; (e) = water surface of sludge deposit; (f) = aerobic zone; (g) = anaerobic zone; (h) = stream bottom.

4. Diffusion of certain soluble products of anaerobic decomposition into the upper layers
5. Turbulence caused by floods and winds or stream obstruction, which may have a scouring effect on sludge deposit

When some anaerobically decomposed matter is brought (by one or more of the above agents) into the aerobic zone, additional oxygen is required for more complete oxidation. When added to the continuous oxygen required of the flowing pollution in the overlying areas, this benthal demand may be extensively significant. Oxygen demand of the sludge may be divided into three separate components:

1. Immediate chemical oxidation of diffused material such as hydrogen sulfide
2. BOD of products of decomposition
3. BOD of the original sludge

When scouring (resuspending or washing away of deposited sludge) occurs, the sludge decomposition becomes aerobic rather than benthal. In that case, an increase in oxygen demand is realized. Benthal decomposition usually makes for grayish-black sludge and a lower rate of oxygen demand.

The course of complete benthal decomposition can be described as occurring in the following three distinctive periods:

I. *Period of Intensive Fermentation*—may cause escape of sediment into supernatant, upper clear water levels, by flotation.
   A. Physical Manifestation
      1. Rapid evolution of carbon dioxide and methane
      2. Internal agitation and flotation of solids
      3. Release of putrefaction odor
   B. Biochemical Manifestation
      1. Lowered pH values, organic acids, reduction of nitrate and sulfate
      2. Large bacteria population; rapid decrease of coliforms
      3. Rapid decrease of BOD[2] but a uniformly high O_2[3] demand

II. *Period of General Consolidation*—determines or controls the reduced rate of transport of oxidizable material to sludge interface
   A. Physical Manifestation
      1. Marked retardation of gas production

---

[2] BOD refers to aerobic condition.
[3] O_2 demand refers to benthal condition.

2. Subsidence and increased density of sludge
3. Surface zone is gray, indicating some aerobicity

B. Biochemical Manifestation

1. Recovery of pH values, release of ammonia and other nitrogen compounds
2. Slow decrease of BOD and relatively high O_2 demand

III. *Period of Quiescent Stabilization*

A. Physical Manifestation

1. Consolidation of sludge is at end
2. Deposition of ferric iron as brownish-red glistening surface film
3. Blackening of lower layers, and tarry odor

B. Biochemical Manifestation

1. Decrease in bacterial population, establishing higher forms of life (worms, larvae, and insects)
2. Nearly sustained BOD, and reduced, but measurable O_2 demand

The rate of benthal decomposition depends primarily on the:

1. Amount of decomposable material [$(L - y)$, (ultimate oxygen demand minus BOD already exerted)]
2. Availability of decomposable material

$$K = \frac{k}{1 + at} \quad \begin{matrix}\text{(velocity constant which increases with turbulence of sludge deposit)} \\ \text{(change in availability)}\end{matrix}$$

where

K = benthal demand
k = reaction velocity constant
t = time
a = coefficient of retardation

Many environmental factors are included in the values of (k) and (a), such as pH, temperature, species composition, etc. (k) and (a) decrease with increasing depth. Thus, the deeper the layer, the harder it is for the diffusible matter to come to the surface of the bed to be oxidized. Resistance to diffusion increases with depth, primarily because of increased density in lower, deeper layers. The L value (ultimate oxygen demand) in benthal decomposition decreases with increasing depth. Greater portions of stabilization are brought about by anaerobic decomposition as the layer gets deeper, thus saving oxygen in the stream. Since methane is the major final product of anaerobic decomposition, the amount of oxygen saved can be measured by the amount of methane formed.

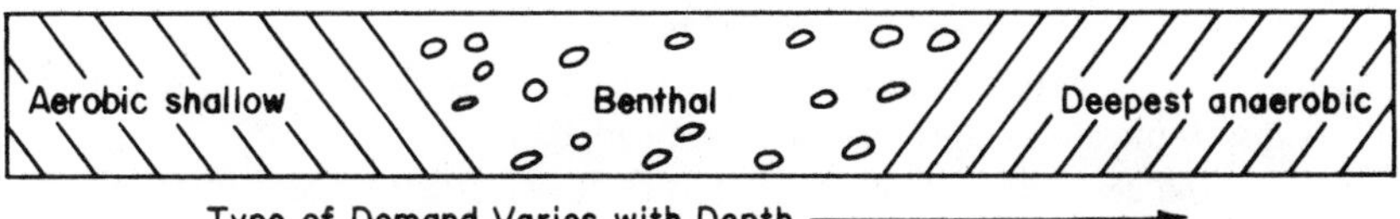

The aerobic phase of sludge not only makes the greatest oxygen demand but also the greatest rate of demand.

Nitrogen in Benthal Decomposition

Slow diffusion in deep beds keeps the aerobic phase from finishing before the anaerobic phase.

Ammonia produced by anaerobic decomposition at the bottom of the sludge bank may exert a delayed O_2 demand on the stream. When it diffuses to the surface, the following occurs:

1. There is loss of nitrogen from the sludge to the flowing body of water.
2. The form of nitrogen lost is organic, ammonia, or nitrate.
3. The percent of these different forms in relation to the total lost depends on the temperature.

Mortimer (1941) found that the reduction of the mud (bottom sludge) surface and the associated increase in supply of solutes to the water may be expected to augment plankton production. The occurrence or nonoccurrence of this event is determined by the balance between (1) the reducing power of the mud and (2) the amount of oxygen supplied to the mud surface. (The latter depends on both the volume of the hypolimnion and the degree of turbulence there.) He further stated that if, as a result of natural (evolutionary) or cultural changes, the reducing power of the mud is increased sufficiently to effect reduction of the mud surface, productivity may be expected to increase relatively suddenly to a higher level. He used redox-potential values and relative amounts of Fe^{+++} and Fe^{++} as indicators of the stage of reduction of the sludge surface.

Lardieri (1954) studied the aerobic and benthal oxygen demand of paper mill waste deposits and found that:

1. The pH values of paper waste sludge deposits studied were found to be under 5.0.
2. The low pH in the deposits is responsible for the low and constant rate of benthal oxygen utilization observed.
3. There are relatively small amounts of gas produced due to this low pH, which retards the activities of the methane producing bacteria.

4. Adjustment of paper deposit sludges to a pH of 7.0 causes an oxygen increase of approximately 100% in both aerobic and benthal decomposition. Gas production is more than doubled after such adjustment.

5. The oxygen demands of bottom and lower strata sludge from a 5-foot deposit were identical per unit weight of volatile solids, evidence that decomposition in these acid deposits proceeds at low and constant rates.

Lardieri stated that sludge from the 1-foot and 5-foot levels showed the same O_2 demand per gram of volatile matter despite the lower percentage of volatile solids in the lower strata. Some of this lower percentage may be due to a greater admixture with inorganic matter (silt) at the 5-foot level, but it seems likely that a greater proportion of the difference in the two levels is due to the destruction of volatile matter. The accumulation of a 5-foot sludge deposit takes a considerable length of time. Despite the fact that the bottom layers in the deposit are composed of older material, its oxygen utilization per unit weight of volatile solids is the same as sludge deposited much more recently.

The benthal O_2 demand of paper waste deposits is much less than the aerobic O_2 demand, and decreases markedly per unit weight of solids deposited with increasing depth. The major portion of the decomposition is probably occurring at the sludge-water interface, where the conditions of O_2 availability and pH are more conducive to biological oxidation.

The importance of including sludge deposits in any stream analysis was brought out by Oldaker (1958) in his report on the Connecticut River Study of 1956 with reference to the effects on water quality of the proposed Enfield Power Project. He states: "indeed it is evident from a plot of the average results of the August of 1956 survey that the low point in the dissolved oxygen concentration of approximately 3.75 parts per million occurred just below the Chicipee River where sludge banks are located. The September 1956 survey did not reveal any pronounced dissolved oxygen sag at all." Presumably no sludge deposits existed in the September survey.

Cairns and Dickson (1971) discuss the use of benthic organisms for the purpose of assessing pollution of the biological life. They describe the commonest types of fauna (bottom-dwelling), their response to pollution, choice of sampling sites, sampling techniques, sorting and preserving organisms, and evaluation of data.

1-5. EFFECT OF DISINFECTION ON BACTERIAL NUMBERS IN STREAMS

The fact that chlorine is required in water disinfection to control bacterial growth is well known. The practice of chlorination has been carried over into sewage treatment for the same purpose. In that case, however, the treatment was designed to prevent contamination of water supplies downstream of sewage effluent discharges. The practice of chlorinating sewage effluents caused many other correlative phenomena to occur in receiving streams. Their importance to us in studying stream reactions to wastewaters is equal to that of controlling disease transmission.

Baity, Merryfield, and Uzzle (1933) were among the first investigators to observe these stream conditions after chlorination. they chlorinated an Imhoff Tank effluent and observed for and analyzed changes in stream conditions. They concluded that the greatest effect was in the improved physical appearance of the stream channel. During periods of nonchlorination, luxurious growths of gray, slimy, filamentous organisms would attach themselves to every possible projecting object on the bed and banks of the channel, and grow to great lengths. In pools and dead areas of the swampy region the microorganisms were characteristic of polluted water and grew in great numbers. Usually the bottom and banks were heavily coated with a murky, black gelatinous deposit, giving to the flowing water an untrue and very disagreeable appearance.

During periods of chlorination, the filamentous growths sloughed off and were washed downstream. The littora and free-floating types of algae practically disappeared, and the discoloring gelatinous growths released their hold upon the bed of the channel, making the water appear clearer and cleaner. There was also a reduction in turbidity during the Cl_2 application. As soon as chlorination of the sewage was suspended, the growths and discoloration quickly returned.

Chlorination also eliminated septic odors at the plant effluent and converted distinct earthy odors in river stations to sweet, grassy ones.

As shown in Fig. 1-8, the investigators found that oxygen sag in the receiving stream was lessened and delayed following chlorination of the treatment plant effluent. For example, with no chlorination, the bottom of the sag was about 25 percent of saturation at 3,000 feet below the outfall. During a period of chlorination, the sag bottom was moved to about one mile below the outfall, and was raised to 40 percent of saturation. Creek flows and temperatures were assumed to be about the same during both periods of testing.

The authors also state that the average stream temperatures during periods of chlorination were higher than during those of nonchlorination. This condition in itself would tend to produce lower concentrations of dissolved oxygen in the stream during chlorination periods. They concluded that the observed stream improvement in oxygen was due to both (1) stabilization of organic matter by chlorine and (2) retardation of bacterial activity until further downstream, when dilution was greater.

BOD values in the receiving stream during these same testing periods are shown in Fig. 1-9. Values were consistently lower at all stream locations during periods of chlorination. Table 1-8 contains bacterial numbers of stream samples —both total and coliforms—per cc with and without chlorination of the sewage

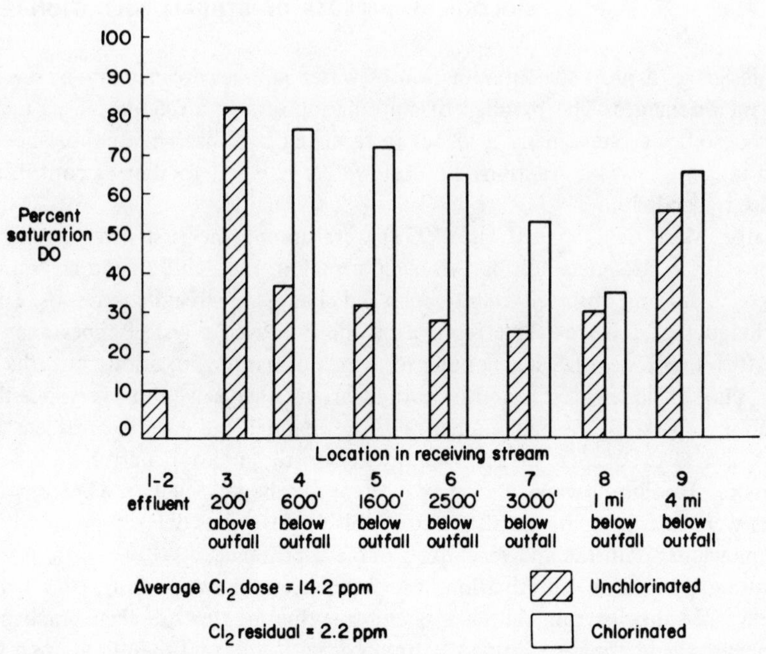

FIG. 1-8. Dissolved oxygen profile downstream from sewage treatment plant.

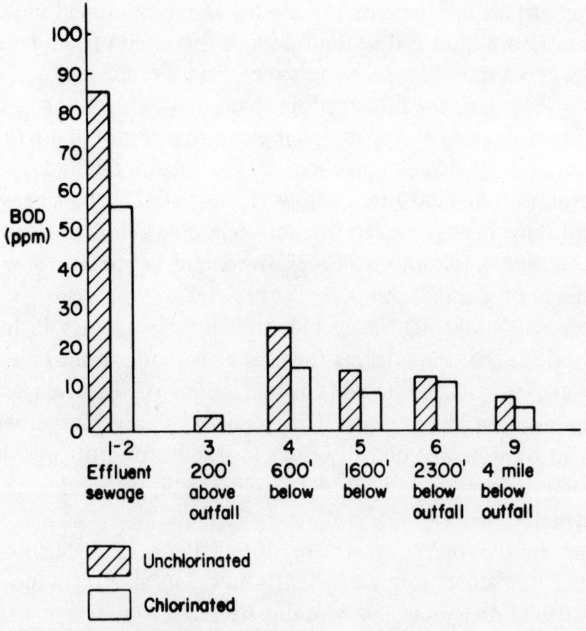

FIG. 1-9. Effect of chlorination of sewage on BOD values in receiving station.

TABLE 1-8 Effect of Chlorination on Bacterial Numbers/cc in Receiving Stream

Sta.	Location	Average Unchlorinated		Average Chlorinated		% Reduction by Chlorination	
		Total	E. Coli.	Total	E. Coli.	Total	E. Coli.
1-2	Sewage Effluent	391,000	40,000	31,500	14,750	91.5	63.1
3	200′ above outfall	104	5	–	–	–	–
4	600′ below outfall	65,000	3,450	6,400	600	90.3	82.6
5	1600′ below outfall	85,000	5,400	2,950	60	96.5	98.9
9	4 mi below outfall	1,200	42	750	3	37.5	92.9

Average Cl_2 dose = 8.13 ppm.
Average Cl_2 residual = 0.12 ppm.

plant effluent. Chlorination, although not killing more than 91.5 percent of the total sewage bacteria and only 63.1 percent of the coliforms, did cause receiving stream reductions of 96.5 and 98.9 percent, respectively, of these bacteria 1,600 feet below the outfall. Bacterial numbers, however, increased at the 4-mile location below the sewage outfall. With proper contact time at STP, chlorination may have been more effective in immediately reducing stream coliform numbers.

It can be seen that chlorine, when applied in dosages just sufficient to kill coliform bacteria, still does not eliminate the organic matter present in a wastewater effluent. It may suppress or delay the degradation of that organic matter until further on downstream. Advocates of chlorination point out that the delay may be sufficient so that, with the added dilution that usually accompanies downstream flow, critical oxygen levels may not be reached. Dissenters maintain that chlorination only puts off the payment of an oxygen debt which must eventually be fulfilled. It appears that both sides have merit in their arguments and that each case should be decided on its own merits.

Heukelekian's (1951) studies in the laboratory measured the effect of chlorination on the aftergrowth of coliform organisms in streams. He chlorinated fresh settled domestic sewage to residuals of 0.1 and 0.5 ppm for 30 minutes. His results showed that coliform organisms could multiply in streamwaters under certain conditions. One of the conditions is the *concentration of sewage* or *degree of dilution. Food* may be the controlling factor. The maximum number attained is affected, in addition, by the residual chlorine in the sewage. *The higher the residual chlorine and the fewer the coliform organisms remaining in sewage, the greater the subsequent multiplication upon dilution with streamwater, although the multiplication may be somewhat more delayed than with a lower residual chlorine.* It remains to be seen whether the relatively few typhoid and dysentery organisms that may be found in untreated sewage multiply similarly when discharged into polluted streams. It is assumed that in properly chlorinated sewage these organisms are not present, and hence their multiplication, even in a highly polluted stream, is remote. The increase (or aftergrowth) has been explained by (1) the dilution of the predatory protozoan population so that they can no longer keep the bacterial population in check, and (2) the fact that the protozoans are eliminated by chlorination to a greater extent than the bacteria.

His conclusions were:

1. Coliforms increase significantly after chlorinated sewage is diluted with stream and distilled waters.
2. The increase is greatest in higher concentrations of sewage or effluents.
3. The increase takes place within one day after dilution.
4. The increase in greater when chlorine residual is greater.
5. No such increase in the number of coliform organisms occurs when chlorinated sewage is diluted with seawater.

Both Baity's and Heukelekian's research points out that chlorination does not sterilize receiving streams. In fact, we can look out for an increase in bacterial numbers at some distance below sewage outfalls despite chlorination of the sewage effluent.

1-6. EUTROPHICATION OF STREAMS

Cultural eutrophication, or the accelerated aging of our streams and lakes by overenrichment, has often been linked to the increased contamination reaching these waters, both through municipal sewers and through underground percolation from soil sewage treatment systems and agricultural irrigation practices. The situation may be easily recognized by the numerous algal blooms which occur in these bodies of water during the hot summer months. It appears that since the world population will continue to increase to the point of doubling in the next 17 years, contaminants reaching the watercourses will also increase (Nemerow and Rand, 1967). The problems this will create can only be surmised, but indications exist that all but the largest, swiftest, and coldest bodies of water may be overloaded with algae in this same period. Algae are simple plants without roots, stems, or leaves. Most algae contain chrlorophyll, which permits them to produce their food by photosynthesis; that is, to use energy from sunlight to convert inorganic substances into organic compounds in their cells. Algae are ubiquitous and resistant to complete elimination. On the other hand, their sudden appearance in numbers great enough to cause a bloom condition defies our precise understanding to date. Algae differ from microscopic animal life in our water bodies in their mode of respiration. Respiration is a process carried on by all plants and animals, and the gaseous exchange of the algae is the opposite of that in photosynthesis; that is, oxygen is absorbed and carbon dioxide is released. However, in algae and other green plants, the maximum rate of their photosynthesis is normally faster than their respiration, or the P/R is greater than 1 (Rand and Nemerow, 1964). These organisms, therefore, release more oxygen during the day than they use, and absorb more carbon dioxide than they release, while animals and other nonphotosynthetic organisms release carbon dioxide and absorb oxygen from their environment. Algae usually react in an opposite manner during the dark, nighttime hours. At this time they actually act simply as dead organic matter and exert an added BOD load. For this reason, the amounts of oxygen and carbon dioxide in water often depend to a large degree upon the relative rates of photosynthesis and respiration being carried on simultaneously by algae, bacteria, and other organisms. As pointed out by the author (Rand and Nemerow, 1964), this is an important point to consider when examining opinions and actions of public-minded groups requesting and demanding the removal of algae from a body of water. The complete removal of algae from a body of water, aside from being a practical impossibility, would often actually be a detriment rather than a benefit. Oxygen supplied by algae during photosynthesis is beneficial to most forms of life in streams. Most researchers agree that the most vital algae limiting nutrient is organic phosphate. Greater populations of people whose homes are connected to a municipal sewer contribute increased quantities of this phosphate to receiving streams. Little or no phosphate is removed from most domestic sewages treated in the conventional manners. Two schools of thought have emerged on ways to combat the problem. The first school has taken the defeatist attitude and concludes that algae growth prevention is impractical. This decision is based partly on the present paucity of information on exact nutrient and environmental requirements for bloom development and partly on the feelings that nutrients are ubiquitous (on land, in the air, as well as in water supplies and wastes) and that natural environmental factors tend to favor blooms. This thinking sometimes suggests control of bloom by physical or chemical removal of the algae after they reach a detrimental level. The other school is more optimistic and proposes to control blooms by limiting nutrients essential for their growth. However, members of this school do not agree on either the proper nutrient to remove (although most are convinced that phosphate is the most logical) or how to remove it practically from the wastewater effluents.

An excellent book concerned with this topic was published in 1969 (Proceedings Symposium NAS, 1969). We do now have some rather effective methods of removing phosphates from domestic wastewaters. It is simply a question of economics and administrative control to enforce this type of treatment.

Mackenthun (1969) gives his considered judgment that to prevent biological nuisances total phosphorus should not exceed 100 μg/l of P at any point within a flowing stream, nor should 50 μg/l be exceeded where waters are more stationary.

In other studies carried out by the author and his associates (Karanik and Nemerow, 1965; Nemerow and Rand, 1964) lime precipitation of raw sewage repressed algae growth, especially of the blue-green type. Percolation of sewage effluents through certain soils has also given some indication of being an effective method of reducing bloom situations. The authors (Nemerow and Rand, 1967) also found that both trickling filter and activated sludge sewage effluents stimulated about twice the growth of green algae as primary settled sewage effluents.

The prevention of algal blooms is important to us in our treatise on receiving waters, primarily from a nuisance standpoint. Blooms are often referred to as "pea soup" and the color they give water is objectionable. Under certain conditions of darkness and warm temperatures these blooms may die, decompose, and render an offensive "sewage-like" odor. Moreover, if the receiving water is used as a raw water supply for some public or private agency, algae may be difficult to remove and hence add certain objectionable tastes to the delivered water. Algae also have the tendency to absorb and concentrate mineral nutrients such as phosphate and nitrogen compounds in their cells. This can always be detected by a decrease in these constituents in the overlaying waters during the growth. At the end of the growing season when the algae die, they settle to the stream or lake bottom (and thereby make the watercourse shallower), from which they release these mineral and organic nutrients at the beginning of the next growing season. In this manner they tend to serve as a form of secondary pollution, contributing organic and mineral nutrients for future biological growth in the same way that fresh wastewater adds these materials. Outside of these objectionable effects of algae (color, odor, taste, and secondary pollution) these plants, in reasonable amounts, may be beneficial to the waters. They serve as a vital part of the ecological food chain and supply oxygen to bacteria, fish, and other animal life.

The stream investigator must comprehend both the detrimental and beneficial effects of eutrophication and cope with their problems when the situation becomes unbalanced in favor of bloom conditions.

Streams or lakes are sometimes treated with copper sulfate to control or to eliminate algae blooms. Care must be exercised in adding this chemical material so that the other biological forms such as fish food and fish are not adversely affected. For example, in New York State one needs a permit when using $CuSO_4$ on watercourses for algae control. Usually doses of less than one part per million of copper sulfate applied at the proper time (usually just prior to bloom conditions) and in the proper manner (usually uniformly dispersed in the entire body) will be sufficient to prevent blooms. In fact, with a total alkalinity of 50 ppm or less, the maximum safe concentration of copper sulfate is only 0.25 mg/l.

1-7. VIRUSES IN RECEIVING WATERS

Although waterborne bacterial infections have been all but eliminated, water resource people find waterborne viral infections much more persistent. Two viral diseases are especially significant: (1) infectious hepatitis, and (2) gastroenteritis. The former is regularly reported and has been at a level of about 55,000 cases per year in the United States. Although gastroenteritis cases are not generally reported, it is estimated that there are hundreds of thousands, and perhaps millions each year. There are more than 100 viruses excreted in human feces that have been reported to be in contaminated water.[4]

The 1970 report of the Committee on Environmental Quality Management, Sanitary Engineering Division of ASCE concluded the following concerning viruses.[5]

1. There is no doubt that the virus of *infectious hepatitis* can be transmitted by drinking water.

2. Although evidence is scanty, it should also be assumed that the enteric viruses and other possible causative agents of viral gastroenteritis can be transmitted through drinking water.

3. There is no doubt that a positive coliform index means that virus may be present; however, absence of coliform does not necessarily mean that virus is absent. Therefore, the coliform index is not reliable for viruses. Better assurance of the absence of virus would be a turbidity of less than 0.1 Jackson Unit, and an $\overline{HOCl}$ residual of 1 mg/l after a contact of 30 minutes.

4. Risk of hepatitis results from the consumption of raw or steamed undepurated shellfish taken from sewage-polluted waters; the PHS coliform standard (70 coliforms/100 ml) is a reliable indication of risk-free, shellfish waters. A high level of protection is provided by activated sludge treatment and chlorination of the effluent to a level of an amperometric chlorine residual of 5+ mg/l after 30 minutes of contact.

5. Virus multiplication in polluted water appears not to be a significant possibility.

6. Viruses are present in certain river waters, and failure to isolate them results presumably from their low concentrations and the relatively ineffective sampling and concentration procedures employed.

7. Enteroviruses and the virus of infectious hepatitis can survive for prolonged periods under conditions prevailing in drinking-water reservoirs. Long detention times, therefore, cannot be considered a safety factor.

---

[4] G. C. White, *Disinfection of Wastewater and Water for Reuse*, New York: Van Nostrand Reinhold Co., 1978.

[5] Committee on Environmental Quality Management, Engineering Evaluation of Virus Hazard in Water, *ASCE J. San Eng. Div.*, (February 1970) 96:111.

8. Enteric viruses differ in resistance to free chlorine. Adenovirus 3 is less resistant than E. Coli, while poliovirus 1 and Coxsackie virus A2 and A9 appear to be more resistant than any of the other enteroviruses studied.

Viruses in Sewage

In the United States, the enteric virus concentration in raw sewage ranges from 2 or 3 to greater than 1000 infectious virus units/100 ml with peak levels occurring in late summer and early fall.[6] Enteric viruses are five to seven orders of magnitude lower than coliform bacteria in sewage. Only one virus infectious unit is needed to produce an infection in humans.

Several workers have detected viruses in water supplies, such as Berg,[7] (Missouri River Water); Coin,[8] (Paris, France water supply); and Nupen,[9] (South Africa drinking water). These waters all have an historic record of some sewage contamination. The most prevalent waterborne disease in the United States continues to be gastroenteritis of unknown etiology.

Of the three groups of viruses (animal, bacterial, and plant) the animal and bacterial types are both present in wastewaters in different concentrations. However, the enteric virus group found in the digestive or respiratory tracts of warm-blooded animals is most important. It includes polio, infectious hepatitis, and Coxsackie viruses. One can also find other enteric viruses in wastewaters and streams, such as echovirus, reovirus, and adenovirus. There are over 100 different enteric viruses of medical interest that could be isolated from polluted waters. They may be transmitted to humans directly through drinking water, swimming water, or from eating shellfish. Viruses tend to clump together, making them more resistant than coliform bacteria to normal doses of and contact times with chlorine. They also may be shielded from inactivation due to chlorine by adsorbing on some suspended matter.

Virus Killing

In the United States, California has done more to study the presence or absence of virus in reused wastewaters than any other states.[10] Their objective was to

---

[6]M. D. Sobsey, "Enteric Viruses and Drinking Water Supplies," *JAWWA,* (August 1975).
[7]N. A. Clarke and P. W. Kalber, "Human Enteric Viruses in Sewage," *Health Lab. Sci.*, (1964), 1:44.
[8]L. Coin et al., "Modern Microbiological Virological Aspects of Water Pollution," Adv. Water Pollution Research 2nd Int. Conf., Pergamon Press, New York (1966), pp. 1–10.
[9]E. M. Nupen, B. W. Bateman, and N. C. McKenny, "The Reduction of Virus by the Various Unit Processes Used in the Reclamation of Sewage in Potable Waters," Virus Symposium, Austin, Texas, April 1974.
[10]M. W. Selna, R. P. Miele, and R. B. Baird, "Disinfection for Water Reuse," Seminar at Annual Conf. A.W.W.A., Anaheim, Calif., May 8, 1977.

find alternative ways to produce a reused wastewater with a median total coliform MPN of 2.2/100 ml or less, and to provide an effluent which would protect swimmers against viral illnesses. The Pomona study revealed several significant inactivation findings:

1. Chloramines do in fact have virucidal efficiency potential equal to that of free chlorine.
2. There is no benefit from free residual chlorination of tertiary effluents to meet California standards where direct potable reuse is called for.
3. The majority of virus removal occurs during disinfection rather than during filtration or carbon absorption prior to disinfection.
4. Virus inactivation in tertiary treatment systems employing combined chlorine residuals of 5 to 10 mg/liter range from 4.7 to 5.2 logs. In experiments using ozonation, virus removal ranged from 5.1 to 5.5 logs; however, attainment of the 2.2/100 ml MPN coliform standard was hampered by water quality variations.

For virus inactivation, reporting destruction as a percent of the original number of organisms is a good comparative method, but may not be compatible with what a regulatory agency is likely to require. One of the major difficulties in the killing of viruses is their variable sensitivity to disinfectants when using minutes of contact time to achieve a 99.99% kill by 0.5 mg/l free Cl_2 residual at pH 7.8 and 20°C; With 16 different viruses, Liu[11] found variations from 36.5 minutes (for poliotype II) to 2.7 minutes (for reovirus I).

For a wastewater effluent we cannot expect much viral destruction unless the effluent is of tertiary quality. For a raw potable water, it is the consensus that a 1 mg/l free Cl_2 residual at the end of 30 minutes contact time at a pH not to exceed 8.0, will destroy nearly all pathogenic viruses.[12]

Chlorine dioxide appears to have the most promise as both a virucide and bactericide.

1-8 UTILITY OF BIOLOGICAL METHODS FOR EVALUATING WATER QUALITY

We have discussed several of the more publicized attempts to indicate the relative condition of streams by observing the living material present. The saprobic system seems to offer the most promise for providing a suitable classification system. Such an approach has many advantages over the more traditional and prevalent chemical analyses. The principal advantages include:

---

[11] "Effect of Chlorination on Human Enteric Viruses in Partially Treated Water from Potomac Estuary," Proc. Congr. Hearings. Proc. Serv. No. 92-94, Washington, D.C., 1973.
[12] G. C. White, *Handbook of Chlorination*, New York: Van Nostrand Reinhold Co., 1972.

1. An analysis of the life forms can be made at any time of the year, usually disregarding current streamflow and momentary pollution.
2. The condition of a stream is ascertained as a result of previous as well as present pollution; it therefore represents an overall resultant of past effects.
3. Any material toxic to biological forms is reflected in the numbers and species of organisms found.

The biological system of statistically evaluating indicator organisms present also has its inherent disadvantages, such as:

1. The analysis does not always reveal what environmental factors are responsible for ecological changes observed.
2. The system (as it is currently developed) cannot be used to predict the quantity of certain wastes that can be assimilated by the receiving stream. Therefore, the degree of pollution abatement required is undecided, which makes stream administration very difficult.
3. Specially trained personnel are required not only to collect, analyze, and identify the various organisms, but also to evaluate the overall biological condition. These scientists do not appear to be as readily available as the more traditional sanitary chemists.

When all aspects are taken into account, an ideal stream classification system should include both biological and chemical analyses.

1-9. BIOASSY TESTING

In order to ascertain the effect of specific concentrations of contaminants on certain biota–usually fish–a bioassay test procedure is used. In principle, this procedure brings together a test solution and a dilution water by dispensing the mixture into an aquarium containing the fishes or other organisms being exposed. The test solution can contain one or more metals, pesticides, ammonia, or special organic compounds. The dilution water may originate from freshwater supplies, or from estuarine or ocean water. The length of time for a given "dose" of test chemical to kill 50 percent of the test organisms (LD_{50}) is usually used as the criterion for toxic level. The bioassy procedure is adapted to any particular problem of contaminant(s) and receiving water.

One such system is the Toxscreen Bioassay System (developed by Enwright Laboratories of Greenville and Summerville, South Carolina). Its system is flexible, with the ability to examine the effects of six different toxicant concentrations during one test sequence. It provides 14 test aquaria which can be adjusted from 0 to 100 ml per minute, and allows for 0 to 10 complete water changes per day in each of the four-gallon aquaria. The system is also able to adjust the ratio of test solution to dilution water needed to attain the desired concentrations for testing.

Bioassay test results should always be verified in actual field experimentation, if possible. Sometimes the characteristic physical, chemical, and biological qualities of a receiving water respond differently than in the laboratory aquaria.

QUESTIONS

1. What is the major basis for Brinley's stream classification?
2. What is the major basis for Heukelekian's stream classification?
3. What is the major basis for Patrick's stream classification?
4. What is the major basis for the Kolkwitz and Marsson stream classification?
5. What is the major basis for Sladecek's stream classification?
6. Describe the effect of benthal decomposition on receiving water biota.
7. What receiving water conditions are ideal for eutrophication?
8. What dangers arise from excessive eutrophication?
9. What are the typical viruses present in streams and lakes containing urban wastewaters? How are these viruses controlled? What dangers develop from the uncontrolled discharge and survival of viruses in receiving waters?
10. What is a bioassay, and why is it important to the stream analyst?

REFERENCES

Baity, H. G., F. Merryfield, and A. Uzzle: *Sewage Works J.*, **5**(3):429, 1933.

Brinely, F. J.: Ohio River Pollution Survey I, *Sewage Works J.*, **14**:147, 1942.

Brinkhurst, R. O., et al.: The Distribution of Aquatic Oligochaetes in Saginaw Bay, Lake Huron, *Limn. Oceanogr.*, **12**:137, 1967.

Butterfield, C. T.: Experimental Studies of Bacterial Death Rates in Polluted Waters, *J. Bact.*, **16**(4):257–267, 1928.

——: Experimental Studies of Natural Purification in Polluted Waters. II. Development of a Suitable Dilute Medium, *Weekly Public Health Reports*, **44**(44):2647–2658, 1929*a*.

——: Experimental Studies of Natural Purification in Polluted Waters. III. A Note on the Relation Between Food Concentration in Liquid Media and Bacterial Growth. Publ. Health Repts. U.S. Publ. Health Serv., **44**(47):2865–2872, 1929*b*.

——: The Selection of a Dilution Water for Bacteriological Examinations, *J. Bact.*, **23**(5): 355–368, 1932.

——: Observations on Changes in Numbers of Bacteria in Polluted Water, *Sewage Works J.*, **5**(4):600–622, 1933.

——: Suggested Procedures for the Presumptive Test in the Determination of the Coli-Aerogenes Group, *Am. J. Public Health and Nation's Health*, **23**(4):343–349, 1933.

——: Some Functions of Bacteria in the Purification of Polluted Water, *J. Bact.*, **39**(5): 527–533, 1940.

——, and W. C. Purdy: Some Interrelationships of Plankton and Bacteria in Natural Purification of Polluted Water, *Indus. Eng. Chem.*, **23**(2):213–218, 1931.

——: and E. J. Theriault: Apparatus for the Study of Biochemical and Other Oxidations in Liquids, *Weekly Public Health Reports*, **44**(38):2256–2267, 1929.

——, and Elsie Wattie: Studies of Sewage Purification. VIII. Observations on the Effect of

Variations in the Initial Numbers of Bacteria and of the Dispersion of Sludge Flocs on the Course of Oxidation of Organic Material by Bacteria in Pure Culture, *U.S. Publ. Health Reports,* **53**(43);1912–1934. Also in: *Sewage Works J.*, **10**(5):815–838, 1938.

——: W. C. Purdy, and E. J. Theriault: Experimental Studies of Natural Purification in Polluted Waters. IV. The Influence of the Plankton on the Biochemical Oxidation of Organic Matter. *U.S.(A.) Treas. Dept. Publ. Health Repts.*, **46**(8):393–426, 1931.

——, E. J. Theriault, and P. D. McNamee: Experimental Studies of Natural Purification in Polluted Waters. V. The Selection of Dilution Waters for Use in Oxygen Demand Tests. *Weekly Public Health Reports,* **46**(18):1084–1115, 1931.

Cairns, J.: Waterway Recovery, *Water Spectrum,* **10**(4):27, 1978.

——, and K. L. Dickson: A Simple Method for the Biological Assessment of the Effects of Waste Discharges on Aquatic Bottom-Dwelling Organisms, *J. Water Pollution Control Fed.,* **43**:722–755, 1971.

Eutrophication: Causes and Consequences, *Proc. Symp. National Academy of Science,* Washington, D.C., 1969.

Fair, G. M., and E. Moore: *Sewage and Industrial Wastes,* Vol. 13, 1941. (I. General Concepts, 270; II. Benthal Decomposition Rates, 279; III. Effect of Sludge-Depths, 288; IV. Effect of Temperature, 756; V. Nitrogen and Iron Changes, 766; VI. BOD Satisfaction. 1209; VII. Effect of Inorganic Mixture, 1212; VIII. Accumulation of BOD, 1221.)

Heukelekian, H.: *Sewage and Industrial Wastes,* **23**(3):273, 1951.

——: Chapter 2 in W. Rudolfs ed., *Industrial Wastes,* Reinhold Publishing Company, New York, 1953, pp. 8–31.

Howmiller, R. P., and M. A. Scott: An Environmental Index Based on Relative Abundance of Oligochaete Species, *J. Water Pollution Control Fed.*, **49**(5):809, May 1977.

Hynes: *The Biology of Polluted Waters,* Liverpool, England, Liverpool Press, 1960, Chap. 13.

Karanik, J. M., and N. L. Nemerow: Removal of Algal Nutrients from Domestic Wastewaters, *Water and Sewage Works,* **112**(12): 460, 1965.

Kehr, R. W., and C. T. Butterfield: Notes on the Relation between Coliforms and Enteric Pathogens, *Publ. Health Repts.*, **58**(15):589–607, 1943.

Kolkwitz and Marsson: "Okologie der tierische Saprobien. Beitrage zur Lehre von der biologische Gewasserbeurteiling," *Int. Rev. Hydrobiol.*, **2**:126, 1908.

Lardieri, N. J.: *TAPPI,* **37**(12):705, 1954.

Liebmann, H.: *Handbuch der Frischwasser und Abwasserbiologie,* Bd. I. Jena, Oldenburg Blg., 588 pp., 1951.

Mackenthun, Kenneth M.: The Practice of Water Pollution Biology, FWPCA, p. 19, 1969.

Mortimer, C. H.: The Exchange of Dissolved Substances Between Mud and Water in Lakes, *J. Ecology,* **21**:280, 1941.

Nemerow, N. L., and M. C. Rand: Algal Nutrient Removal from Domestic Wastewaters, *Proc. First Annual Water Quality Research Symp., N.Y. State Health Dept.,* Albany, N.Y., pp. 37–58, 1964.

——, and M. C. Rand: Algal Growth Affected by Degree and Type of Wastewater Treatment, *Algae, Man, and the Environment,* Chap. 18. A Text of the Proceedings of an International Symposium at Syracuse University, New York, June 18–23, 1967.

"Oekologie der Saprobien. Über die Beziehungen der Wasserorganismen zur Unwelt," *Schr. Riehe Ver Wasserhyg.*, **4**, 1969, 64 pp.

Oldaker, W. H.: Bottom Deposits in a River and Their Potential Effects on Dissolved Oxygen Concentrations, in M. A. Churchill and N. L. Nemerow, eds., *Am. Soc. Civil Engrs. San Eng. Div. Research Rept. No. 20,* paper 1779, SA5:9, 1958.

Palmer, C. Mervin: Algae in Water Supplies, *Public Health Serv. Bulletin No. 657,* 88 pp., 1962.

Patrick, Ruth: *Some Diatoms of Great Salt Lake, Bull. Torrey Bot. Club,* **63**(3):157, 1936.

——: The Diatoms of Linsley Pond, Connecticut, *Proc. Acad. Nat. Sci., Philadelphia,* **95**:53, 1953.

——: Estudio Limnologico e Biologico das Lagoas da Regiao Litoranea Sul-Riograndense, *Bol. Mus. Nac., Rio de Janeiro Bot. Ser.,* **2**:1, 1944.

——: A Proposed Biological Measure of Stream Conditions, Based on a Survey of the Conestoga Basin, Lancaster County, Pennsylvania, *Proc. Acad. Nat. Sci., Philadelphia,* **101**:277, 1949.

——: Biological Measure of Stream Conditions, *Sewage and Industrial Wastes,* **22**(7):926, 1950.

——: A Proposed Biological Measure of Stream Conditions, *Proc. Int. Assn. Limn.,* **11**: 299, 1951.

——: A Study of the Number and Kinds of Species Found in Rivers in Eastern United States, *Proc. Acad. Nat. Sciences, Philadelphia,* **113**(10):215–258, 1962.

——, Matthew Horn, and John Wallace: A New Method for Determining the Pattern of the Diatom Flora, *Notuale Naturae,* Philadelphia, **259**:1–12, 1954.

——, and B. W. Dickerson: Research on Recovery of a Polluted River, *Water and Sewage Works,* **102**(5):214–215, 1955.

Purdy, W. C., and C. T. Butterfield: The Effect of Plankton Animals upon Bacterial Death Rate, *Am. j. Publ. Health,* **8**:499–505, 1918.

Rand, M. C., and N. L. Nemerow: Removal of Algal Nutrients from Domestic Wastewater, I. Literature Survey Report to New York State Health Dept., Syracuse University, New York, September 1964.

Rao, R. S. V., V. P. Singh, and L. P. Mall: Pollution Studies of River Khan (Indore), India. I. Biological Assessment of Pollution, *Water Research,* **12**:555–559, 1978.

Sladecek, V.: The Future of the Saprobity System, *Hydrobiologia,* **25**, Dr. W. Junk Publishers, The Hague, 1965.

2 HYDROLOGICAL CONSIDERATIONS IN STREAM ANALYSIS

2-1. WATERCOURSE SAMPLING FOR EFFECTIVE EVALUATION

The writer recommends that in each instance of stream analysis the investigator begin with a thorough study of his real objectives. It might be helpful to list these overall specific objectives prior to any planned sampling program. Then, later, as the sampling program unfolds, the investigator can refer to these stated objectives and decide whether certain samples are actually necessary or whether other samples must be added to the list.

Consideration must be directed toward three main operations: (1) collection of data; (2) analysis of data to determine specific self-purification characteristics and degree of contamination, if any; and (3) projection of estimates to other situations–primarily the evaluation of expected stream quality at various drought severities and ranges of pollution load. The present and future uses of a watercourse–whether for water supply, bathing, fish and shellfish culture, recreation, etc.–have important bearing on the types of analyses to be made, and hence, upon data collection. Once these have been decided, collection of data takes on meaning, and a sampling procedure can be designed.

The scientific collection of data for stream analysis has been divided by Velz (1950) into three major categories: (1) hydrologic factors; (2) sources of pollution; and (3) watercourse sampling.

Hydrologic Factors

A considerable number of past streamflow records (the author recommends records for the last 10 years) is necessary to define the characteristics of the daily hydrograph. The physical characteristics of channels contribute to the control of deoxygenation, bacterial death rates, benthal decomposition, and reaeration. These physical characteristics in relatively small streams are best determined by making channel cross sections every one-half mile in noncritical areas, and every one-quarter mile in critical[1] areas. Critical areas may be defined as areas of unusual water deterioration and/or usage. These determinations need not be precise but can be carried out by making depth soundings from a boat or bridge, and by the use of a rope to determine the width. The data and time should be noted so that the surface of the water can be referred to a gage height corresponding to the prevailing river discharge. These measured flows can then be adjusted to other streamflows at the same site by use of rating curves as shown in Fig. 2-1. The main purpose of channel cross sections, which should be made at different river stages in order to establish and/or verify the curve of Fig. 2-1, is to determine the occupied channel volume associated with a particular steady streamflow. It is then possible to compute the steady streamflow at any gage height to provide the all-important travel time or "time of passage" of pollution from reach to reach. (A reach is generally assumed to be a distinct and

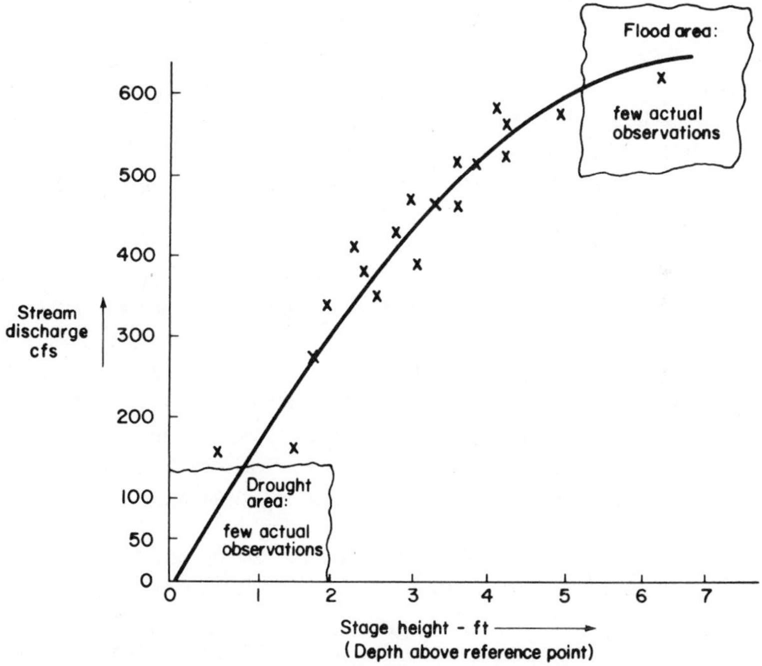

FIG. 2-1. Typical rating curve for a particular stream station.

[1]Refers to significant change in flowing stream characteristics or in its use by society.

separate volume of streamwater in which the characteristics remain mainly the same.) Integrating the various loads with this time of passage affords a measure of the pollution load satisfied or of the bacterial survivals. In addition, cross sections provide the mean channel depths applicable to each reach, which are important in computing reaeration rates and extent of aeration (as you will learn in Chapter 5). They also provide the variation in mean velocity through various reaches, from which it is possible to evaluate the potentialities of sludge accumulation and scour. Other means of computing average velocity, such as dye additions, floats, salt additives, etc., are also useful in certain situations.

Sources of Pollution

Sources of pollution should be determined from an onshore survey. Municipal pollution is readily determined from tributary population but is better established by actual sewer flow measurements. The industrial waste load is usually the most difficult, and entails a separate water survey of major industries. Industrial waste is the subject of an entire textbook by the author (Nemerow, 1978). Septic tank and drainfield effluents represent nonpoint discharges and are often quite significant. Their effect on the stream is usually quite constant in any event, and is often considered as a background level of pollution. Agricultural drainage yielding mineral and organic nutrients should be recognized and accounted for in the sampling. The same is true for temperature rises, which often follow the discharge of cooling waters from power plants or industry.

Immediate oxygen demand, or the oxygen used up in only 15 minutes of incubation, should be differentiated from BOD. The fraction of BOD in suspended, dissolved, and colloidal forms should be ascertained. In addition to 5-day BOD values (see page 1 of Chapter 3 for further explanation of BOD), 20-day values should be run for the major industries that are suspected of having unusual wastes. This 20-day or long-term BOD value increases in significance as streams are delayed, by impoundments and reuse, in their passage to the seas. It may also be necessary to employ different dilution waters to measure accurately the effect on the stream during critical conditions, since mineral water content may be vastly different at different streamflows. A reasonable measure of pollution load can be obtained from a direct survey at the sources, using the BOD of rivers only as a check against the onshore load computations.

Urban Runoff. During and following rainstorms, urban runoff contains significant quantities and types of contaminants. Because of the great variation in land use and population density, rainfall happenings can affect a watercourse's pollutant level from dilution to concentration of specific contaminants. Wanielista (1977) et al. suggest that these wide variations in pollutant types and quantity from nonpoint sources reduce predictability of hydrological effects of runoff. Their experimental results do show significant quantities of heavy metals (iron, lead, zinc, and manganese) coliforms, phosphorus, and nitrogen in urban stormwater. "First flush" analyses also show an increase in suspended solids, BOD, COD, coliforms, and nitrogen in the runoff. As expected, they also found a general decrease of pH, and of total dissolved solids, specific conductivity, alkalinity, hardness, and inorganic carbon, during rainy periods compared to dry periods.

Stream Sampling

In fresh water the types of samples collected are usually tested for both short- and long-term BOD, dissolved oxygen, pH, temperature, and (if bacterial quality is under consideration) coliform density. More recently, stream analysts have become more concerned with the "total" stream picture and have included such analyses as toxic chemicals, persistent, organic or refractory matter, phosphates, alkyl benzene sulfonate (ABS), etc. The factors are more completely discussed in Chapters 3 and 7. It naturally follows from Chapter 1 that further biological parameters be included. However, some aspects of sampling are presented here.

Location, Number, and Intensity of Sampling. Rivers are dynamic systems and are subjected to much variation. A few locations with sufficient numbers of samples to define the results in terms of statistical significance are much more reliable than many stations with only a few samples at each. Velz (1950) believes it is better to concentrate collections during a relatively short interval (about one week) with intensive sampling when the river regime is stable, that is, during a steady hydrograph, rather than to attempt to sample all conditions by grab samples here and there, now and then, during several stages of river flow—which usually defines no condition, requires a longer study period, and may be very misleading. Averaging of such noncomparable results is a dangerous procedure. A week of intensive sampling is usually adequate to provide 16 to 20 samples at each station only when a steady flow is established. The most probable season for the occurrence of such periods of steady flow can be ascertained from statistical analysis of past flow records. However, one must also be certain that the normal onshore point pollution sources are also being discharged during the intensive sampling period. In addition, steady flow periods of streamflow are not easily identified until after they have passed.

Sampling in Relation to Source of Pollution and Tributaries. The investigator should establish a station some distance downstream from the point of pollutant entrance, where dispersion throughout the section has been reasonably complete. "Sampling stations should not be located immediately below a junction of a tributary. It is better to locate the station on the main stream above the junction and to establish a secondary station on the tributary just above its mouth. Otherwise the sampling station on the main watercourse should be located sufficiently downstream to ensure dispersion through the cross section." (Velz, 1950, p. 670) The actual distance downstream varies with each situation. A "test run" or "onshore" observation beforehand will usually be ample to indicate the correct location of the station(s).

Sampling in Relation to Physical Characteristics and River Developments. Rivers which are either wide, deep, or nonuniform in flow and waste distribution must be sampled at different points transversally in the channel, which greatly adds to the number of samples required to obtain a reliable mean for the section. The common practice has been to select as many as three equidistant points across the river and sample each of them at the .2 and .8 depths if they are over 10 feet deep (otherwise one sample at .6 depth would be sufficient). These depths were chosen to obtain an average velocity which would truly represent the velocity in a vertical cross section. (See diagrams in Fig. 2-1A and Fig. 2-1B and further discussion in Chapters 2 and 3.) Because of density differences, currents, and the physical peculiarities of certain wastes, multidepth sampling should be encouraged. Certain sampling points can be eliminated after the first few collections, when the results show no significant difference from another sampling under all conditions of streamflow and onshore discharges.

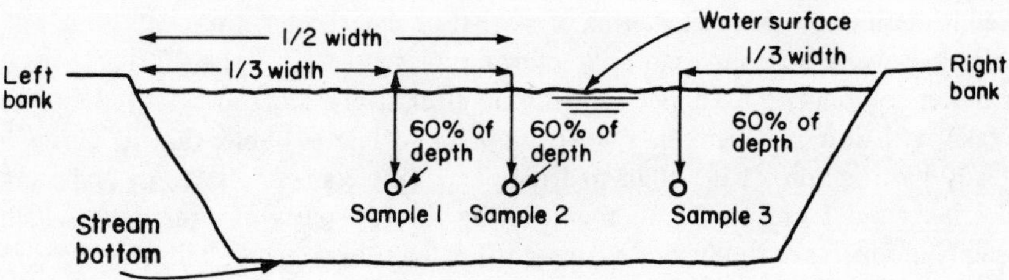

FIG. 2-1A. An example of selected sampling points for a deep, wide, stream with a uniform depth.

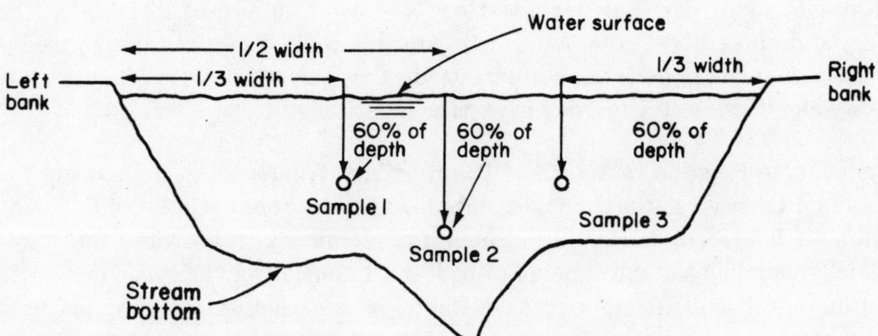

FIG. 2-1B. An example of selected sampling points for a deep, wide, stream with a nonuniform depth.

The construction of various types of hydraulic structures in rivers can have significant effects on the quality of water downstream of the structure. A task committee (1978) listed various types of these hydraulic structures along with their water, land, and air environmental effects. These are shown in Table 2-1A.

The committee concluded that impoundments seem to improve conditions of downstream water quality. Some exceptions noted were the adverse effects of temperature, dissolved nitrogen, and lower dissolved oxygen, on recreation and fisheries.

Sampling in Relation to Waste Discharge Practice. Consideration should be given to operating cycles of industries which are located above sampling points. Some industrial plants operate 7 days a week and 24 hours a day, constantly repeating the same manufacturing processes; on the other hand, other plants only run 10 hours each day and shut down for repairs and "cleanups" during the weekends. Stream samples will reveal vastly different conditions depending upon the type and operational characteristics of the industrial plant discharging above the sampling points. Some industries discharge wastes according to streamflows and/or company production cycles of predetermined policies. Sampling points should be selected sufficiently below industrial waste entrance to afford an opportunity for adequate mixing but not far enough for biological or chemical

TABLE 2-1A. Construction of Example Matrices
[After Task Committee, 1978.]

	Environmental effects		
Hydraulic structure (1)	Water (2)	Land (3)	Air (4)
Reservoir	Salinity	Salt	Evaporation
Selective outlet (epilimnion)	Phosphate	Sedimentation	Humidity
Outlet energy dissipator	Nitrogen	Erosion	
Spillway intake channel	Iron	Slides	
Spillway energy dissipator	Manganese	Terrestrial animals	
Power plant tailrace	Dissolved oxygen	Plants	
Bridges	Supersaturation		
	pH		
	Turbidity		
	Temperature		
	Sedimentation		
	Fecal coliform		
	Phytoplankton		
	Zooplankton		
	Fish		
	Benthic organisms		

degradation. If complete mixing does not occur, several samples should be taken at each downstream point to reveal the complete contamination picture.

Abnormalities in Natural Purification Characteristics. Unusual and discontinuous stream conditions often influence the location of stations and number of samples to be collected. A test run will usually determine the general characteristics before the location of sampling stations is decided upon. Four major abnormalities may be encountered: (1) immediate oxygen demand (IOD); (2) sludge deposits; (3) biological adsorption on the streambed; and (4) algae.

Immediate oxygen demand (IOD) (caused by chemically reduced compounds such as H_2S) would be detected if an industrail waste survey were made prior to the sampling, and the location of sampling stations could be planned to reflect the abnormality. Such a situation is shown graphically in Fig. 2-2. In this case the IOD of the waste causes a steeper, rather immediate oxygen sag with a more rapid subsequent recovery. The curves are somewhat similar in shape, since both wastes contain oxygen-demanding matter; however, the one with IOD exerts its demand sooner and over a much shorter period. The exact location of the bottom of both oxygen sag curves would depend upon many stream and wastewater variables, such as concentration and quantity of oxygen-demanding matter and streamflow.

Sludge deposits may produce similar abnormalities. The greatest abnormality is associated with pools of detention periods of a few hours with a flow-through velocity sufficiently low to permit deposit and accumulation. The difference between what is normally expected in oxygen sag with the short time of passage and that associated with the demand from sludge accumulation is then pronounced. The sag curve is similar to that of IOD, except that recovery is usually slower and doesn't saturate as rapidly (see Fig. 2-3).

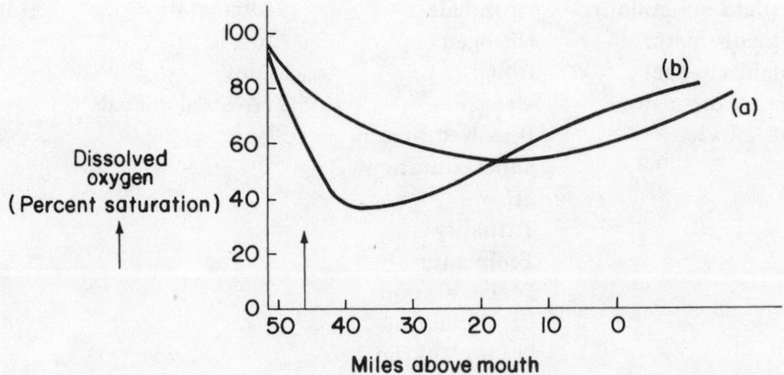

FIG. 2-2. Effect of BOD on oxygen sag below a source of waste entry. (a) = computed profile with normal demand; (b) = computed and observed profile with a portion of total load as IOD.

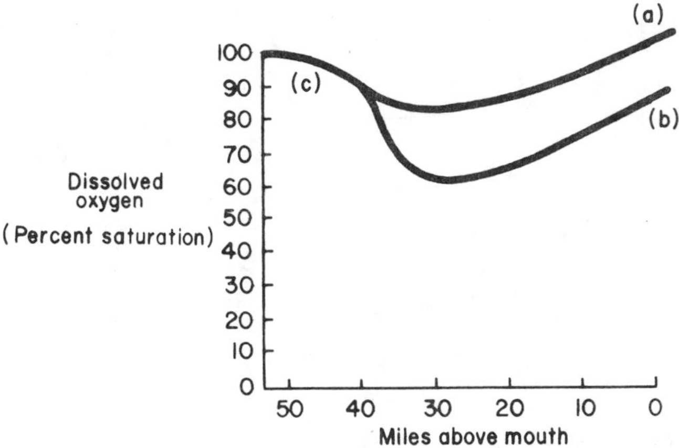

FIG. 2-3. Effect of sludge deposits on oxygen sag. (a) = no sludge deposit; (b) = sludge deposit; (c) = wastewater added here.

Growths of the bed extract and accumulate organic matter as BOD from the overflowing water. The action is similar to that taking place in a trickling filter. This accumulation produces a very sharp drop in dissolved oxygen far beyond what can be accounted for by the time of passage through the reach. The demand in excess of the normal rate results from a slime growth accumulation in excess of one day's flowing load. Usually biological adsorption on streambeds does not occur when depths are more than three or four feet. In deeper streams any bottom growths would tend to be anaerobic and nonadhering. The abnormality of biological absorption may not always be detected, as sludge deposits are, as it is influenced by excessive scouring stream velocities which may remove the growths completely. When the BOD and DO curves show rather steep drops below a source of pollution, this abnormality can be suspected (see Fig. 2-4). Both BOD and DO reduce more than expected, since growths adsorb organic matter and create areas of intensive biological activity, which in turn extract more dissolved oxygen than normal from the flowing water.

Algae during the daylight hours give off oxygen as a by-product of photosynthesis; during the night they do not and may even utilize oxygen in respiration. When algae are involved, the problem of stream analyses becomes highly complicated. Sampling is essential several times during the 24-hour daily cycle. Variations due to algae production are most pronounced on sunny days; cloudy, dark weather may result in conditions which can be expected without algal contribution of oxygen. In deep rivers the temporary surface effect of algae may be avoided by taking samples at lower depths. The photosynthetic effect of algae is most noticeable from around noon to 3:00 p.m., and least influential or detectable just prior to daybreak (however, algae respiration may add to the deoxy-

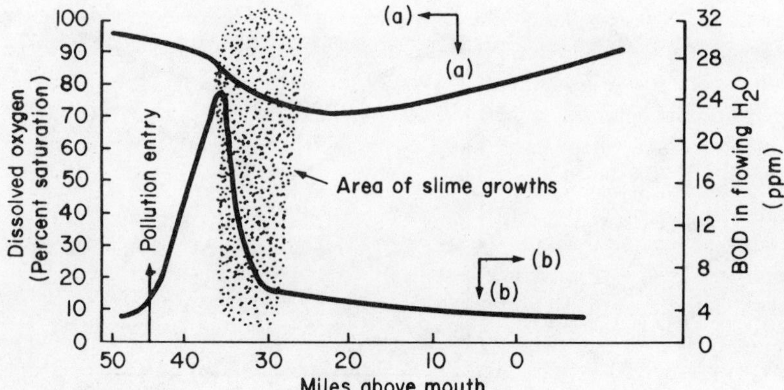

FIG. 2-4. Effect of slime growths in shallow streambeds on oxygen sag and BOD. (a) = typical oxygen sag with slime growths; (b) = typical BOD with slime growths.

genation effect of bacteria; see Fig. 2-5). During the day, photosynthesis by algae ($+O_2$) exceeds respiration ($-O_2$). At night, respiration ($-O_2$) exceeds photosynthesis ($+O_2$). Both occur to some extent during both day and night periods.

Sampling Tidal Waters. Sampling tidal waters is becoming much more important today, since greater quantities of pollution are moving toward the mouths of larger streams. Many of these larger streams terminate in the oceans, where they are subjected to the effect of tidal currents. Such rivers as the Delaware,

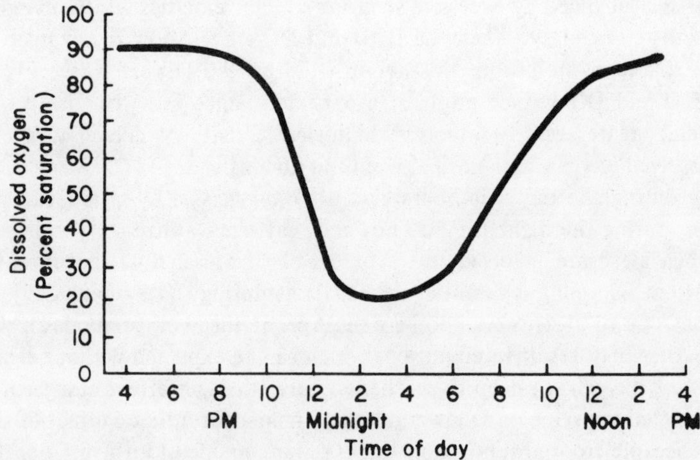

FIG. 2-5. Effect of algae growth in oxygen sag at one location in stream.

Hudson, and San Joaquin, which flow into estuaries, are subject to the moment-to-moment variations in tide throughout any given day. These variations are associated with the movement of the moon around the earth. In addition, Velz (1950) describes the day-to-day variations which depend upon the position of the moon relative to the earth and the sun. The most noticeable variation from day to day is that related to the moon's phase, which has a cycle of $29\frac{1}{2}$ days. During the new and full moons, the range (level above mean low water) is considerably above the average; during the first and third quarters, tide levels are substantially below the average. To add to the complexity of predicting the effects of pollution in tidal estuaries, the moon's declination, which has a cycle of $27\frac{1}{3}$ days, induces diurnal inequality between the morning and evening tides. There are other long-term factors, such as wind and physiographic features, which can radically alter the normal tide pattern. The dominant influences, however, are the daily or semidaily cycle, the cycle of the moon's phase, and the cycle of the moon's declination. Because of all the complications of mixtures of fresh and salt water, random sampling in tidal waters is of little value. Controlled sampling, taking into consideration the major variables, is required. Salinity, the key to the position and division of land and seawater, is a basic measurement in sampling tidal waters. Runoff now has a double significance, because land–water variation also affects ebb–flood translation and the extent of seawater intrusion. It becomes doubly important, therefore, to control these influences by confining sampling to a period of steady runoff, allowing adequate time to ensure establishment of a steady-state regime. The second variable to control is the daily variation associated with the moon's phase. Fortunately, this is a periodic phenomenon which is predicted and obtained in the tide tables; thus sampling definitely can be planned to coincide with one distinct phase. Figure 2-6 illustrates the importance of such concentrated sampling in polluted waters, as the high-range phase in tide produces a distinctly different quality level from that of the low range. If samples were collected on only one tide phase, the true variation in streamwater quality would not be obtained. Obviously, samples from these two phases cannot be lumped, in a statistical sense–it being much better to define the two extremes, and from these to draw conclusions about transitional phases. In comparing quality of such waters, results are comparable only when samples are taken during comparable ranges in tidal phase. For a more detailed treatise of tides as they affect estuary analysis, see Section 2-5.

Depth of Receiving Stream. The depth of the receiving stream may be extremely significant in the location and extent of oxygen sag below a source of pollution. It had been assumed that the rate at which oxygen dissolves in water per unit area of surface per unit of time is independent of the depth, but in fact, on the contrary, a shallow stream is likely to be more turbulent than a deep one. In that case, the rate of solution would be greater in a shallow stream, owing to

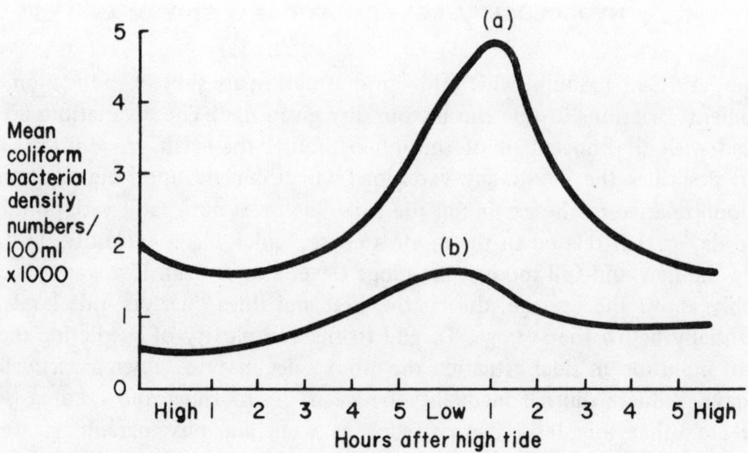

FIG. 2-6. Composition of samplings at large range and small range in tidal phase. (a) = large range in tidal phase; (b) = small range in tidal phase.

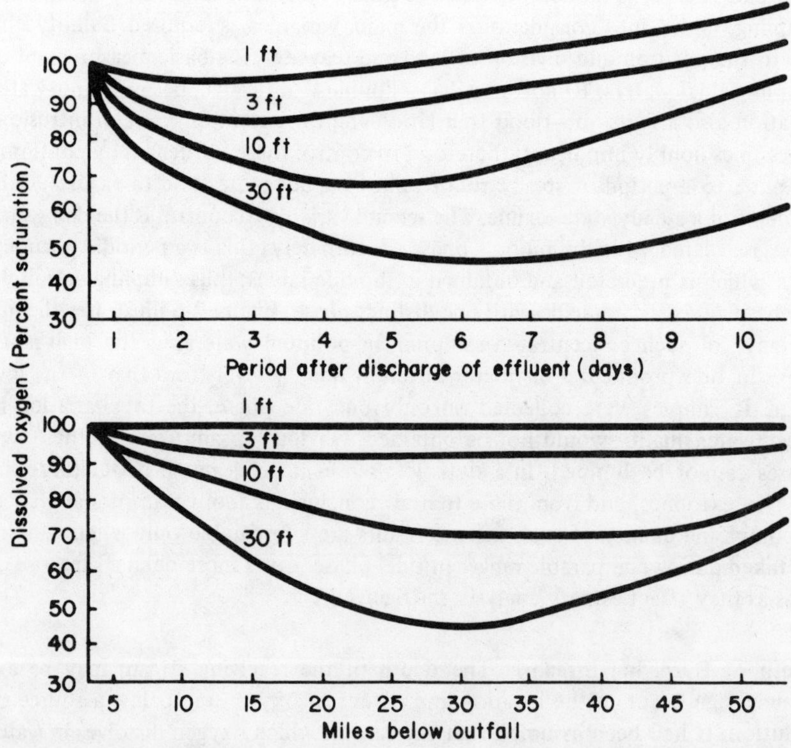

FIG. 2-7. Effect of depth of a stream on the distribution of dissolved oxygen brought about by discharge of a given effluent to it. (Depths shown are those of a hypothetical given river when DO is measured. All curves are obtained from a single river, with constant width and discharge; only the velocity varies.)

the effective increase in surface area caused by turbulence, thus accentuating the effect of depth on the dissolved oxygen found in the water. Since a deep river must flow more slowly than a shallow one with the same width and discharge ($Q = D \times W \times V$), the effect of depth is more pronounced when dissolved oxygen is plotted against distance from the point of effluent discharge[2] (see Fig. 2-7). Naturally, in shallower streams the velocity and reaeration are generally greater, and hence influence the oxygen concentration positively.

2-2. FACTORS AFFECTING STREAM "SELF-PURIFICATION"

Velz (1947) states that "fundamental to any stream pollution survey and abatement program is the evaluation of the factors influencing self-purification. No two streams being the same, the individual characteristics of drainage area, topography, climatic setting and runoff pattern must be determined, which in turn are collectively reflected in the natural self-purification capacities of the stream."

Velz relates the self-purification phenomenon to the business of finance. BOD, being the oxygen required for stabilization, may be considered as liabilities/outstanding debts, and DO as assets. The solvency or condition of the stream, therefore, depends on the relation of liabilities to assets.

To quote Velz,

> The amortization schedule of the BOD debt is peculiar, but very sound. Mother Nature is an exacting banker, insisting upon a short amortization (payback) period with no postponement of the evil day by putting off payments until the end. She will not tolerate equal daily installments; rather, she insists upon heavily unbalancing the schedule, with the larger payments coming due in the early periods. And then, to make doubly sure of payment, when temperature rises and business accelerates, she shortens the amortization period. And if anyone should be so foolish as to go into debt, beyond the capacity to pay, she sends in the wrecking crew, the anaerobes, who literally tear block from block, extracting her payments and reducing the stream to a foul putrescible ruin.
>
> Similarly, there are characteristics peculiar to the income or asset side of this business. There are but two dependable sources of income, oxygen contributed from increments of stream runoff along the course of the river and oxygen absorbed by reaeration from the atmosphere.

Velz omitted algal production of DO because it was relatively insignificant in 1947, our lakes and impoundments not having been overly fertilized with algal

[2] "Discharge" is understood to be a given volume of water or wastewater flowing in a given period of time.

nutrients at that time. Also, our awareness of this factor and its meaning, as well as some of the causes, has increased in the last decade. There is, however, a limit on the assets, which is the saturation level of water exposed to the atmosphere. This level decreases as temperature increases, thereby resulting in a maximum strain on the stream during periods of low flow and high temperatures.

The liability factors are temperature, time of passage, pollution load, and effects of sludge deposits. At normal summer temperatures (20°C) practically the whole oxygen demand (99%) is paid in 20 days; at 10°C the amortization period is extended to about 32 days, whereas at 30°C debts must be paid in about 13 days.

The time of passage of pollution is a liability in that it is dependent upon volumetric displacement of river water by runoff. Alteration in the natural flow, such as a dam or a storage pool in the zone of pollution, will increase and alter the time of passage, and hence increase and shift the depth amortization burden along the river course. While impoundment increases the volume of the debt paid it also increases the volume to be reaerated. The net result of this will depend upon which effect is greater, additional deoxygenation or additional reoxygenation.

Artificial storage above the zone of pollution, operated to augment low runoff, may increase the critical dry weather flow and thereby shorten the natural time of passage, and thus reduce payments required for debt amortization.

The debt itself, or the pollution load, is a liability that is easily understood. The effects of sludge deposits are determined by the hydrological characteristics of the river. For deposits to occur, river velocity and turbulence must be reduced to a point where settling is possible. When sludge deposits are very shallow and evenly spread, one notices very slight effect on the normal amortization rate. If, however, the runoff and velocity conditions are such that heavy localized deposits occur, the amortization may be affected appreciably by causing a heavy oxygen demand, especially in periods of transition from spring to summer. This effect is due to a sudden increase in temperature combined with unamortized sludge accumulated during the winter, which, however, seldom occurs because of the scouring action of the spring floods.

One can list the assets as dilution, reaeration, algae (also a liability), nitrate oxygen, and negative DO (oxygen-requiring compounds which can be satisfied by oxygen of the air rather than the water). The first two are the most important.

Dilution by runoff can appreciably increase the "cash on hand." The reaeration depends upon the physical and hydrologic characteristics of the river course, which can be grouped as mean water temperature, mean depth, occupied channel volume, and stream turnover. In Chapter 5 we discuss reaeration in more detail.

The three remaining assets are less important, but must be included. Algae can add oxygen but may also be a liability when they start dying off. Stored DO in the form of nitrates is a reserve fund which should not be drawn from but considered as a safety factor in the event of the complete exhaustion of available oxygen. Bacteria will utilize the oxygen in nitrates after dissolved oxygen is exhausted and before they use the oxygen from sulfates to produce odors. Negative DO is a liability in that hydrogen sulfide results from sulfate reduction, which, if not accounted for, remains in the dissolved state in the water and may cause an extra burden on resources downstream. If, however, the hydrogen sulfide is lost to the atmosphere where the debt is paid by the oxygen of the air, such a loss of negative DO might be considered as described by Velz, as "bankruptcy proceedings where at least the operation in the red is wiped from the books and the river can start again from scratch." In Chapter 3, Section 4, we consider all of the oxygen donors and users in considerable detail.

It is evident that an important part of a stream pollution survey consists of obtaining all the pertinent hydrological features of the stream in question.

When toxic effects of certain industrial wastes are absent, the natural purification of a stream can be defined as its natural ability to stabilize organic matter without seriously affecting its dissolved oxygen. This ability, so important in maintaining the biodynamic cycle of a stream, is a complex function of many variables, one of them being the hydrologic characteristics. The volume and rate of flow, the type of flow, the physical features of the stream channel, and the drought probability are some of the factors affecting a stream and its natural self-purification.

Some of the other specific factors affecting the self-purification of a stream include streamflow, sunlight, temperature, sedimentation, currents, and other physical properties. The currents, when present in degree, will afford a complete mixing of the sewage with the water, thus dispersing the organic matter and preventing sludge deposits; in addition, they will speed up the aeration process. It should be noted, however, that a current may reduce the recovery time, but only at the expense of affecting a longer stretch of the stream.

The behavior of a shallow and turbulent stream in its process of self-purification was investigated by Kittrell and Kochtitsky (1947) using a stream 42.7 miles long. About 25 percent of this mileage consisted of rapids, and 75 percent of shallow pools with an average depth of 3.60 feet at low water stage. The average surface slope at low water stage was 2.96 feet per mile, and the average velocity during the study was 0.70 feet per second. The one major tributary increased the stream discharge in the section about 10 percent. Many smaller streams contributed to the additional increase in discharge from 6.20 cfs at Mile 42.7 to 880 cfs at Mile 0.0, while control of streamflow at a small upstream dam caused the discharge to fluctuate nearly 40 percent in 24 hours.

During the investigation of long-term BOD, Kittrell and Kochtitsky obtained results suggesting that the utilization of organic material as bacterial food may become more rapid as the concentration of food increases; the investigation of the BOD in the stream showed an 80 percent reduction in one day. The investigators believed that sedimentation alone did not account for the rapid rates of purification observed and that the biological slimes found attached to the stream bed, acting in a manner similar to that occurring in a trickling filter, were an important factor in the rapid decline of BOD. The turbulence of the water, offering frequent contact with the biological slimes, caused and enhanced the adsorption of the organic matter, as well as increased oxygen adsorption from the atmosphere.

In studying the dissolved oxygen, the investigators (Kittrell and Kochtitsky, 1947) calculated reaeration coefficients for various stream sections and found that these coefficients in similar sections of the stream lead to the conclusion that large portions of the bottom deposits decompose and escape without utilizing dissolved oxygen from the water (an acceptable possibility, especially in the case of rapid flow). The authors support the logic of this suggestion by citing results obtained by Symons and Fair, Moore and Thomas (1941), noting that one of "the chief constituents of the major industrial waste in this stream was an organic compound that could decompose anaerobically to carbon dioxide and water without the use of any oxygen other than that contained in the compound" (Kittrell and Kochtitsky, 1947).

The results obtained in the turbulent stream were compared with those obtained by a United States Public Health Service for the Ohio River, using this river as a prototype of a hypothetical sluggish and deep stream. In the turbulent stream, the BOD showed a reduction of 109.5 percent of the added BOD in one day, in contrast with a reduction of 22.5 percent in one day and 53.4 percent in three days in the hypothetical sluggish and deep stream. For the dissolved oxygen, the turbulent stream reached its minimum content of 21 percent saturation in 0.25 day, recovered to 58 percent saturation in one day, and obtained its maximum degree of saturation of 82 percent in 2.25 days. In contrast, the hypothetical stream descends to 7 percent in one day and to total depletion in 1.5 days.

Total depletion continues for 1.5 days and recovers to only 10 percent saturation in 3.5 days. This process indicates that a turbulent stream has a much greater assimilative capacity for organic pollution without excessive dissolved oxygen depletion, due to its greater aeration capacity. However, its ability to reduce rapidly high concentrations of BOD by a possible adsorption of organic material on the slime beds, tends to counteract its favorable reaeration capacity. Deposits on the streambed tend to concentrate the oxygen demand in those sections where they occur, adding their DO requirements to those of the BOD present in the water.

Velz (1949), one of the country's most respected stream analysts and statisticians, describes rivers as individualistic, with natural purification capacities that vary radically from stream to stream and from reach to reach along a river course. He sees a river not as static but rather as living, dynamic, and constantly responsive to the laws of biological change and the vagaries that apply to living things. Further, each river is sensitively responsive to hydrologic fluctuations, which follows the laws of chance and probability.

Natural self-purification is not a fixed quantity, but rather a range in variability of capacity associated with biological and hydrological changes. Failure to recognize this dynamic character of natural purification results in rigid stream standards which cannot possibly be maintained under the normal patterns of stream variability.

Streams cannot be generalized: hence, stream standards cannot be applied wholesale to all rivers without regard to their widely varying natural self-purification capacities. Furthermore, standards cannot be rigidly applied to a particular river, as though self-purification capacity remained constant along the course of a stream. And worse than rigid stream standards is the practice of arbitrarily specifying quality requirements of the effluents of sewage and waste treatment works, regardless of the varying natural purification capacities of different bodies of receiving water. No generalization and no short-cut will solve this problem. One approach is to apply rigid standards to be on the safe side, rather than overtaxing the stream. The essence of stream control and sewage and waste treatment design is the evaluation of factors influencing self-purification in each specific case. In Chapter 7 we consider some of the arguments for and against both stream and effluent standards.

Measures of stream condition, to be of value, must be expressed in terms quantitatively related to water quality, reproducible under similar conditions, and related to a measure of efficiency of sewage and waste treatment. Velz (1949) felt that the best such single measure was the quantity of oxygen required to stabilize pollution by natural biological processes of decay, or biochemical oxygen demand. Where pollution is discharged into a stream, the oxygen requirements to satisfy this demand are obtained from the oxygen dissolved in the river water. If the demand for oxygen is excessive in relation to that available, the DO is exhausted and the stream goes into complete bankruptcy. Thus, the BOD of pollution may be regarded as liabilities—outstanding debts—and the DO of the river taken as assets. As mentioned previously, stream solvency depends on the relation between rate of amortization of liabilities and the rate of income from assets, where oxygen instead of dollars is the medium of exchange.

Velz (1949) defends the practice of considering each stream on an individual basis in order to evaluate self-purification. Two factors, sludge deposits and drought probabilities, constitute major reasons for variations from stream to stream and day to day of the oxygen profile. We have already described the biological phenomenon of benthal decomposition in Chapter 1, Section 4. Suspended solids subject to deposit along the course of a stream or in pools behind dams can produce a wide variety of effects upon deoxygenation and, hence, upon resultant DO. In a natural stream the limiting factor which determines the occurrence and accumulation of sludge deposits is the velocity of flow at which scour commences. Shields (1939) has defined this critical velocity as

$$V_c = \left[8\beta g(s - 1) \frac{D}{f} \right]^{1/2}$$

where V_c is the mean channel velocity at which scour commences; β is a constant for a particular sludge situation; g is the gravity constant; f is a friction factor similar to the Weisbach-Darcy value; s is the specific gravity of the particle; and D its diameter.

From what is known of the performance of grit chambers, β for fresh depositions of organic matter is 0.06, and a critical velocity of about 0.6 fps is required to scour organic matter granular in shape with diameters up to one millimeter. Under natural stream conditions, greater turbulence due to a higher value of f would allow an increase in β up to 0.22 with comparable results. However, in the event channel velocities fall below 0.6 fps over extended periods, organic settleable solids will deposit and, in the process of accumulating, will cohere and compact, resulting in an increase in β. If β is assumed in such instances to increase to 0.8, the velocity required to induce scour of the accumulation would increase to 1.15 fps, or practically double that where accumulation and compaction did not occur. Also, during accumulation, deposited particles no longer remain discrete, since they coalesce to form larger aggregations; hence, the size of sludge particles to be scoured increases and the velocity required to induce scour will be further increased. To offset these tendencies, however, gases produced during digestion of organic deposits will agitate the accumulation by ebullation and lift some particles into the flowing stream. Thus, fresh organic deposits undergoing digestion may readily be scoured, probably at velocities between 0.6 and 1.0 fps.

For resettled partially or wholly digested material for which active gasification has subsided, the previous tendencies adverse to scour hold, and higher velocities are required such as occur during freshets or flood discharges. Taking an extreme situation where β is 1.0 and organic deposits are stabilized material, a velocity to scour a 1 mm particle would be approximately 1.3 fps.

A unique opportunity to observe such critical balances between channel velocity and deposit and scour was afforded in studies of the Kalamazoo River made for the National Council for Stream Improvement (Velz, 1949). Two deposit areas separated by an intervening clean stretch were observed. A study of river hydrology and channel cross sections revealed the following relations. For the reach in the river in which no deposits occurred, a channel velocity of 0.6 fps would not occur until the runoff declined to about 130 cfs.

From probability studies it was determined that a runoff as low as this could be expected once in 20 years; hence, for practical purposes sludge deposit and accumulation, according to theory, were rarely permitted. The cross-sectional soundings through this reach revealed a complete absence of organic matter and a clean sand and gravel bottom.

For the upper deposit area located below the city of Kalamazoo, a channel velocity of 0.6 fps or less is attained at a runoff below 700 cfs, which occurs frequently each year; consequently, it can be expected that at times, sludge of organic character will accumulate. It will be further noted, however, that with runoffs between 700 and 1,500 cfs, which also occur frequently each year, velocities would range between 0.6 and 1.0 fps, sufficient to scour fresh, actively decomposing sludge. An examination of cross sections taken during a period when the runoff was around 835 cfs indicated the presence of organic sludge, but no large accumulations.

Further downstream in a pool area, a large accumulation of relatively stabilized material was observed. For this area, channel velocities of 0.6 fps or less occur when the discharge is below 1,500 cfs. Flows below this level occur for protracted periods, giving opportunity for extended accumulation. Thus, this zone receives the deposit by-passing or scoured from the upper area. Scour velocities in the pool area occur only during freshets and the flood season.

Such a critical balance of sludge deposits and scour with river runoff gives rise to a very sensitive dissolved oxygen profile with periods of very sharp, deep drops, when runoff is such as to permit deposit and accumulation.

Accumulation and Demand of Sludge Deposits. To experience oxygen depletion in the overlying water in excess of that which is normally expected, by demands satisfied in the flowing stream, sludge deposits not only must be able to occur, but they must be able to accumulate. The extent of the demand exercised is primarily a function of the level of accumulation. The level of this storage, in turn, is a function of the rate of biological decomposition and the length of period during which deposit may take place without scour. The rate of the former was discussed in Chapter 1, Section 4, but, in practical application, the latter is controlling, and the exact rate of decomposition is of less importance. Unless interrupted by scour, sludge accumulation continues until the daily BOD demand exerted by the accumulation equals the BOD of the daily deposit added. A means for practical evaluation is afforded by Streeter's unimolecular equation

$$L_d = \left(\frac{P_d}{2.3k'}\right)(1 - 10^{-kt})$$

where L_d is the cumulative BOD in pounds; P_d is the BOD added to the deposit in pounds per day; k' is the specific rate of decomposition of the deposit; and t is the time of accumulation in days.

Depending upon hydrologic conditions, the daily demand from sludge deposits may vary from a small fraction to more than the BOD of the daily deposit added. The extent of the distortion to the DO profile from these sludge deposits, in addition to varying demand, depends upon the time of passage of the flowing water through the zone of deposit. If the time of passage is long—10 days or more—the difference in deoxygenation is slight, because a large proportion of the BOD would be satisfied in the flowing stream, regardless of the occurrence or nonoccurrence of deposits. On the other hand, where time of passage is much shorter, the deoxygenation effect of sludge deposits greatly exceeds that which takes place in the flowing stream in the short period, and a sharp drop in the DO profile occurs. The benthal demand itself is not greater during shorter passage times, but its proportion of the total oxygen demand is greater than during longer passage times. However, in such instances, the zone of deposit serves as a natural treatment device, reducing the debt load for downstream reaches and thus raising the downstream DO profile above that which would occur if the sludge were deposited later.

Velz (1949) states that no generalized expression such as the oxygen sag equation is applicable for refined analysis. Obviously, the effects of deposit cannot be lumped as an increased deoxygenation constant k. Not only is it necessary to compute deoxygenation and reaeration separately, but it is also essential to deal with separate fractions of the load, integrating the colloidal and dissolved BOD fraction remaining in the flowing water, then separately integrating the suspended solids BOD fraction, dropping out the residuals in the areas where velocity permits deposit. The deposited fraction in turn must be integrated at its decomposition rate over a considerable period prior to the data under investigation in order to obtain the quantity of accumulated storage of BOD.

Drought Probability. Runoff plays a still larger role, for it affects assets as well as liabilities. Drought flows coincident with high temperatures produce the greatest strain on self-purification, as amortization demands are increased while income declines. Since droughts follow the laws of probability, no evaluation of self-purification is complete which does not include a statistical analysis of droughts for each stream. An excellent means of determining one extreme (drought probabilities) is afforded by the application of the theory of extreme values developed by Gumbel (1941) for another extreme (flood flows). The full explanation of the use of the method would require a separate chapter. The following example is offered, however, to illustrate the results of an application and to emphasize how radically the DO profile varies with different levels of drought severity, and how essential it is to recognize and consider drought probability in determining a stream improvement program.

Figure 2-8 contains a set of typical DO profiles that would result from a certain treatment scheme at summer droughts ranging from 1,000 to 337 cfs. Probability studies disclosed that monthly average droughts of severity of 1,000, 750,

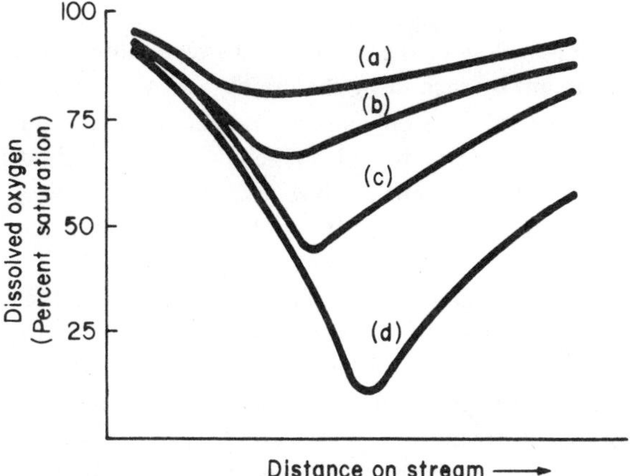

FIG. 2-8. Drought flows affecting oxygen sag. (a) = 1,000 cfs (once in 2 yr); (b) = 750 cfs (once in 5 yr); (c) = 396 cfs (once in 10 yr); (d) = 337 cfs (once in 20 yr).

and 396 cfs may be expected yearly, once in two years and once in five years, respectively. It is to be noted that a drought flow of 337 cfs which occurs once in 20 years will produce a sag which almost depletes the oxygen in the stream. It is noted that each drought flow produces its corresponding DO profile so that obviously, as drought severity increases, the DO profile will decline; but these steeper declines are expected less frequently as drought severity increases. Further, it must be recognized that under the laws of probability there will inevitably occur, on rare occasions, a drought of such severity as to cause stream conditions to fall below a desired level. These are the tolerances which must be taken into consideration in any rational approach to stream control. We usually design pollution abatement facilities for the one-in-ten-year drought flow (or the low flow which will occur in 10 percent of the years). In New York State, the minimum seven-day flow likely to recur once in 10 years is used to establish treatment plant design criteria.

2-3. STREAMFLOW MEASUREMENT

Since biological organisms and the concentration of chemical constituents in streams–including dissolved oxygen–depend to a great degree upon the flow, the measurement of river flow becomes an important part of any survey. Although this information is generally available from the United States Geological Survey, the stream analyst should understand the basic general methods involved in streamflow measurement. Furthermore, many, if not most, of the flows of streams in this country are not continuously measured and recorded. In many cases, the stream analyst must project base flows from other similar drainage basins on a square mile basis to the unmeasured stream under study in order to obtain an approximate flow of the unmeasured stream. Computing the flow at a given point in a stream by measuring the drainage area contributing and average rainfall and runoff conditions is only an approximate method, and then can only be used with "rough" accuracy for small catchment areas. Although the student should consult any of the excellent textbooks on hydrology, or manuals prepared by the U.S. Geological Survey for detailed procedures, a general method is given here.

Briefly, streamflows are computed by accurately measuring the cross-sectional area of a stream section and the average velocity of the water flowing past the section. A current meter placed at a depth of 60% of the way from the surface to the bottom (0.6 depth) is usually recommended for determining the velocity at a given location on a stream.

The specific instrument relation of the speed of rotation of the wheel to the velocity of the water which causes the rotation must be established–a process involving "the rating" of the current meter before using it. In the United States, this normally is done by the National Bureau of Standards in Washington, D.C. A graph plotting the revolutions per second versus the velocity of the stream in feet per second is supplied with the meter. The average stream horizontal velocity thus obtained is multiplied by the average water cross-sectional area to obtain average stream volumetric flow in cubic feet per second (or, meters per second).

Current meters are of two general types, differing essentially in the form of the meter wheel: (1) cup type, differential action, vertical axis of rotation; and (2) screw type, direct action, horizontal axis of rotation. With either type, the speed of rotation of the meter varies with the velocity of the water that strikes it, and the rotation is the measure of the velocity of the water in which it is placed. The Price current meter is an ideal example of the cup type, and Haskell and Ott meters are outstanding examples of the screw type. On the other hand, fluorescein dye, radioactive carbon, salt, or another suitable tracer is used to ascertain an average velocity in an entire stream stretch. The product of the area and velocity yields a streamflow. However, since stream channels are seldom uniform at any section, and since the velocity varies in both a horizontal (from bank to bank) and a vertical (from surface to streambed) direction, incremental subsections are usually used, and true (integrated) cross-sectional areas and velocities are determined. A typical cross section is shown in Fig. 2-9, and data recording and computation are given in Table 2-1B. Subsections of 10-feet widths were selected such that at least 10 equal sections were sufficient to cover the entire cross section from the left to the right bank. Depths of each section were measured at each end of the subsection. Streamflow horizontal velocities were also measured in each subsection at one depth (0.6 of the total depth) or in the case of deeper streams at both the 0.2 and 0.8 depths (Fig. 2-9A). Flow discharges were computed either by the mid-section or mean-section method. The

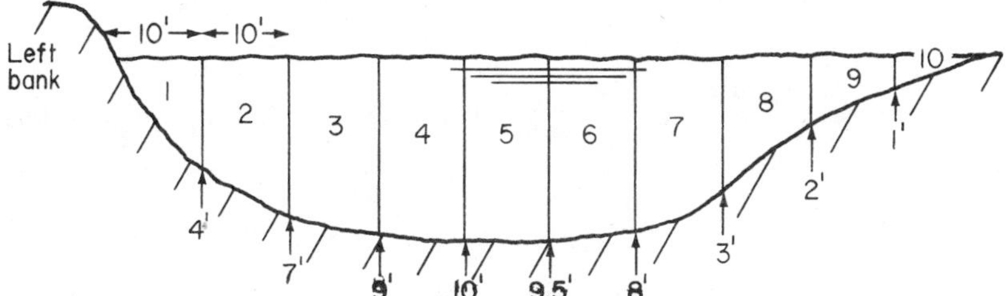

FIG. 2-9. Typical river cross-section for flow measurement.

summation of the product of areas of each subsection and the average velocity at 0.2 depth and 0.8 depth equaled the volume of flow.

$Q_i = A_i V_i$, where Q_i = flow in cubic feet per second of a section, i; A_i = area in square feet of a section, i; V_i = velocity in feet per second of a section, i; and $Q_t = Q_i + Q_x + Q_y + Q_z + Q-$; where Q_t = total flow in cfs, and Q_i, Q_x, Q_y, Q_z, $+Q-$ equal all incremental flows of all sections.

Computation of discharge measurements by the mid-section method is not greatly different from that by the more similar mean-section method. The main difference is that in the mid-section method a discharge is computed directly from each set of observations of depth and velocity and entered on the same line as the station designation, whereas in the mean-section method, sets of observations of depth and velocity for adjacent stations are averaged to obtain a discharge which is entered on the line between the station designations. In the mid-section method (see Table 2-1) the mean velocity in vertical is also the mean velocity in section, so that the column headed "Mean in section" is left blank. Likewise, the depth of observation is also the mean depth, so that the column headed "Mean depth" is left blank. The area is the product of the observed depth times the width, and the discharge equals this area multiplied by the mean velocity in the vertical. Incidentally, with the adoption of the mid-section method of computation, the form for recording discharge measurement notes will be revised to arrange for proper columns and headings.

In the mid-section method, the width is computed as half the total distance from the preceding to the following station. That is, subtract the preceding station (distance from initial point) from the following station and divide the result by two. If there is no preceding station, such as when beginning the computation at the edge of water or at a pier, the width becomes half the distance to the following station. Similarly, if there is no following station, such as when the computation ends at the edge of water or at an abutment, the width becomes half the distance from the preceding station to the station at the edge or abutment.

At an edge section, multiplying a width by zero velocity will result in a zero discharge for the section. Therefore, observations should be taken as closely as practicable to the edges of water. If care is taken in spacing observations, these

TABLE 2-1B. Sample of Discharge Measurement Notes Computed by Mid-Section Method.

Dist. from initial point*	Depth	Observation depth	Revolutions of rotating cups	Time in seconds	Velocity: At point	Velocity: Mean in vertical	Velocity: Mean in section	Width	Mean depth	Area	Discharge
0	0		0			0		3		0	0
6	2.6	.6	7	52	0.30	0.30		7		18.2	5.5
14	2.9	.2	20	54	.81	.68		8		47.2	32.1
		.8	15	60	.55						
22	7.1	.2	25	47	1.16	.95		7		49.7	47.2
		.8	20	59	.74						
28	8.2	.2	30	52	1.26	1.11		6		49.2	54.6
		.8	25	57	.96						
34	9.0	.2	40	57	1.53	1.28		6		54.0	69.1
		.8	25	53	1.03						
40	9.2	.2	40	54	1.61	1.40		5.5		50.6	70.8
		.8	25	46	1.18						
45	9.3	.2	40	52	1.67	1.52		5		46.5	70.7
		.8	40	63	1.38						
50	9.4	.2	40	48	1.81	1.64		5		47.0	77.1
		.8	40	59	1.47						
55	9.5	.2	40	49	1.77	1.70		5		47.5	80.8
		.8	40	53	1.64						
60	9.7	.2	40	44	1.97	1.78		5		48.5	86.3
		.8	30	41	1.59						
65	9.7	.2	50	52	2.09	1.82		5		48.5	88.3
		.8	40	56	1.55						
70	10.0	.2	40	43	2.02	1.82		5		50.0	91.0
		.8	40	54	1.61						
75	10.3	.2	40	44	1.97	1.72		4		41.2	70.9
		.8	30	44	1.48						
78 est.	10.0		est. 0.9 of sta. 75			1.55		1.5		15.0	23.2
PIER		9:50 a.m.									
		10:00 a.m.									
92 est.	6.5		est. 0.9 of sta. 95			.97		1.5		9.8	9.5
95	6.8	.2	30	51	1.28	1.08		4.5		30.6	33.0
		.8	20	50	.87						
101	7.1	.2	30	47	1.39	1.23		6		42.6	52.4
		.8	25	51	1.07						
107	8.1	.2	30	44	1.48	1.34		6		48.6	65.1
		.8	30	54	1.21						
113	8.3	.2	30	43	1.52	1.38		6		49.8	68.7
		.8	25	44	1.24						
119	8.5	.2	40	56	1.55	1.38		6		51.0	70.4
		.8	30	54	1.21						
125	8.1	.2	30	44	1.48	1.32		6		48.6	64.2
		.8	25	47	1.16						
131	7.1	.2	30	47	1.39	1.20		7		49.7	59.6
		.8	25	54	1.01						
139	6.7	.2	25	45	1.21	1.05		9		60.3	63.3
		.8	20	49	.89						
149	6.3	.2	20	43	1.01	.90		8		50.4	45.4
		.8	15	42	.78						
155	5.9	.2	20	48	.91	.68		3		17.7	12.0
		.8	10	49	.45						
	Vertical Well under Water										
155	2.2	.6	20	54	.81	.81		4.5		9.9	8.0
164	1.7	.6	7	56	.28	.28		8.5		14.4	4.0
172	0		0		0			4		0	0
								158		1006.5	1423.2

*Left edge of water at 9:05 a.m.; right edge of water at 10:45 a.m.

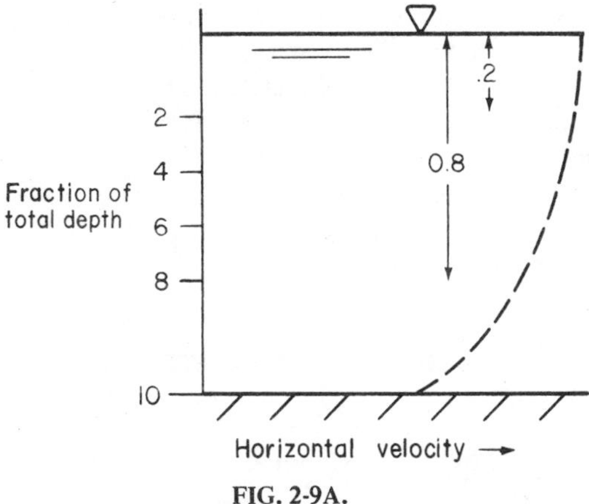

FIG. 2-9A.

small edge sections will be very small proportions of the total measurement. At piers, after a set of observations is taken as closely as is practicable, the depth and velocity at the edge of the pier may be estimated and these estimates entered as regular observations in the same manner as is done in the mean-section method.

If a vertical drop is encountered where a large change of depth takes place, it is treated in the usual manner by listing the station twice and obtaining depth and velocity observations just before and just after the change. This method, in effect, assumes a pier of infinitesimal width at this point in the cross section, and prevents observations on one side of the vertical drop from being used beyond the drop. In the mid-section method these sets of observations are treated as separate stations: one set of observations is used for an effective width halfway back to the preceding station, and the other for a width halfway to the following station.

Field practices for the mid-section method are essentially similar to the mean-section method except that, for most accurate results, any significant changes in velocity should be, theoretically, nearly midway between the selected observation points. The importance and necessity of this exact selection of observation points decreases as the number of sections in a measurement is increased. For instance, in the mean-section method, observation points, theoretically, should be selected *at* any changes in velocity. However, it is not practical to adhere strictly to this exact location of points, and in most cases, with a satisfactory number of observations, velocity variations are compensating. It should be remembered that for both methods the same figure for area will result with a given set of soundings. Accordingly, observation points should be selected to obtain the necessary definition of the river bottom as in the past, with proper attention and emphasis being given to any irregularities in velocity.

A sample of measurement computed by the mid-section method is shown in Table 2-1 and includes measurements at water's edge, at piers, at abrupt changes in profile, and with varying widths of sections. This sample is furnished for the benefit of those unfamiliar with this method of computation, and may be used as a guide, if so desired.

Barret (1975) found that cyclical stream temperature changes can trigger alterations in riverbed form, thus altering the stage-discharge, rating curve relationship of alluvial streams.

2-4. EFFECT OF IMPOUNDMENT ON WATER QUALITY AND QUANTITY

Stream Pollution Caused by Artificial Impoundment Reservoirs for Power Generation

The storage of water within flowing streams (impoundment) is generally considered beneficial for handling larger amounts of pollution downstream, but it can also be quite damaging to the water quality. The procedure involves the storage of water behind large dams, where the water is used mainly for power generation, municipal water supply, and flood control. Large storage reservoirs with controlled discharge provide, in general, a larger amount of water for periods of drought or low flow in the streams, so that the minimum streamflow is likely to be much higher below a large storage reservoir than it was in the stream without or before storage. However, as Ingols (1956) points out, there are at least four reasons why the water quality downstream is markedly affected by storage during part of the year.

1. Power generation usually involves taking water from the bottom of the water reservoir, where the water may be colder but completely devoid of DO, for periods during the summer months, when the total amount of oxygen that can dissolve is less than during the winter. If pollution is added in the stretch of the stream below the dam where the dissolved oxygen has not yet reached the level that is needed to carry the pollution load, the one discharging pollution below the power dam may be penalized if the amount of oxygen depletion is excessive, a consequence which has been created by the reservoir on the stream. If a downstream user is a municipality, the water supply may contain shock loads of stratified layers of precipitated manganese and iron released from the impoundment upstream in the water selected for use in cooling by the power plant. This can be very troublesome to a water treatment plant operator. Iron may be naturally present in streamwater or artificially added in wastewaters, but storage provides an opportunity for oxidation, precipitation, and concentration in layers.

2. A further degeneration in water quality below power dams is brought about by sudden changes in the rate of flow caused by the use of the power

dams for "peaking" operations. While a river flows at a low level, organic matter added as pollution induces the growth of clumps of fungi or biological slimes on the surfaces of stones at the bottom of streams. However, when the flow is suddenly increased, these same fungi are sloughed off from the stones and are distributed throughout the water with an accompanying decrease in the water quality. A rise in water quantity usually means a rise in pounds of dissolved oxygen available and a higher degree of dilution, which is provided for a discharge of waste. However, when flow increases as a result of power generation practice, the water quality may decrease markedly, and the dissolved oxygen then drops. The decrease in quality is usually a result of algal growth which occurs during storage and the warmer temperature of the discharge during high power demands. The sewage fungi are capable of creating a very severe BOD because of the presence of food and the extreme abundance of the biota, indicated by the fungi floating in the water. In general, one thinks of water storage on a flowing stream as improving the quality of a water, but this is not necessarily correct. One must consider also the degeneration in water quality brought about by the power-generation practice in the various government agencies and private power companies.

3. An increase in pollution may accompany a sharp fall in water level. When a tributary stream or pipe discharges large volumes of wastes, a rise in the water level in the large stream may cause the wastes to accumulate temporarily in the tributary. When the level of water in the river drops, there is a large relative increase in the quantity of waste discharged into the main stream. This drop in water quality during a drop in volume of river flow again becomes important in the control of severe damage below a waste outfall.

4. Another type of pollution that is created by electric power generation occurs when power is generated from fuel and rivers must be used to cool the condensers in order to get the most efficient use of the energy produced as heat. This causes a rise in temperature in the receiving water. When large volumes of water are needed, and when the stream, during periods of drought, does not provide this normal flow, reuse of the water by all industrial users may occur, and the temperatures rise to the point where the operation of the plant becomes inefficient and may have to be suspended, or the temperature in the river rises to a point (85°F-90°F) where fish life cannot be maintained. The rate of exertion of the BOD is hastened by the increase in temperature, and where the pollution exists, a higher temperature causes a more rapid drop in the DO. This process again creates a more serious problem for the industry discharging into a stream with a higher temperature than that which is normal to the streams in the area. In addition, waters at higher temperatures are not capable of holding as much oxygen as colder waters, because of the lower saturation levels.

Kittrell (1969) experienced years of the effects of impoundments, both while with the Tennessee Valley Authority and with the United States Public Health

Service. He believes, in connection with storage of water in streams, that "when man alters nature for his own good, he often produces unanticipated results which may or may not be beneficial."

The type of impoundment exerts an important effect on the dissolved oxygen resources of its waters. The main types of impoundments are storage impoundments, main stream reservoirs, and pumped storage reservoirs. Under certain conditions storage reservoirs may function as main stream impoundments, so that the separation by type is not absolute.

Storage reservoirs are usually located on a tributary stream, which frequently has a steep slope. The dam is usually 100 to 125 feet high, and stored water spreads well beyond the former river channel. Cross-sectional areas are large in relation to streamflow, and flow velocities are negligible. Water in this type of reservoir may be retained for several months. This reservoir is used to store water from high surface runoff for release when runoff is low.

The typical main stream impoundment alters the configuration of the stream much less than the storage reservoir. The dam is 60 to 80 feet in height, and may be less. The increase in surface area is small in comparison with that of the storage reservoir. The flow velocities are less than those of the uncontrolled stream, but remain appreciable. Most of the water is replaced in relatively short periods.

The highest quality natural surface water contains enough organic matter to support biological organisms that use dissolved oxygen. Reaeration in the natural stream tends to balance this biological depletion. As the pollution by organic matter increases, reaeration becomes increasingly important.

Reaeration is a function of stream velocity and depth. Impounded waters have a great effect on reaeration. Both increased depth and decreased velocity significantly reduce the reaeration capacity.

Thermal stratification in the storage reservoir resembles the three-strata pattern of natural lakes. After the winter season, the water in the reservoir is cold and has a high density. The warmer water coming into the reservoir has a lower density, and the difference in density between the top and bottom strata is sufficient to prevent wind-induced mixing from bottom to top. Summer stratification persists from April to November. The thermocline, a stratum which separates epilimnion (upper stratum) and the hypolimnion (bottom stratum), results from the nonmixing of these two layers.

Both the physical changes in configuration of the water mass and the thermal stratification following impoundment reduce reaeration rates drastically. The effects vary with the type of impoundment.

In storage reservoirs the increased time of water travel through the epilimnion compensates for the reduced aeration. The relatively quiet waters encourage increased algae growth, which in turn generates oxygen during daylight periods, making up to some extent for the loss of natural stream reaeration oxygen.

The hypolimnion, however, shows very low oxygen content. The depth of this stratum inhibits light penetration, and therefore photosynthesis is negligible.

Oxygen removed by decomposition of organic matter is not replaced. The storage time of water in the hypolimnion may be several months and may permit total oxygen depletion.

Before the development of large valves that could be operated under high pressures, most water was drawn from the oxygen-rich epilimnions. With the development of valves that operate under high heads, the outlets were placed at increased depths. These deep intakes withdraw water containing little or no dissolved oxygen, and discharge it below the dams.

In spite of the absence of oxygen-demanding materials, the recovery of dissolved oxygen in streams receiving impounded waters from storage reservoirs is usually very slow.

Main stream reservoirs lose reaeration capacity but not to the extent found in storage reservoirs. The stratification characteristics of main stream reservoirs are not as well known as those established for storage reservoirs. Limited data suggest that it may not be realistic to expect reaeration in main stream reservoirs to provide large portions of the dissolved oxygen needed to stabilize organic wastes.

Numerous methods for correcting the reduction in dissolved oxygen resources caused by impoundments have been tried, either experimentally or in actual practice on small reservoirs.

The use of multilevel intakes would permit withdrawal of water from the surface stratum regardless of the water surface level. High level penstock weirs increase reaeration. Special design of the tailrace structure to increase turbulence would probably increase the dissolved oxygen picked up at this point. The application of compressed air to the discharge might increase the dissolved oxygen, but would be costly. Stratification may be greatly reduced by circulation of the water. If we could maintain a high quality of water entering an impoundment, we would be able to reduce the algal effect that normally occurs in shallow areas of highly mineralized impounded waters.

Some practical method of preventing or eliminating stratification of large reservoirs should solve a substantial portion of the existing problems of low dissolved oxygen caused by storage impoundments.

Kittrell's text (1969) contains much excellent information of relevance to the hydrological factors under consideration in this chapter. His chapters on sampling frequency, mixing, and waste sources are especially useful.

2-5. STORMWATER MANAGEMENT

Because of changes in storage characteristics of a watershed, urbanization changes the timing of runoff. McCuen (1979) observes that the most obvious change is the decrease in the time-to-peak. If one utilizes a stormwater management basin to store the first portion of storm runoff, one can delay the time-to-peak. This delay may appear to be desirable; however, the basin runoff characteristics may not ever return to those existing prior to stormwater management. In fact, the runoff holding basin may actually create flooding conditions downstream. McCuen illustrates this effect with his studies of Crabbs Creek in Maryland as shown in Fig. 2-9B. He shows here that the stormwater management basin has even less effect on the runoff from a 100-year storm than on the 10-year event. However, the peak outflow from the stormwater basin is 64% greater than the peak discharge before development. The time-to-peak is only four minutes later than the time-to-peak before development. These, according to McCuen, are further indications that storage in stormwater management basins does not exactly replace the natural storage that existed prior to development. There are also indications and some evidence that stormwater holding basins to manage flood flows also increase downstream bed sediment. In addition, there are quality implications and problems with stormwater management basins. For example, basins could be compartmented for separate release at certain times to ameliorate the peak discharge effect, but water qualities from these compartments may differ widely, and some may even be highly contaminating to downstream users.

McPherson (1979), in talking about the substantial synergistic benefits of the use of detention storage for both quantity and quality management, brings up some of the questions remaining.

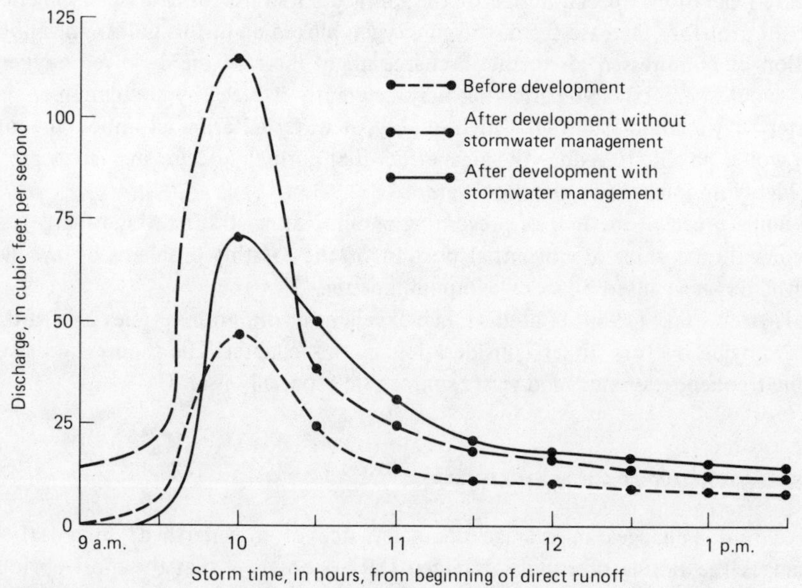

FIG. 2-9B. Crabbs Branch simulated flow at cross section 2 (100-yr storm). [*After McCuen, 5, 1979*].

a. Since available evidence indicates that the effect of urbanization on peak runoff may decrease markedly with increasing rarity of events, why should we expect benefits from storage designed to contain a 10-year event to be equally useful during a 100-year event?

b. Since some studies have indicated that the peak flow attenuation effect of detention storage distributed over a catchment diminishes as the catchment tributary area becomes greater going downstream, how do we reconcile storage shares for pollution abatement and flood control in the benefit expected?

c. And since storage basin design requirements for water quality control differ in several important respects from those for flood containment alone, how do we allocate costs between those services?

2-6. GROUNDWATER POLLUTION

All groundwater which does not remain stored underground, evapotranspired, or withdrawn for surface uses will migrate to surface water streams, lakes estuaries, and eventually, to oceans. This groundwater contributes volume and contaminants to our surface waters. The amounts of water and pollutants are extremely variable and depend mainly on original inputs to the ground, geology of the area, and relative piezometric heads of ground and surface waters. Contributions of groundwater to surface waters must be ascertained on an individual (case-by-case) basis.

Pollutants entering surface waters from exfiltration of groundwater are termed "nonpoint" sources. Nemerow (1978, Chapter 28, pp. 722-723) defines these contaminants and describes their contributions to the quality of surface waters. Not all nonpoint contributions are from groundwater; some run off directly from land surfaces but are diffused over wide areas.

It is not the purpose of this textbook to assess the precise part played by groundwaters in the reaction of surface waters. The reader must be aware, however, of these contributions. As Viessman et al. (1977, p. 296) advise, linkage between surface and groundwaters should be investigated in all regional studies so that adverse effects can be noted, if they exist, and opportunities for joint management understood. Although surface streamflows are sustained by the groundwater resource, groundwaters may also be replenished by infiltration from surface runoffs, and sometimes by surface streams. You are directed to Viessman's excellent textbook for a better understanding of the entire subject of groundwater flow. In most cases, groundwater's contribution to surface waterflows will be an increase in volume, serving as additional dilution of the contaminants in the latter. This is especially true in the case of dissolved organic matter and BOD. Added dilution enhances the assimilative capacity of surface waters.

Groundwaters generally contain higher dissolved inorganic salts than surface waters. The latter are enriched by these groundwater contributions and result in harder, more mineralized waters. Such waters may support denser crops of plankton and higher forms of biological life. Hard waters may be troublesome for water users such as industrial and municipal consumers. See Chapter 7, Section 2 for recommended water quality limits.

2-7. THEORY OF TIDE FORMATION AS RELATED TO ESTUARY[3] ANALYSES

Since the amount of dilution water (whether fresh or salt) affects the quality of the receiving water, we are required to understand the effect of tides on these waters. The general result of tides is to "hold" the land drainage (outflow) water for longer periods of time before ultimately discharging all of it into the oceans. The length of time the contaminated river outflow remains in an estuary is variable but predictable. Although we will not consider estuary analysis until Chapter 8, we will consider here the tidal phenomenon as a hydraulic system.

As Defant (1958) pointed out, "the tides are the heartbeat of the ocean, a pulse that can be felt all over the world." Tides occur with great regularity and uniformity in all coastal areas of the earth. When the seawater currents rush inland, the water level rises; this is called the *flood tide.* When the channels drain the water to the sea and the seawater currents are away from the land, the water level drops; this is called the *ebb tide.* The regular recurring interval between two successive high (flood) or low (ebb) tides is *12 hours and 25 minutes.*

Our time on earth is determined by the sun. It takes a point on earth 24 hours to make one revolution and return to the same position again facing the sun. The moon, which rotates about the earth, takes slightly longer, or, 24 hours and 50 minutes (known as a lunar day). Therefore, the moon passes the same meridian 50 minutes later each solar day, which corresponds to the average lag in the tides (see Fig. 2-10).

Copernicus (1473-1543; 1960) believed that the *sun*, not the earth (as previously believed by Ptolemy and his followers) was the center of the solar system. He showed that the planets of the solar system, including *Earth*, keep revolving about the sun.

The sun revolves around the center of the giant galaxy of about 100 billion stars at a speed of 175 miles per second. The earth and all the other bodies of the solar system tag along with the sun at this speed.

The force of gravitation holds the moon in its path around the earth. The *centrifugal force* caused by the moon's speed exactly balances the *centripetal force* of gravity. So the moon revolves around the earth.

The sun does move. It moves in two ways. First, it is speeding along at about

---

[3] For the purpose of this discussion an estuary shall be considered as any body of water which is affected by or mixed with seawater.

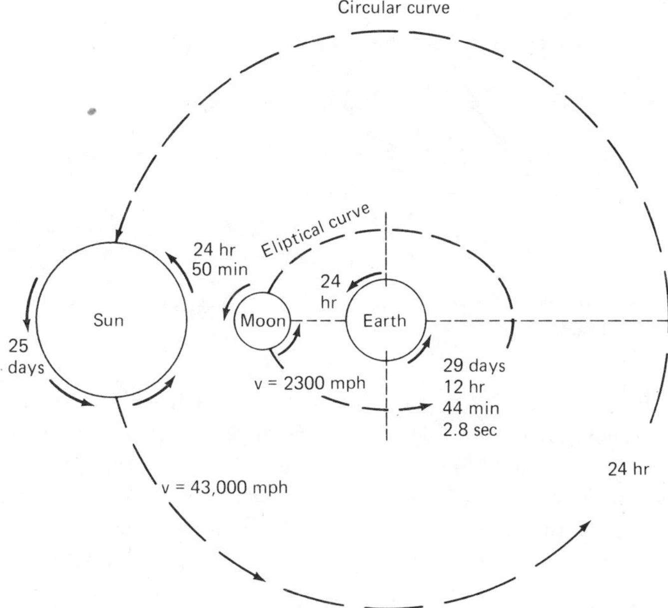

FIG. 2-10. Relative movement and positions of moon, sun, and earth.

43,000 miles per hour among the billions of other stars that make up the Milky Way. It carries with it the earth and all the rest of the solar system. *The sun also rotates*, or spins about, like a giant flaming top. But, instead of turning around once in 24 hours as the earth does, it takes about 25 days for one rotation. And because the sun is made of gas, parts of it spin around in less time than other parts.

The moon is the only natural object in space that revolves around the earth. It goes around the earth in an elliptical, or oval, path at an average speed of 2,300 mph.

The position of the moon affects not only the time of the tides but also the height and the mass of water involved in the tidal current.

During *full moon* and *new moon* (syzygies) the difference between high and low tide water, known as the range or amplitude of the tides, is at its maximum—highest high water and lowest low water. These are known as the *spring tides.*

During the first and last quarters of the moon, the range is particularly small and we have *neap tides.* The difference between spring tides and neap tides is significant and is shown graphically in Fig. 2-11.

The range (difference in elevation between high and low tides) or *phase inequality* not only depends upon the relative positions of the moon and the earth, but is also affected by the position of the sun, since the phases of the moon de-

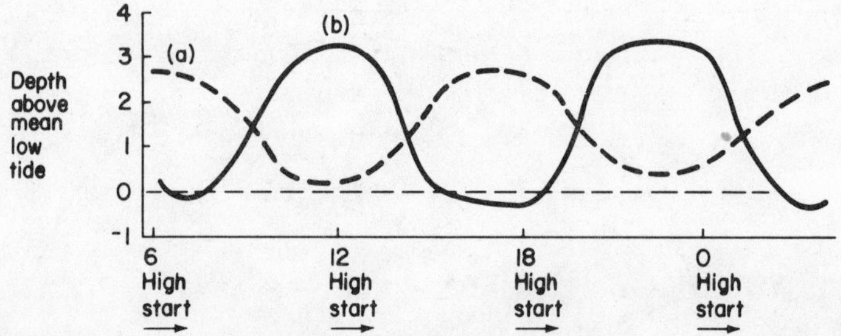

FIG. 2-11. Water depth difference between spring and neap tides. (a) = mean neap tide; (b) = mean spring tide.

pend on the position of the earth relative to both (sun and moon). The tidal range sometimes varies from tide to tide; the morning high water may be higher than the following afternoon high water, or vice-versa. This phenomenon is known as the *diurnal* (*daily*) *inequality*. The *phase* and *diurnal* are the major basic inequalities, the former caused by the relative positions of the moon and the earth, and the latter caused by the position of the earth in relation to the sun and moon.

The range in tides differs from one coastal town on the earth to another from a few inches to as much as 60 feet and more.

The regular course of the tides is often altered by winds. Seaward breezes drive water towards the coast, and the water level rises; it drops when water is driven away from the coast. These variations do not depend upon the tides. Coupled with tides, particularly with high spring tides, they can become extremely dangerous.

To explain the tide-generating forces of the moon and sun, we must be familiar with *Newton's Law of Gravitation*, which states that "the force of attraction between two bodies is proportional to the square of the distance between them."

Without this force of attraction, the earth and the moon would not move *annually* about their common center of gravity of the earth. Therefore, if the earth and moon were not in motion, the force of attraction would cause them to collide.

Now, the motion about the common center of gravity–like all circular motion–produces *centrifugal forces* which just balance the force of attraction. The fact that these two forces are equal and opposite and are at the center of the earth and the moon is the basis for the entire stability of our universe as we know it.

Despite the fact that the overall earth-moon system is stable and in equilibrium, separate material on earth is not. Equilibrium depends upon centrifugal forces being exactly equal and opposite to attractive forces. While the centrifugal forces on all parts of the earth are equal, attractive forces at all locations on earth, which depend upon the distance between the earth and the moon, are not. The individual residual forces (centrifugal–attractive) are the tide-generating forces. These residual (tide) forces can be visualized schematically be referring to Fig. 2-12. At Z (the zenith position on earth which directly faces and is closest to the moon)

> the centrifugal force is $C = -3.38$ mg weight
> and attractive force is $A = +3.49$ mg weight.
> Therefore, the residual or tide force is ($T = +0.11$ mg)

At N (the nadir position on earth which faces directly away from the moon and is farthest away)

> the centrifugal force is $C = -3.38$ mg weight
> while the attractive force $A = +3.27$ mg weight (since it is farther away).
> Therefore, the residual or tide force ($T = -0.11$ mg weight)

At E (the center of the earth) the force of attraction (A) draws *each kilogram* of the earth's mass towards the moon, the very small force of +3.38 mg weight C being exactly equal and opposite here and the net tidal force equaling 0. Only at points Z, N, B, and C does the tide force act perpendicularly (either upwards or downwards from the center of the earth).

The resultant tide forces are really very small, causing man, for example, actually to weigh 1/9,000,000 less than he does at points Z and N only.

Except at points Z, N, B, and C the tide-generating force also has a horizontal component–the attractive force–acting along the surface of the earth. Although the attractive force is very small, it is more important than the vertical compo-

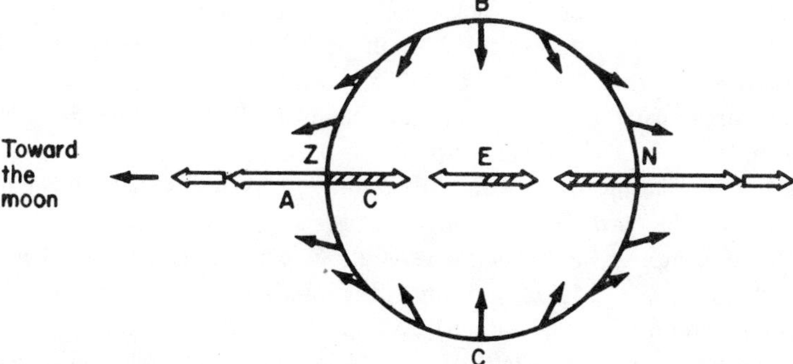

FIG. 2.12. Distribution of resultant tide forces along the meridian. Reproduced with permission. From *Ebb and Flow–The Tides of Earth, Air and Water*, by Albert Defant. © 1958 by The University of Michigan Press.

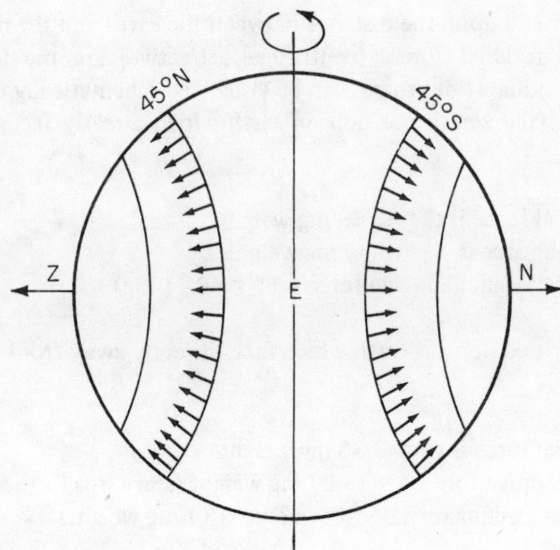

FIG. 2-13. System of horizontal components of tidal forces. Reproduced with permission. From *Ebb and Flow–The Tides of Earth, Air and Water*, by Albert Defant. © 1958 by The University of Michigan Press.

nent, since other horizontal forces acting along the earth's surface are usually also small. The horizontal component of the tide-generating force is 0 at *Z* and *N* and strongest along the two circles at 45° to *Z* and *N* (see Fig. 2-13).

This system of tide-generating forces is determined by the position of the moon (see Fig. 2-14).

Now the earth itself is not at rest but rotates about its own axis once a day. The attractive forces will continue to increase as a point near *Z* or *N* leaves it (gets farther away from the moon) and will reach a maximum as it reaches either 45°N or 45°S from *Z* or *N*. Then it begins to reduce again until it reaches *E*, where it is zero. Twice during the 24-hour period a point on the earth's surface will pass through maximum tides at 45° from *Z* and *N*. Thus the earth's rotation produces semidiurnal changes in the tide-generating force both in direction and is magnitude.

The tide-generating force of the moon must necessarily vary with the moon's distance from the earth. This distance changes slightly and continuously in the course of one month, and the tide-generating force is affected by these small alterations (see Fig. 2-14). In addition, the moon keeps changing in declination (from a maximum of 28.5°N and S), and the resulting asymmetry of the system of the tide-generating forces with respect to the earth's equator causes the diurnal inequality in the tide-generating force; therefore, the greater the declination, the greater the inequality.

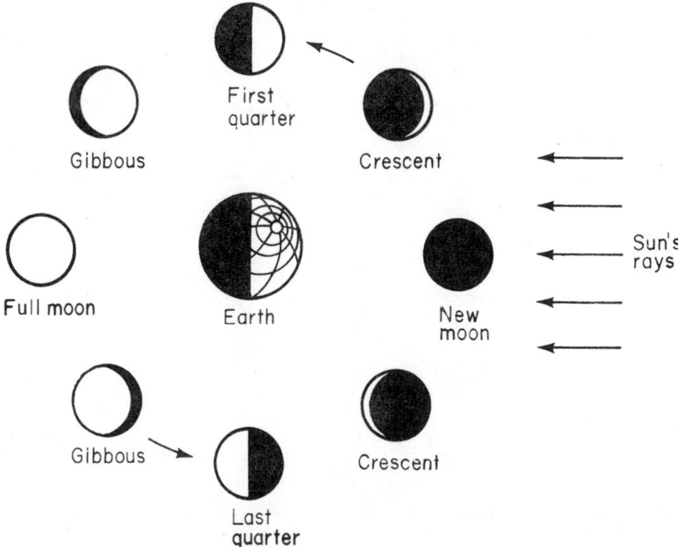

FIG. 2-14. Phases of the moon. The age of the moon is the time elapsed since new moon. The age to full moon is about $14\frac{1}{2}$ days. Reproduced with permission. From *Ebb and Flow—The Tides of Earth, Air and Water*, by Albert Defant. © 1958 The University of Michigan Press.

The *sun* too produces a system of tide-generating forces on earth. However, the semidiurnal tidal effect of the sun is only about 50% of the lunar tidal effect, because the sun is so much farther away from the earth (even though its mass is greater than that of the moon). The sun is 93,000,000 miles away from the earth (or about 300 times as far as the moon).

The sun is a huge globe of incandescent gases measuring 865,380 miles in diameter (about 109 times larger than the earth) and containing more than a million times the bulk of the earth. Its surface temperature is 11,000°F, and its interior temperature has been estimated to be 40 million degrees.

The *twice monthly inequality* is the result of the joint effects of the moon and the sun. During the new moon phase, the sun, moon, and earth are in one straight line. The moon's tidal effect is then added to the solar effect, and the resultant attractive force is increased in the ratio of 3 : 2. The same happens during full moon when moon and sun are in opposition. During the first and third quarters the moon has moved 90° away from the sun, and the resultant attractive force is roughly one-half of the lunar force alone (see Fig. 2-15).

Tide generation, like all gravitational forces, is proportional to the mass it attracts. In comparison with equal volumes (1,000 cc) of the three main masses on earth we see that

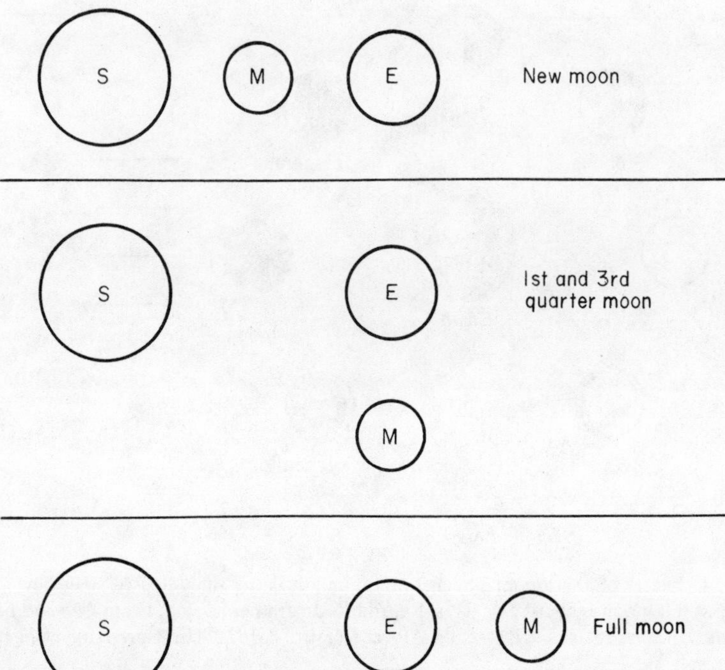

FIG. 2-15. Relative positions of moon and sun with respect to earth. Reproduced with permission. From *Ebb and Flow–The Tides of Earth, Air and Water,* by Albert Defant. © 1958 by The University of Michigan Press.

1,000 cc of rock from the earth's crust weighs ~2.5 kg
1,000 cc of sea water from the earth's surface weighs ~1.0 kg.
1,000 cc of air from the earth's surface weighs ~1 g.

Therefore, the tide-generating forces are in the proportion of 2.5 : 1 : 0.001. However, solids are very inelastic when compared to water, and gases of the air are so low in density that the tide-generating force has its main effect on the oceans.

Facts About Our Moon. The moon is our nearest planetary neighbor (satellite) to remain always about one-quarter of a million miles away (mean distance actually is 238,857 miles. Range is 222,000–253,000). The moon moves in an orbit around the earth at 2,300 miles per hour.

The moon has a diameter of about 2,160 miles, or a little more than $\frac{1}{4}$ that of the earth.

The moon's surface is about $\frac{1}{14}$ of the earth's.
The moon's volume is about $\frac{1}{49}$ of the earth's.

The moon's density is $\frac{6}{10}$ of the earth's, or $3\frac{1}{3}$ times the density of water.
The gravity at the surface is $\frac{1}{6}$ of that on the earth.
All the light from our moon is reflected sunlight.

QUESTIONS

1. What is the best method of sampling a stream in order to ascertain all of the changes in physical conditions of a stretch of stream?
2. What is a rating curve and how can it be used for our objective?
3. Why are stream depths so important? Why are stream velocities so important?
4. How are stream velocities normally measured?
5. Of what importance is a correlation of the survey with onshore pollution?
6. What are the advantages and disadvantages of using actual streamwater for dilution water in BODs?
7. Discuss the pros and cons of a large number of samples from a few stations as opposed to a small number of samples from a large number of stations.
8. What is meant by intensive sampling at a steady flow?
9. Name some of the stream situations that may call for sampling stations.
10. Discuss the assets and liabilities of a stream.
11. What factors have the most important influence in self-purification?
12. Of what importance is an accurate knowledge of the drought flow of a stream?
13. Describe how we measure a streamflow.
14. Discuss effects of an upstream impoundment on a downstream user.
15. How do we attempt to overcome problems caused by dams?
16. Why is a knowledge of tides becoming increasingly important in stream pollution analysis?
17. How is water quality affected by ebb flows and flood flows? How much time elapses between ebb tides and flood tides on the eastern coast?
18. Describe the relative positions and rotations of the earth, moon, and sun; what effects do the movements and positions have on our tides?
19. What causes the tide-generating forces?

REFERENCES

Barret, C. B.: Quantitative Prediction of Seasonal Rating Curve Shifts, *Water Research*, 9(4):526, August 1975.

Copernicus, Nicolaus: *Concerning the Revolutions of the Celestial Spheres* (1543), World Book Encyclopedia, Chicago, Ill., 1960, Vol. 3, p. 822.

Defant, Albert: *Ebb and Flow–The Tides of Earth, Air, and Water*, University of Michigan Press, Ann Arbor, 1958.

Environmental Effects of Hydraulic Structures, Task Committee on Environmental Effects of Hydraulic Structures, *Journal of Hydraulics Division, ASCE*, HY2:203, February 1978.

Fair, G. M., E. W. Moore, and H. A. Thomas: The Natural Purification of River Muds and Pollutional Sediments, *Sewage Works J.*, **13**:1209, 1941.

Gumbel, E. J.: The Return Period of Flood Flows, *Annals of Mathematical Statistics*, **12**:163, 1941.

Ingols, R. S.: Surface Water Pollution and Natural Purification, *Municipal South*, p. 31, January 1956.

Kittrell, F. W.: Effects of Impoundments on Dissolved Oxygen Resources, *Sewage and Industrial Wastes*, **319**:1065–1081, 1969*a*.

——: A Practical Guide to Water Quality Studies of Streams, 1969*b*.

——, and E. Kochtitsky, Jr.: *Natural Purification* Characteristics of a Shallow Turbulent Stream, *Sewage Works J.*, **19**:1032, 1947.

McCuen, R. H.: Downstream Effects of Stormwater Management Basins, *Journal Hydraulics* Division (A.S.C.E.) HY II; November 1979.

McPherson, M. B.: Challenges in Urban Runoff Control, *Proc. 14th Annual Henry M. Shaw Lecture Series in Civil Engineering*, Raleigh, N.C., March 1979.

Nemerow, N. L.: *Liquid Wastes of Industry–Theories, Practices and Treatment*, Addison-Wesley Company, Reading, Mass., 1970.

—: *Industrial Water Pollution*, Addison-Wesley Publishing Co., Reading, Mass., 1978.

Shields: An Analysis of Sediment Transportation in the Light of Fluid Turbulence, by H. Rouse, U.S. Dept. of Agriculture, Sedimentation Division, Washington, D.C., 1939.

Velz, C. J.: Factors Influencing Self-Purification and Their Relation to Pollution Abatement, *Sewage Works J.*, **19**:629–644, 1947. (Presented at 19th Annual Meeting, New York State Sewage Works Association; New York City, January 17, 1947.)

Velz, C. J.: Factors Influencing Self-Purification and Their Relation to Pollution Abatement II. Sludge Deposits and Drought Probabilities, *Sewage Works J.*, **21**(2):309–319, 1949.

Velz, C. J.: Sampling for Effective Evaluation of Stream Pollution, *Sewage and Industrial Wastes*, **22**(5):666, 1950.

Viessman, W. Jr., J. W. Knapp, G. L. Lewis, and T. E. Harbaugh: *Introduction to Hydrology*, Harper & Row Publishing Co., New York, 1977.

Wanielista, M. P., Y. A. Yousef, and W. M. McLellon: Nonpoint Source Effects on Water Quality, *J. Water Pollution Control Fed.*, **49**:441, 1977.

3
BIOCHEMICAL FACTORS INVOLVED IN STREAM ANALYSIS

3-1. MEANING AND DETERMINATION OF BOD

Although many contaminants, both organic and inorganic, enter our watercourses each day, no measurement is more all-encompassing, utilitarian, or more generally accepted and used than the biochemical oxygen demand (BOD). The BOD value represents the amount of oxygen required by bacteria to decompose aerobically an amount of organic matter in a given period of time at a stated temperature. In the laboratory, the standard oxidation or incubation time is usually five days, and the temperature is 20°C. To eliminate nonuniformity and to provide optimum conditions for degradation of organic food, mineral nutrients and bacterial seed are usually added to the incubation bottles in the standardized laboratory determination (Standard Methods, 1970). The general point of the first four sections of this chapter is to comprehend the BOD and its related effect on dissolved oxygen in the watercourse. In addition, the rate of exertion of the BOD is the sole subject of Chapter 4. It is imperative, then, that the reader study and assimilate carefully, step by step, the meaning, derivation, and determination of BOD.

Determination of BOD

Dissolved oxygen (DO) of streamwater is the chemical constituent most frequently determined in observing the effect of organic pollution on streams. This parameter alone sometimes provides an adequate check on existing conditions. When it is necessary to predict changes in stream conditions that will follow proposed reductions in plant wastes, or to predict the effects of new industries or treatment plants discharging to streams, the biochemical oxygen demand (BOD) determination must be used as an adjunct to the DO determination.

Some workers feel that the BOD test probably is the most misunderstood and the most abused of all the analytical tools used in stream analyses. They recognize that the test is not perfect, but realize also that some of the apparent difficulties with the tests are caused by an imperfect and incomplete knowledge of some of the fundamental reactions involved rather than by imperfections in the method of analysis itself.

To determine the BOD, measured amounts of wastewater or receiving waters are added to a standard 300 ml laboratory BOD bottle. The bottle is then filled completely with standard laboratory dilution water. The latter contains seed (microorganisms) and mineral nutrients to ensure optimum biochemical oxidation. The bottle is incubated at 20°C for five days, after which its oxygen content is determined and compared to a "blank" (similar bottle sample not containing any wastewater). The oxygen difference between samples and blanks when multiplied by the appropriate dilution factor yields what has become known as the standard laboratory, 5-day, 20°C, BOD value.

In any event, the BOD test is the best method we have for evaluating the biochemical demand for oxygen from the stream. More precise chemical methods are available, such as the chemical oxygen demand (COD) or total organic carbon demand (TOD), but what do they mean if these values of organic material are not equivalent to the organic matter actually being utilized by microorganisms?

The COD is measured by chemically oxidizing the organic matter with carefully measured amounts of strong oxidizing agents such as potassium dichromate. The oxidizing chemical used up is equivalent to the chemically oxidizable organic matter. The TOC is obtained by igniting a sample of dried solid residue at over 600°C. The amount of carbon dioxide given off is equivalent to the total organic carbon present. Microorganisms may not utilize either the total COD or TOC completely for food in a relatively short period of time (usually five days) and at 20°C. Therefore, ratios of COD and TOC to BOD 5, 20°C are often as high as two or three to one.

Most moderately contaminated streams have BOD values of between 1 and 8 ppm; BODs can be determined on the undiluted river water samples. The dilution water available greatly affects the resulting BOD. Toxicity, concentration of microorganisms, inorganic nutrient salts, and other factors will greatly influence the rate of reaction (Standard Methods, 1970). For this reason, many onshore laboratory BOD tests of wastes which utilize the standard dilution water will not agree with the river tests employing natural streamwater for dilution. The former is a standard technique which can be compared in any laboratory in the world, while the latter is applicable only to that particular wastewater being discharged into that particular river. Because of this, I would suggest that both natural and artificial dilution water be used in determining onshore BOD of sewages and wastes.

Streeter and Phelps (1944) made the following generalization: "The rate of the biochemical oxidation of organic matter, k_1, is proportional to the remaining concentration of unoxidized substance, measured in terms of oxidizability."

$$\int_{L}^{L_t} \frac{-dL}{dt} = KL \text{ or } \frac{dL}{dt} = -KL$$

integrating to

$$\log_e \frac{L_t}{L} = Kt$$

or since K for $\log_{10}$ = .434K for $\log_e$

$$\text{Log}_{10} \frac{L_t}{L} = -.434Kt = -kt$$

and since $e^{-Kt} = 10^{-kt}$, whence

$$\frac{L_t}{L} = 10^{-kt}.$$

k = an empirical constant, but in reality, a statistical average
L = oxidizability at time zero
L_t = the corresponding value of oxygen demand remaining at time t
$\therefore$ L_t/L is the fraction remaining

If time, t, is taken in days, k is found experimentally to have a value of 0.1 at 20°C for a normal domestic sewage,

then $10^{-.1} = .794$ (exactly the value adopted for the rate of reaction on basis of the relative stability tests [an entirely different procedure]).

The law may now be stated in simpler numerical form. The biochemical oxidation of organic matter of sewage origin in the presence of an excess of oxygen at 20°C proceeds at such a rate that it is 20.6% completed $(1-10^{-.1})$ at the end of the first day, and during each succeeding 24 hours the remaining unoxidized organic matter is further reduced by 20.6% of its value. The BOD remaining at any time = 0.794 × BOD ultimate. The reaction rate of organic matter originating from industrial wastes may be slower, equal to, or greater than this rate for domestic sewage.

The typical monomolecular curve described by Streeter and Phelps may also be formulated in the following manner:

$$\frac{dc}{dt} = K'C \quad \text{a first-order reaction similar to that of the decay of nuclear energy}$$

where

C = concentration of organic matter
t = time (in days)

When the initial concentration of organic matter = L, the ultimate BOD, and y = the BOD at any time, t, the above equation can be expressed as

$$\frac{d(L - y)}{dt} = -K'(L - y) \qquad \text{BOD left at any time, } t$$

or rewritten

$$\frac{d(L - y)}{L - y} = -K'dt.$$

This must be integrated over the entire oxygen utilization period

$$\int_{y=0}^{y=y} \frac{d(L - y)}{L - y} = \int_{t=0}^{t=t} -K'dt$$

and obtaining

$$\ln \frac{L - y}{L} = K'dt$$

or

$$\frac{L - y}{L} = e^{-K't}$$

or

$$L - y = Le^{-K't}$$

$$y = L - Le^{-K't} = L(1 - e^{-Kt})$$

$$y = L(1 - e^{-K't})$$

since

$$e^{-K't} = 10^{-k't} \quad y = L(1 - 10^{-k't})$$

The monomolecular relationship between the amount of BOD exerted and the time of reaction is shown over a 20-day period in Fig. 3-1. Many factors affect

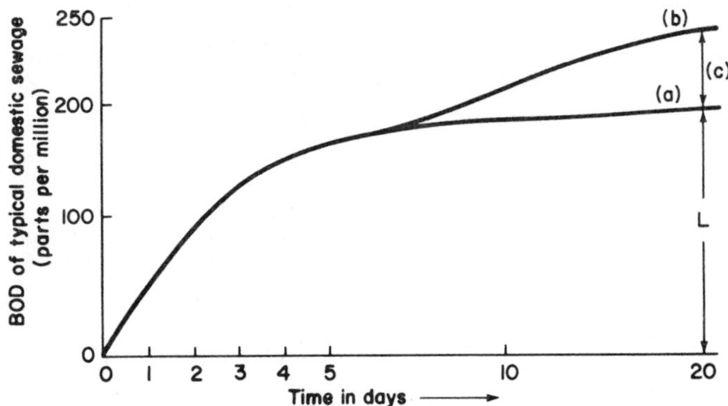

FIG. 3-1. Schematic presentation of typical BOD curve of domestic sewage. (a) = first stage (L) of BOD reaction; (b) = second stage of BOD reaction; (c) = nitrogen oxygen demand (NOD). (See Chapter 4, Section 7, for more complete discussion of NOD.)

the shape of the general curve shown in Fig. 3-1. In the laboratory test only the nature of the organic matter, seed, temperature, and nutrients can alter it drastically. Since the standardized laboratory test uses given amounts and similar type seed, temperature, and nutrient material, only the nature of the organic matter itself varies, and hence, will affect the shape of the curve. Individual laboratory BOD values may vary as much as ±17%, however, due to the nature of biological reactions and the accuracy of the chemical analyses. Stream conditions may not agree with the laboratory reaction rates because of differences in nutrients and reaction environmental conditions.

3-2. EFFECT OF TEMPERATURE ON BOD

The rate of the BOD reaction has been found to be a function of temperature, just as biological reactions are speeded up (to a certain point) by increased temperature.

A simple temperature function that applies is

$$\frac{k_1}{k_2} = \theta^{(T_1 - T_2)}$$

where k_1 and k_2 are the velocity constants at temperatures T_1 and T_2, respectively.

θ is known as the temperature coefficient. Its best mean value from a comparison of several independent studies, as adopted by Streeter and Phelps, is 1.047. This means that the velocity of the reaction is increased by 4.7% for each 1°C rise in temperature. Temperature rise increases both bacterial activity and contact of molecules reacting. Since this function is exponential, an increase of

TABLE 3-1. Relation of Time and Temperature to 5-Day BOD of Sewage. [*After Phelps and Streeter, 1944. By permission of publishers.*]

Time (Days)	Temperature °C 10	14	20	28
0.5	.08	.106	.16	.26
1	.15	.20	.30	.48
2	.29	.38	.54	.81
3	.42	.53	.73	1.06
4	.51	.64	.87	1.25
5	.60	.76	1.00	1.38
10	.91	1.07	1.33	1.65
20	1.12	1.26	1.46	1.71

10°C in temperature increases the velocity, not by 10 times θ, but by its 10th power, or 1.584-fold (see Table 3-1).

Although there are some differences of opinion as to the *exact* effect of temperature, it appears to be a fact that the ultimate oxidizability increases with temperature and, as shown by Theriault (1927), in the region of 20°C, at approximately 2% per degree.

G. W. Fair (Streeter and Phelps, 1944, p. 72) introduced a temperature characteristic C, because he found that a rise of 1°C not only increased the rate in the ratio of 1.047 but also simultaneously and independently increased the ultimate demand upon which the ratio is based in the ratio of 1.02: then the overall increase became 1.068(1.047 × 1.02).

$$\theta = e^c; \; C = \frac{\log \theta}{.434 \leftarrow} \quad \left(\frac{1}{2.303}\right)$$

$$L_T = L_{20°\mathrm{C}}[1 + .02(T - 20)]$$

Solution of the General BOD Equation

Combining the general equation

$$\mathrm{Log}\,\frac{L_t}{L} = -kt$$

with the temperature relation of L and kt

$$L_T = L_{20°\mathrm{C}}[1 + .02(T - 20)]$$

$$kt = .1 \times 1.047(T - 20)$$

where

t = actual temperature of the BOD reaction observed
$L_{20°C}$ = ultimate BOD at 20°C.

The above expression denotes the accepted standard time and temperature conditions for carrying out the laboratory test for the BOD of sewage or wastes. The 5-day, 20°C BOD value is of only limited value in the discussion of stream pollution, merely providing a datum for which, with the help of the combined time/temperature equation, it is possible to evaluate the BOD of the wastewater in question under any stated conditions.

In Table 3-1 we present relative BOD values of domestic sewages at varied temperatures between 10°C and 28°C, and from various times of from one-half day to 20 days. The reader must keep in mind that these values were developed over 40 years ago when domestic sewage was vastly different from that of today.

The values are still useful, however, to illustrate the relationship of the BOD reaction of one temperature to another. For example, few workers question the validity of the five-day values showing that at 10°C the reaction is only 60 percent complete as compared to the reaction at 20°C.

Influence of Temperature on k_t

Bacteriological processes are accelerated as the temperature is increased, and hence the rate of oxygen utilization. The relationship of the slope of k_t (as derived on page 110) to temperature is illustrated in Fig. 3-1A, from which k_t at any temperature is readily obtained. For example, k_t at 20°C is read from the graph as .10, and k_t at 10°C, as .06.

Churchill (1961) found that the reaeration rate, k_2, also increases with water

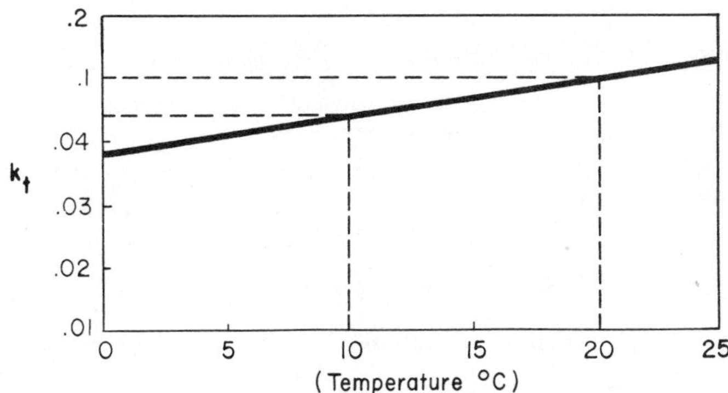

FIG. 3-1A. Biological activity and deoxygenation rate versus temperature of reaction.

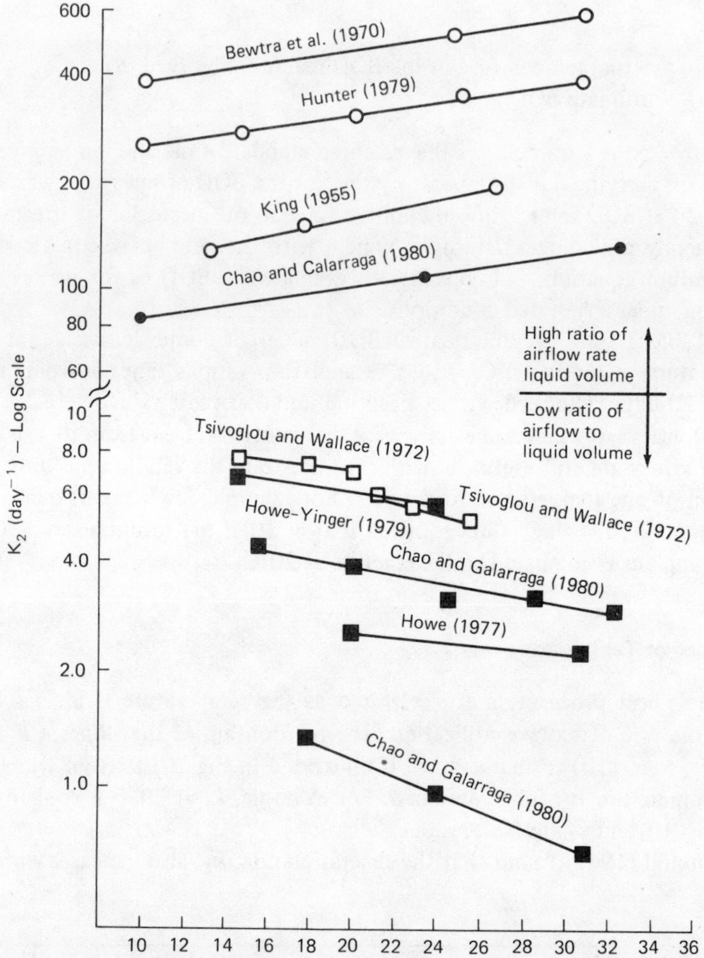

FIG. 3-1B. Influence of Temperature on K_2. (*From Chao, et al., 1980*)
⊙ Increasing K_2 vs. Temperature for Aerators–Literature data
● Increasing K_2 vs. Temperature–Author's data
⊡ Decreasing K_2 vs. Temperature for Streams–Literature data
■ Decreasing K_2 vs. Temperature–Author's data

temperature at the geometric rate of 2.5 percent per degree centigrade, or

$$k_2(T^\circ) = k_2(20^\circ) \times 1.0238(T - 20)$$

where T = temperature in °C.

Presumably, reaeration proceeds faster as the temperature is raised because of an increase in molecular activity and surface exchange of oxygen. This will be useful in studying the net effect of temperature changes on BOD reactions and stream reactions as shown in Chapter 6.

However, in 1977, Howe proposed that the k_2 value decreases as the liquid temperature increases. Since his modification is counter to the well-recognized one of increasing the oxygenation coefficient with higher temperatures, Chao et al. (1980) presented a more detailed summary (shown in Fig. 3-1B), verified in further studies. They concluded that their preliminary results showed that the K_2-temperature relationship is dependent on the magnitude of k_2 observed. If the k_2 value is less than 40 day^{-1}, it has a tendency to decrease for increasing liquid temperature. Based upon their experiments, they believe that Howe's modification is valid–and since most stream k_2 values are below 40 day^{-1}, Howe's suggestion is recommended by Chao for calculating the k_2 value at 20°C for stream modeling and simulation. When using Howe's modification, the oxygen transfer rate of streams will be greatly reduced at the critical temperature level. A review of Chao's work also seems to indicate that the ratio of air flow rate to liquid volume may also be involved in the discrepancies of researcher values of k_2 as the temperature increases. The ratio is normally much lower (as is the k_2 value) for streams than for that used in aeration of wastes in experiments.

3-3. REACTION RATE *k*; ITS MEANING IN STREAM WATER QUALITY

At any given moment, there is a momentary deoxygenation rate expressed as a constant fraction of the remaining unoxidized substance in any arbitrary time unit, and so decreasing progressively as the concentration of the residual organic matter decreases.

The normal experimental value of k_1, 0.1 in common logs, becomes 0.23 in natural logs. Therefore, if this value is put into the differential equation, the equation then reads, in words: "The rate of change at any instant of time is such that if continued uniformly for 24 hours, the continuous reaction rate would be 23% (per day implied), which becomes a daily rate of 20.6%, this being the sum over one day, of the small 23% fractions of the diminishing residue. If one would avoid the calculus, it is exactly the same thing to say that the daily rate is 20.6% of the residual at the beginning of any day."

Another measure of the rate of reaction has also been explained as the "half life" or time required for a certain percent completion.

Velz (1953) describes nature's rule of oxygen requirements in a stream as really quite simple:

At any particular temperature, the percent of the total biochemical debt paid per interval of time is always a constant percentage of that remaining unpaid from the previous interval.

It is evident that, for the mixed organics of municipal sewage and the major-

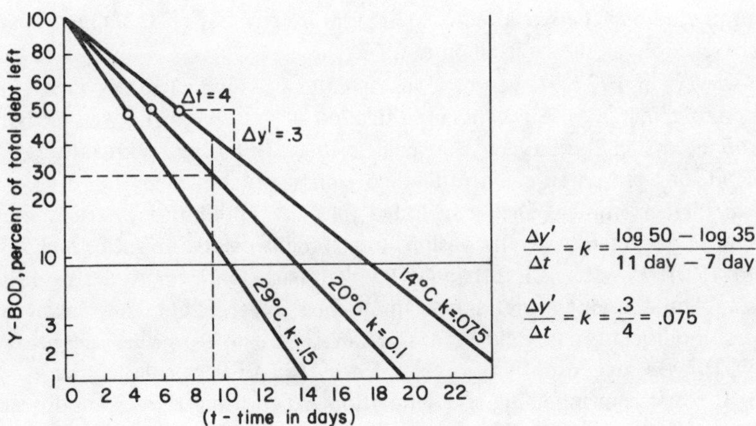

FIG. 3-2. Relationship of BOD remaining at various temperatures over a 24-day period of reaction.

ity of industrial wastes, 50 percent of the ultimate organic debt is amortized (eaten, oxidized) in three days at 20°C (Fig. 3-2). In accordance with the basic law, $\frac{1}{2}$ of what is left is oxidized in the next three days, etc., with some fraction always remaining. This plan is decidedly unbalanced, requiring heavy payments on the debt in the early periods; smaller payments only can be postponed to a later day. This imbalance Velz likens to the interest-principal relationship on home mortgages.

It will also be noted from Fig. 3-2 that as the temperature increases from 14°C to 29°C the amortization schedule shortens; each succeeding 50 percent reduction in outstanding debt is accomplished in a two-day interval with less than 1 percent outstanding at the end of 14 days at 29°C. As the temperature decreases to 14°C, the amortization period, however, is lengthened, each succeeding 50 percent reduction taking place in a four-day period. If the data in Table 3-2 are plotted on semilog paper, the percentage of the total BOD remaining against time in days, perfect straight lines develop as shown in Fig. 3-2. Each straight line has a specific slope k depending upon the temperature.

TABLE 3-2. Percent of Total BOD Remaining at Various Temperatures and Oxidation Times

% of total (ult.) BOD left	at 14°C	at 20°C	at 29°C
	Time required		
100	0 days	0 days	0 days
50	4	3	2
25	8	6	4

In other words, if we take the log of the percent BOD remaining in a stream sample and plot this against the time of flow in the stretch, the slope of the line will be k, the rate of debt amortization.

3-4. CREDITORS AND BENEFACTORS IN OXYGEN BALANCE

Oxygen in a receiving water has been likened to an extremely limited bank checking account. Since the amount of money in the account is very small at any time, the deposits and withdrawals play a vital role in maintaining a surplus. The objective of this section is to focus our attention on the delicate oxygen balance of streams and lakes, as affected by each definable factor, in somewhat more detail than found in Section 2-2.

Creditors[1] demanding resources from our oxygen account consist of:

1. organic matter in the continuously flowing water
2. slime growths on attached rocks, debris, and other surfaces (shorelines) over which the water flows
3. primary organic bottom sludge deposits (benthal demand)
4. secondary organic bottom deposits (dead algae)
5. temperature rises causing oxygen vapor loss and increased microbiological metabolism
6. fish and other aquatic organisms' respiration needs
7. organic contamination in branch streams
8. salinity
9. time (detention)

Benefactors contributing oxygen resources include:

1. reaeration due to the physical reaction of air and water
2. photosynthesis
3. temperature decrease increasing oxygen saturation potential and decreasing microbiological activity
4. dilution from uncontaminated streams

If the summation of the demands of the eight creditors is greater than the contributions of the four benefactors, the oxygen account is deficient. The solution to the problem is to control creditors' demands so that they will not exceed, and in fact, will be considerably less than the contributions from benefactors. The normal procedure used to accomplish this has been waste treatment to

---

[1] Usually these creditors will represent about 90% of the total pollution load. Other sources are nonpoint sources difficult to detect.

reduce the demands of the first creditor. Little concern or effort has been expended in behalf of the other creditors because of the lack of simple mathematical tools for predicting either the exact quantity or time of the other creditor demands. A discussion of all the demands and contributions in one section seems in order here, so that we may gain better insight into the total picture of oxygen resources.

Creditors

1. Organic matter in continuously flowing wastewater will demand oxygen at a rate generally proportional to the reaction rate k_1, which equals

$$\frac{1}{\Delta t} \log \frac{L_a}{L_b}$$ (See page 99)

Stated in another way, the rate of oxygen demand of this type is roughly proportional to the logarithm of the BOD remaining to be oxidized at any time, t. At constant temperature, this demand rate for oxygen decreases as the distance (flow time) from the point of contaminant injection increases, since the concentration of organic matter remaining decreases accordingly. Of importance to us when considering this demand is the type and character of the organic matter itself. Just as one creditor may be more "demanding" than another, so may one type of organic matter, such as sugar waste from a candy manufacturing plant, be more demanding of immediate oxygen payment than another type of organic matter, such as a pulp mill waste containing cellulose. The nutrients and bacterial seed present in the flowing water will also affect the demand rate of this type of organic matter. Since all these factors may affect the oxygen demand so as to alter materially the "account balance" at any given time, laboratory and actual receiving stream reaction rates for each waste should be determined prior to formation of treatment design decisions.

2. Slime growths, such as *Sphaerotilus natans* (see page 29) for more discussion) attached to physical surfaces along the stream channel accelerate both the removal of flowing dissolved organic matter and oxygen. This fact is overlooked in the determination of the laboratory reaction rate. As we have seen in Chapter 2, Section 2, the slime growths act in much the same manner as trickling filter slime, zoogleal growth. They cause an accelerated rate of payment of the outstanding debt, resulting in a rather sudden demand for oxygen resources in stream areas that are particularly conducive to slime growth. A physical examination of the stream, when coupled with a determination of the stream reaction rates, may reveal this phenomenon. The oxygen demand of slime growth is often confused or at least coupled with that resulting from benthal demands. Slime growth demand cannot be depended upon in long-term computations, since it varies with the level of streamflow and the type of organic matter in the flowing water, as well as with the season of the year. Since the demand is likely to be

greatest during low flow, high temperatures, and with carbonaceous-type wastes, its effect under these circumstances becomes extremely significant.

3. Primary organic bottom sludge deposits represent a persistent creditor demand on the oxygen resources of the moving stream or quiescent reservoir above. Much work, some of which has already been presented in Chapter 1, Section 4, has been done on phenomena associated with benthal demands. Most researchers have found that this type of demand represents only a small fraction of the aerobic demand of creditor 1. For example, Lardieri (1954) found that the benthal oxygen demand of paper mill waste deposits is much less than the aerobic oxygen demand, and decreases markedly per unit weight of solids deposited with increasing depth. Products of anaerobic decomposition within the sludge diffuse to the surface of the deposit, where the environment tends to be aerobic. It is here that a great percentage of the actual benthal demand for oxygen is satisfied. When scouring occurs, the sludge decomposition tends to become aerobic, and hence the demand for oxygen is increased. Naturally, the k_1 value in such stream areas would appear to be greatly affected.

4. Secondary bottom deposits occur when plant growth takes place in the quiescent overlaying waters, such as pools and stagnant stream areas, and subsequently dies owing to adverse environmental conditions such as lack of sunlight, prevalence of cold weather, and presence of toxic agents. The dead algae, when deposited on the stream bottom, begin to decompose in a manner similar to that of other benthal sludge. This decomposition results not only in demands of oxygen from the stream resources, but also in the release of mineral nutrients such as nitrates and phosphates upon which additional plant life can feed. In highly eutrophicated (mineralized or fertilized) water, the secondary sludge deposit creditor demand can represent a significant portion of the total oxygen being utilized.

5. As shown in Section 2, an increase in temperature causes a decrease in the saturation value of oxygen in water. For example, at 15°C, a normal winter temperature of many of our eastern streams, the water will hold 10.03 ppm oxygen at 100 percent saturation. On the other hand, at 25°C, a normal midsummer temperature, the same water will only hold 8.18 ppm at 100 percent saturation. This represents a decrease of 18.4 percent. When this 18 percent loss in our oxygen account is coupled with the increase in reaction rate k_1, (creditor demand rate) of from 0.792 to 1.260, the loss of oxygen becomes even more critical. The increased deoxygenation reaction rate is somewhat countered by an increase in reaeration rate at these same temperatures, from .924 to 1.082. However, when one considers that the temperature increase usually occurs at a time when the dilution water is at its lowest volume, our already minimal oxygen resource account can be diminished rapidly and extensively. Investigations can predict this critical oxygen level with some degree of reliability by carrying out the stream survey during the critical period described above.

TABLE 3-3. Oxygen Consumption of Some Fish and Representative Animals

Water living	cm^3 O_2/gram/hour of wt. of animal	Nonwater living	cm^3 O_2/gram/hour of wt. of animal
Jellyfish	.0034–.005	Sparrow	6.7
Starfish	.03	Horse	.25
Leech	.023	Sheep	.34
Mussel	.0549	Dog	.83
Crayfish	.04	Butterfly (at rest)	.6
Eel	.04		
Goldfish	.07		
Trout	.22		

6. Fish require oxygen to live, and even more to propagate. Although in some cases the total demand by fish creditors is not very great, it seems to this writer that this creditor as such has been overlooked entirely. The metabolic rate, usually measured in terms of the rate of oxygen consumption per unit of biomass, differs considerably with different species. Generally, this respiratory rate is more closely related to the surface area of the organism than to its weight; thus smaller organisms generally have a higher metabolic rate; for example, the rate for nematode worms (often present in the sludge deposits in streams) is about 10 times that of the larger earthworm. In assessing the relative importance of different species populations in a given stream, it is not their numbers, or their biomass, but their total metabolic activity that is important. Those skeptics who may still feel that oxygen consumption by fishes is insignificant may be referred to Table 3-3, which was compiled by Heilbrunn (1952).

Although fish are cold-blooded, their metabolic rate at normal temperatures is not too much lower than that of warm-blooded animals. Most of the data in Table 3-3 apply to the state of rest, and the respiration rate increases with activity. The oxygen consumption of man during rest, for example, is about 1.6 to 3.3 cc/gram/hr,[2] (Drinker, 1954).

Haskell (Haskell, Davis, and Neckahn, 1960) computed that 1.7 pounds of BOD are contributed by 1,000 pounds of fish per day, which, he reported, resulted in the utilization of 0.22 parts per million of dissolved oxygen in one particular instance.

7. Organic matter entering a main stream or lake from a tributary also reflects a creditor demand which cannot be overlooked. This type of demand, although similar to creditor No. 1, usually represents a sudden and usually continuous surge in oxygen resource requirement which may upset an otherwise

[2]Computed from Drinker's (1954, p. 32) basic air consumption rate of mean of 600 liters per hour at rest.

smoothly operating amortization of oxygen debts in the main stream. It is especially important that the engineer determine the pattern of this type of organic matter inflow as well as its reaction rate after it mixes with the main flow.

8. Salinity in a flowing water or estuary causes a depression in the amount of oxygen which the body of water is able to carry. For example, at 25°C, during the midsummer critical conditions when the salt concentration is also increased to 10,000 ppm, the oxygen saturation value is reduced to 7.56 ppm from 8.18 ppm, or about 8 percent at 100 percent saturation. This phenomenon occurs rather generally in tidal streams. It may also occur to a limited degree in waters receiving certain industrial wastes such as oil field brines or cucumber brine wastes from pickle processing. Agricultural practices (such as those in irrigation) can render a stream quite salty. For example, the Colorado River has been found to contain 760 ppm of salts at Hoover Dam and 1,264 ppm by the time it reaches the Mexican border (Ripley, 1972). Runoff from salted streets in northern states has also contributed much salinity to receiving waters. The eighth creditor, although usually relatively insignificant or totally absent, should be considered as a further potential drain on oxygen resources. Salt water has been argued by some to prevent or at least deter algal blooms, with its associated good and bad ecological effects.

There are some optimistic and encouraging aspects of the oxygen balance. All is not on a downhill slide from saturation to suffocation. Fortunately, we are blessed with at least four kind benefactors who assist in "refilling the till" with oxygen.

1. Reaeration occurs in the receiving stream almost in direct proportion to the deficit in oxygen that exists in the water. Although we are grateful to this benefactor, we observe that its effect is not fully felt by the water until the deficit has reached the danger level. Since the reaeration rate K_2 depends both on the result of deoxygenation and on the existing oxygen deficit, it may be computed by the following equation

$$K_2 = K_1 \frac{\bar{L}}{\bar{D}} - \frac{\Delta D}{2.3 \Delta t \bar{D}}$$

where $\bar{L}$ and $\bar{D}$ are the average of the upstream and downstream ultimate BODs and dissolved oxygen deficits, respectively, and ΔD and Δt are the change in deficit and flow time between upstream and downstream stations, respectively. K_1 is the stream deoxygenation rate per day. This and other relationships used to obtain K_2 are the entire subject of Chapter 5.

However, as early as Streeter's work in 1925, it was proposed that K_2 depended upon many physical variable factors of streams such as velocity, depth, slope, and channel irregularity.

Therefore, an equation of the general form

$$K_2 = \frac{CV^n}{H^2}$$

was utilized. O'Connor (O'Connor and Dobbins, 1956) has modified the formula slightly to read

$$K_2 = \frac{D_2 U^{1/2}}{H^{3/2}}$$

where V and U are horizontal velocities of the water and H is the depth of flow; D_2 is the coefficient of molecular diffusion throughout the liquid film. Churchill (1961) found that the following equation was applicable to controlled uncontaminated sections below dams in the Tennessee Valley:

$$K_{2_{20°C}} = 5.026 \frac{V^{0.969}}{H^{1.673}}$$

It is apparent from the above that K_2 can be quite accurately estimated in any stream in which the velocity and depth are known, or computed in those streams that are uncontaminated. In the majority of situations, however, where a certain amount of organic contamination is encountered, K_2 will be about the same or slightly higher in contaminated streams, and can be approximated by measuring it in an uncontaminated stream. Empirical data yielding the deoxygenation rate, BOD, and dissolved oxygen levels as well as the temperature and flow time in the reach must be collected and used in these instances to obtain actual K_2.

2. Photosynthesis—the growing of generally green plant life in watercourses from sunlight, carbon dioxide, and other stream nutrients—while active, produces oxygen to help replenish the oxygen resources. In a recent experiment with a blue-green algae, *Microcystis aeruginosa*, it was found that it produced 62.5 cc of oxygen per hour per milligram of dry algae in a raw sewage environment. Odum (Odum and Odum, 1959) reports algae production rates as high as 60 grams per square meter per day in bloom situations, or as low as 0.5 gram per square meter per day in deep lakes. Oswald and Gotaas (1956) found that for every pound of algae cell synthesized 1.6 pounds of oxygen were produced. Therefore, under bloom conditions we could expect that up to 96 grams of oxygen could be produced by photosynthetic action per square meter per day. One can estimate, then, the oxygen produced by algae in any given receiving water by either (1) measuring the algae mass and surface area of the watercourse, or (2) actually measuring the photosynthetic rate of algae samples in a Warburg Respirometer in the laboratory. The sanitary engineer should keep in mind that the oxygen produced by algae is a net production figure which is composed of a predominantly negative rate at night and a potentially-positive production rate

TABLE 3-4. Maximum Effect of Temperature on Oxygen Solubility (Saturation)

Temperature (°C)	Dissolved Oxygen Solubility (mg/l)
30	7.44
25	8.18
20	9.02
15	10.03
10	11.27
5	12.79

during the daytime. An algal cell "produces" oxygen only because it makes about 1.5 times the oxygen it uses. If the oxygen produced by algae results in supersaturation during the day, oxygen may be lost to the atmosphere. Under such conditions, the negative rate at night could result in rather serious oxygen depletion.

3. Temperature decreases, which occur nightly, and normally during all seasons, but especially during spring and fall, "unlock" the oxygen bank account and allow more oxygen to be deposited. This is in accordance with the natural law that defines solubility of a gas in a liquid as proportional to the type of gas, temperature of the liquid, the equilibrium partial pressure of the solute gas in the gas phase, and the concentration of the solute gas in the liquid phase. If the type of gas (oxygen) is known, and its existing partial pressure in the air and concentration in the water are constant, its solubility varies only with the liquid temperature. This is shown in Table 3-4, as proposed by Churchill in 1961. Some idea of the magnitude of the "cold temperature" benefactor is obtained by observing the effect of going from an extreme summer temperature of 30°C to an extreme winter temperature of 5°C–a gain of about 72 percent in oxygen saturation. When this is coupled with the greatly decelerated rate of bacteriological activity at 5°C, it is small wonder that seldom does oxygen depletion in streams occur during cold temperature periods. A temperature profile of the receiving stream, during all seasons, is therefore a vital and valuable tool for the stream analyst.

4. Any dilution of our main receiving water with cold, clean tributary waters will bolster our oxygen resource account. The effect of this benefactor may be one of keeping the main stream in a "solvent" state, even when the bulk of oxygen demands comes due. Engineers should always be on the lookout for and consider the effects of this type of oxygen contribution, which may dilute the effect of organic wastes so as to reduce the resultant concentration.

Summary

Although oxygen represents only one type of stream resource susceptible to depletion, it is usually considered by both the public and by sanitary engineers as the most important. The oxygen balance is influenced by at least eight commonly known creditors and four benefactors; each must be carefully observed and evaluated by the engineer in order to assess the assimilative capacity of the stream.

3-5. WATER QUALITY—CHEMICAL AND ELECTROCHEMICAL

The biological factors significant in stream analysis include many organic and inorganic chemical constituents other than BOD and dissolved oxygen. Since most of these affect our use of water resources for drinking purposes, most criteria have been established with this use in mind. Water quality for other uses is discussed in Chapter 7. Three major considerations for this water quality include:

a. Raw water characteristics
b. Quality standards
c. Pollutants in water supply

Raw Water Characteristics

In 1963 the nation used about 355 billion gallons of water per day. By 1980 the total dependable (uncontaminated) fresh water supply that would be available in the United States was projected to be about 515 billion gallons per day. But, by the time we can achieve this fresh water supply of 515 billion gallons per day, our required needs for fresh water should have climbed to 600 billion gallons per day. These figures and facts are presented at the beginning of this section primarily to inform the reader that water reuse and more utilization of contaminated raw water will be the general rule rather than the exception during the next decade. We are not running out of water, but we are rapidly depleting our reserve of uncontaminated water. The Ohio River, for example, while once supplying its inhabitants and industries with an abundant supply of previously unused water, now is used 3.7 times before it reaches the Mississippi.

One can appreciate this accelerated demand on water when one realizes that our population of over 200 million is increasing at the rate of 2 percent per year, a rate which, if continued, will approximately double our present population in the relatively short period of 35 years. In 1972, however, the population increase slowed down, and some doubt is being expressed that the 2 percent rate of increase will hold true over the long term. Industrial production may also slow down its rise, and the per capita water use may also rise. Both of these latter contributions may overshadow any trend towards a lower population growth.

Each community, when it seeks to expand its water supply, is faced with the problem of using sources that may not be quite as pure as the present supply. This is especially true because the dwindling fresh waters usually exist in upland areas in insufficient quantity to justify economical development. Municipalities must then consider the larger, more dependable sources, which usually contain considerable contaminants. This does not necessarily signify that these supplies are polluted, for a polluted state is one in which the contamination is excessive for the desired best usage.

Quality Standards

Because of the increased contamination in our raw water supplies, the United States Public Health Service revised its drinking water standards in 1962 (U.S. Public Health Service, 1962). It serves our needs best to take careful note of these recommended standards as presented here in Table 3-5. Legislation being considered in Congress would make these recommended standards the maximum allowable values in all states; stricter laws already existing in states would be enforced as is.

In areas in which the nitrate content of water is known to be in excess of the listed concentration, the public should be warned of the potential dangers of using the water for infant feeding. Nitrates should not be acceptable in a water supply in excess of the listed concentrations where, in the judgment of the reporting agency and the certifying authority, other more suitable supplies are or

TABLE 3-5. U.S. Pulbic Health Service Recommended Drinking Water Standards

Item	Concentration (mg/l)
Alkyl benzene sulfonate (ABS)	0.5
Arsenic (AS)	0.01
Chloride (Cl)	250.0
Copper (Cu)	1.0
Carbon chloroform extract (CCE)	0.2
Cyanide (CN)	0.01
Fluoride (F)	1.0
Iron (Fe)	0.3
Manganese (Mn)	0.05
Nitrate (NO_3)	45.0
Phenols	0.001
Sulfate (SO_4)	250.0
Total dissolved solids	500.0
Zinc (Zn)	5.0

TABLE 3-6. U.S. Public Health Recommended Drinking Water Standards

Item	Concentration (mg/l)
Arsenic (As)	0.05
Barium (Ba)	1.0
Cadmium (Cd)	0.01
Chromium (Cr^{+6})	0.05
Cyanide (Cn)	0.2
Lead (Pb)	0.05
Fluoride (F) 2 times that of Table 3-5	(about 2.0)
Selenium (Se)	0.01
Silver (Ag)	0.05

can be made available. The presence of the substances shown in Table 3-6, in excess of the concentrations listed, shall constitute grounds for rejection of the supply. In addition, other contaminants which are becoming more significant and are considered dangerous by the U.S. Public Health Service include (1) radioactivity and (2) coliform group bacteria. Limits of these contaminants are found in Table 3-7. Also, the U.S. Public Health Service states that drinking water should contain no impurity which would cause offense to the sense of sight, taste, or smell. These characteristics are known as physical, and are given in Table 3-8. Under general use, these limits should not be exceeded. The pres-

TABLE 3-7. Other Significant Contaminants

Radioactivity	Concentration
*Radium 226	3 μμc/liter
*Strontium 90	10 μμc/liter
†Gross Beta radiation	1000 μμc/liter
Coliform bacteria (10 ml portions of water examined)	10% of total portions (10 ml sample) examined in any month showing presence of coliform bacteria

*When these concentrations are exceeded, a water supply shall be approved by the certifying authority *if* surveillance of total intakes of radioactivity from *all* sources indicates that such intakes are within the limits recommended by the Federal Radiation Council for control action.

†In the known absence of Sr^{90} and alpha emitters, 1,000 μμc/l shall be grounds for rejection of the supply except when more complete analysis indicates that concentrations of nuclides are not likely to cause exposures greater than the Radiation Protection Guides as approved by the President on recommendation of the Federal Radiation Council.

TABLE 3-8. Physical Characteristics

Turbidity	5 units
Color	15 units
Threshold odor number	3

ence of coliform groups in three or more 10 ml portions of a single standard sample shall not be allowed if this occurs in

1. two consecutive samples
2. more than 1 sample/month when less than 20 samples are examined/month
3. more than 5 percent of the samples when 20 or more samples are examined/month

When using the Membrane Filter Technique, the arithmetic mean coliform density of all standard samples examined per month shall not exceed *one per 100 ml.* Coliform colonies per standard sample shall not exceed $\frac{3}{50}$ ml, $\frac{4}{100}$ ml, $\frac{7}{200}$ ml, or $\frac{13}{500}$ ml in

1. two consecutive samples
2. more than 1 standard sample when less than 20 are examined/month or
3. more than 5 percent of the samples examined when more than 20 are examined/month

Contaminants in Water Supply

In this text we cannot discuss the causes of these contaminants in the water supply and the implications they may have on our health and economic life. However, we can study some of the newer contaminant limits–as indications of the changes that are being made in quality control in water, and some that we may expect. The following contaminants have been selected for a brief discussion, since each originates from wastewaters which are reaching our watercourses in startlingly increasing amounts. This list is not complete but is intended only to include a few of the more significant contaminants.

1. Turbidity
2. Alkyl benzene sulfonate (ABS)
3. Chloride and dissolved solids
4. Carbon chloroform extract (CCE)
5. Fluoride
6. Phenols
7. Chromium
8. Radioactivity
9. Coliform bacteria
10. Chlorinated hydrocarbons
11. Phosphates
12. Fossil oil

Turbidity—Turbidity is a general cloudiness of water resulting in an increased reflection of light directed into it. In natural stream conditions is is usually caused by silt originating from soil erosion. However, many municipal and industrial wastewaters will also produce turbidity in streams. The U.S. Public Health Service (1962) states that requirements for physical characteristics can be met during general use by properly designed and operated treatment plants and distribution systems. Further, it signifies that failure to meet these requirements is an indication of either inadequate treatment facilities or improper operation of the system. However, many of our raw water supplies are becoming more turbid owing to increased erosion of the naked soil and the presence of increased industrial contamination of this type. These contaminants may not be so readily removed in even a properly designed and operated water plant. In addition, many plants will require redesigning to remove this type of contamination from their water supply. It may represent an entirely new type of contaminant for a given plant.

ABS—Alkyl benzene sulfonate (ABS) is a typical anionic surfactant, as shown in Fig. 3-3, a synthetic organic chemical having high residual affinity for organic matter at one end of its molecule and low affinity at the other. The U.S. Public Health Service recommends that the concentration of ABS in drinking water be limited to 0.5 mg/l, inasmuch as higher concentrations may cause the water to exhibit *undesirable taste and foaming.* Concentrations of ABS above 0.5 mg/l are also indicative of questionably undesirable levels of other sewage pollution such as hardness, salts, bacteria, etc.

Since sewage normally contains about 10 ppm of ABS, this means that raw sewage or its equivalent would be diluted only 20 times in the receiving stream at 0.5 ppm. There apparently is no question here of ABS toxicity directly. One study (National Research Council, 1956) evidently shows "that the interfacial tension existing in the digestive tract of a *healthy human* is so low that it will not be further lowered by the small amounts of synthetic surfactants which may be present in food."

Since this limit is set so low, and since the detergent manufacturers are changing over to production of a more biodegradable surfactant, the visible effect of this "indicator contaminant" will no longer be useful to the water engineer. In other words, the sewage pollution may still exist—and as far as coliform bacteria are concerned, may even be in greater concentration—although the limits for ABS are being met satisfactorily. Biodegradability, however, is being measured by oxidation in an activated sludge treatment plant. All detergent wastes not

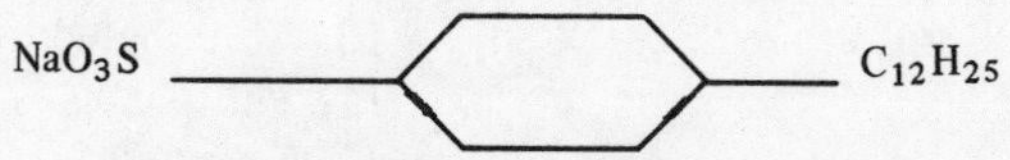

NaO_3S — (benzene ring) — $C_{12}H_{25}$

FIG. 3-3. Typical structure of ABS.

subjected to this type of treatment may not exhibit the standard degree of degradation expected.

Chloride and dissolved solids–The importance of chloride, sulfate, and dissolved solids as they affect water quality hinges upon their taste and laxative properties. There is evidence that excessive amounts of these constituents cause consumer reactions which may result in individual treatment or rejection of the supply.

The Public Health Service limits of 250 mg/l for chloride are largely based upon taste results presented in a study by Richter and MacLean (1939). However, there may be a great difference between a detectable concentration and an objectionable concentration of the neutral salts. The factor of acclimatization is particularly important. The U.S. Public Health Service points out that more than 100 public water supplies in the United States provide water with more than 2,000 mg/l of dissolved solids, even though the standard of 500 ppm is recommended.

Carbon chloroform extract–This method of CCE analysis provides a safeguard against the intrusion of excessive amounts of potentially toxic material. It is a generalized procedure (Middleton, 1961) for measuring the organic chemical content of drinking water. Analysis of data indicates that water supplies containing over 200 micrograms CCE per liter of water represent to the water consumer an "exceptional and unwarranted" dosage, with ill-defined chemicals. The Public Health Service is primarily concerned here with insecticides, herbicides, and other agricultural chemicals, as well as accidental spills from industrial operations. The procedure in general calls for passing a large quantity of water (usually 5,000 gallons) through an activated carbon filter, drying the carbon material in clean air, and extracting the organic contaminants from the carbon with chloroform for about 35 hours.

Increased emphasis is being given, and will continue to be given, to production of more food from a given acreage of farmland in order to feed the increasing population of the world adequately. In order to do this, insecticides may be used more extensively and may find their way into water supplies. However, recently more restrictions have been placed upon DDT and other related insecticides, so that the effect of these compounds upon our watercourses in the future may not be as extreme as anticipated. New industrial processes and products will continue to create an increase of the exotic type of chemicals in wastewaters. The use of antibiotics, diethylstilbestrol (DES), and other chemicals to fatten cattle in feedlots is an example, since as much as 50 percent of the injected or ingested material is excreted. Because of their characteristic resistance to biological degradation in our watercourses, water supply and treatment plant engineers will be faced with the difficult task of removing these from the finished water. Our general treatment practices today do not provide satisfactory solutions to this problem.

Fluoride—This is a rather unique problem; we are advised to add this chemical to water supplies in which it is deficient in order to prevent dental cavities in the human consumer. Although about 30 percent of the public water supplies today are supplemented with fluoride for this purpose, there still exists a considerable resistance to fluoridation. The objections to even small (1 ppm) doses are primarily raised by the following groups: (1) "pure water" advocates; (2) religious groups or anti-forced medication supporters; (3) medical personnel who seriously question the safety of even small quantities of fluoride taken internally; and (4) persons who continually vote against any positive actions. We should be primarily concerned with group (3).

According to the Public Health Service information presented by Shaw (1954), bone changes will occur only when the water containing 8–20 mg fluoride per liter is consumed over a long period of time. Stained teeth, however, can occur with continued use of water below this fluoride level. That it prevents dental cavities in concentrations of 1 mg/l is almost uncontested.

Phenols—(Assumed to be the presence of single, double, or triple benzene rings with one OH group attached to each.) This term also includes cresols and xylenols. Although the limit of 1 part per billion or 0.001 mg/l remains unchanged from the older 1946 drinking water standards, it would be advisable for us to consider its implication. This contaminant arises largely from industrial sources such as synthetic textile plants or steel mills. Both types of industries, as well as others producing phenols, are increasing production. The level of .001 mg/l is extremely low, and any slight increase in the present stream concentration levels will constitute contravention of standards. The phenol limit is set because of the undesirable taste that often results from chlorination of waters, a practice which is also on the increase because of increased bacterial contamination of our water supplies. Concentrations of phenol which are injurious to health are several thousand times that of the standard established for taste.

Chromium—This limit of .05 mg/l for hexavalent chromium was originally established in the 1946 standards, based upon "the lowest amount analytically determinable at the time it was established." The U.S. Public Health Service admits that at present, the level of chromate ion that can be tolerated by man for a lifetime without adverse effects on health, is unknown. We do know that when inhaled, chromium is a known carcinogenic agent for man, but that the trivalent form apparently exhibits none of the toxicity of the hexavalent form.

We can look for increased chromium in water supplies because of the emphasis being placed upon corrosion resistant metals in all types of terrestrial as well as extraterrestrial equipment. Once again, little of this contaminant is removed by existing treatment methods in general use. Since both radioactivity and coliform bacteria (items 8 and 9 in the list on page 101) have already been considered in this section, no further mention of them will be made here.

Chlorinated hydrocarbons—These are known to be carcinogenic in test animals. States sometimes limit them to a concentration of 0.5 parts per billion in treated drinking water supplies. They are produced either by direct discharge of these compounds into water supplies, or by chlorination (during water treatment) of hydrocarbons in the supply.

Phosphates—Water supplies from runoff from fertilized land areas or from wastes containing detergents reaching these supplies cause phosphates to arise. Phosphates stimulate algae growth in exposed reservoirs, resulting in tastes and odors in drinking waters.

Fossil oils—These oils can reach water supplies by mining wastes or by their inadvertent discharge into ground areas and the subsequent seepage into surface waters. They add taste, odor, and an unpleasant appearance to the water.

Other Contaminants

Lead is a cumulative poison, which makes it difficult to establish definite safe limits. The limit was reduced by 50 percent in 1962 to .05 ppm.

Mercury, especially when present in the form of methyl mercury, is also a cumulative poison about which there are still many unknowns. A limit of 0.005 mg/l has been proposed by the U.S.S.R. Its greatest effect appears to be on the fish food chain. The Food and Drug Administration has set a fish pollution limit of 0.5 mg/l mercury that may reach a stream or lake from many sources such as natural, agricultural, and industrial (chlorine production, mercury cells, and plastics), as well as from individual homes, laboratories, and hospitals. This limit applies only to the fish and not to the concentration of mercury in the water.

Japan has recently (Oberdorfer, 1972) announced the first restrictions on the amount of fish which citizens should consume in relation to meat and vegetables. For example, the allowable limits announced for a week, based primarily upon mercury hazards, include 47 slices of tuna, or 2.3 medium-sized squid, or 10.3 small sea bass, etc. This is not very much fish when one considers that the average Japanese meat diet consists of over 90 percent fish and less than 10 percent beef.

The U.S. Public Health Service states that polluted sources should be used only when other sources are economically unavailable, and then, only when the provision or personnel, equipment, and operating procedures can be depended upon to purify and otherwise protect the drinking water supply continuously. This leaves the door open to the public to use contaminated water supplies when necessary or economical, but no one as yet has defined the term economical, except to say that it is determined by what the people are willing to pay. But are the people willing and/or able to pay the true economic worth for water? Some aspects of this question are considered in both Chapters 8 and 9.

The Public Health Service also recognizes that surface waters may be sufficiently protected to warrant their use as a supply without coagulation and filtration, although they are becoming rare. When not using complete treatment, it recommends intensive surveillance of the quality of the raw water and the disinfected supply in order to ensure constant protection. During times of unavoidable and excessive pollution of a source already in use, it may be necessary to "provide extraordinary treatment such as exceptionally strong disinfection, improved coagulation, or special operation." If the pollution cannot be removed satisfactorily by treatment, the Public Health Service recommends that use of the source should be discontinued until the pollution has been reduced or eliminated.

It appears to the writer that even these measures are not adequate to treat excessively contaminated water, and that more effective and new treatment methods should be developed. For example, little if anything is recommended by the Public Health Service concerning the subject of algae contamination, virus removal, or excessive mineral contaminants such as phosphates. More information on these and other modern day contaminants will be forthcoming in the next few years.

To add to our water quality problem, the water supplies of our country are warming up, owing in part already to increased reuse. The elevated temperatures may encourage chemical and biological contaminants to develop and warrant consideration by water engineers.

In addition, the United States Environmental Pollution Agency in 1980 published a list of chemicals which are classed as toxic in the water environment. The reader can refer to page 147 and Tables 7-15A and 7-15B for specific chemicals and concentrations in the list.

QUESTIONS

1. What is meant by BOD?
2. How would you predict the decomposition rate of an unknown wastewater?
3. What factors determine the shape of the BOD curve?
4. How does the temperature affect the deoxygenation rate of BOD?
5. What is the real meaning of the value of "k"?
6. How is the oxygen utilization rate comparable to home mortgage payment rates?
7. Enumerate the various creditors of the oxygen account in any stream.
8. What are the few benefactors of the oxygen account?
9. Discuss the significance of each contaminant tested in Table 3-5, page 99, for drinking water use. Should any of these be monitored in streams used for purposes other than drinking? Tables 3-7 and 3-8 characteristics?
10. Relate the importance of the 12 contaminants listed on page 101 as pollutants. Are there any others not listed which would qualify as potential, significant stream contaminants?

REFERENCES

Churchill, M. A.: Effect of Water Temperature on Stream Reaeration, *J. San. Eng. Div.* SA6:2997, November 1961, 31st Progress Report of Committee on San. Eng. Res.

Drinker: *Industrial Dust*, 2nd ed., McGraw-Hill, 1954.

Elmore, H. L., and W. F. West: Effects of Water Temperature on Stream Reaeration, *J. San. Eng. Div., Proc. ASCE*, **87** (Sa6):59, 1961.

Food Protection Committee, *The Relation of Surface Activity to the Safety of Surfactants in Foods, Nat. Acad. Sci., Nat. Res. Council Pub. No. 463*, Washington, D.C., 1956.

Haskell, Davis, and Neckahn: Fish Hatchery Pond Design, *U.S. Fish and Game J.*, July 1960.

Heilbrunn.: *An Outline of General Physiology*, W. B. Saunders Co., Philadelphia, 1952.

Howe, R. H. L.: K_2-Temperature Relation, *J. Env. Eng. Div., Proc. ASCE*, **103**(EE4):729, 1977.

Hunter, J. S. III: Accounting for the Effects of Water Temperature in Aeration Test Procedures in Proceedings Workshop Toward an Oxygen Transfer Standard, EPA-600/9-78-021, p. 85, 1979.

Lardieri, N. J.: *TAPPI*, **37**(12):705, 1954.

Metzger, I.: Effect of Temperature on Stream Aeration, *J. San. Eng. Div., Proc. ASCE*, **94**(SA6):1153, 1968.

Middleton, F. M.: Nemenclature for Referring to Organic Extracts Obtained from Carbon with Chloroform or Other Solvents, *J. Am. Water Works Assoc.*, **53**:749, 1961.

Oberdorfer, Don: In Fish-Eating Japan, the People Are Told to Cut Their Consumption, *International Herald Tribune*, Paris, France, June 26, 1972, p. 1.

O'Connor, D., and W. Dobbins: The Mechanisms of Reaeration in Natural Streams, *J. San. Eng. Div. (ASCE)*, SA6:1115-1-1115-30, December 1956.

Odum, E. P., and H. T. Odum: *Fundamentals of Ecology*, 2nd ed., W. B. Saunders Co., Philadelphia and London, 1959, 546 pp.

Oswald, W. A., and H. B. Gotaas: *Ind. Eng. Chem.*, **489**:1457, 1956.

Richter, C. P., and A. MacLean: Salt Taste Threshold of Humans, *Am. J. Physiol.*, **126**:1-6, 1939.

Ripley, Arthur: Salt in Colorado River Up, 7 States Want U.S. to Act, *New York Times*, February 21, 1972, p. 45M.

Shaw, J.: Fluoridation as a Public Health Measure, *AAAS Pub. No. 38*, Washington, D.C., 1954, pp. 79-109.

Standard Methods for the Examination of Water Sewage and Waste. 13th ed., 1970, Fed. Water Poll. Control Assoc., American Water Works Assoc., and U.S. Public Health Service, Washington, D.C.

Streeter, Harold, and Earl Phelps: *Stream Sanitation*, John Wiley and Co., New York, 1944.

——, and E. Phelps: A Study of the Purification of the Ohio River, *U.S. Publ. Health Service Bulletin No. 146*, Washington, D.C., 1925.

Theriault, E. J.: Public Health Bulletin 173, United States Public Health Service, 1927.

U.S. Public Health Service, *Drinking Water Standards,* U.S. Dept. Health, Education, and Welfare Pub. No. 56, Washington, D.C., 1962.

Velz, C.: Recovery of Polluted Streams, *Water and Sewage Works*, **100**(52):495, 1953.

4 MATHEMATICS INVOLVED IN PREDICTING STREAM POLLUTANT DEOXYGENATION RATES

As we have seen in Chapter 3, the computation of the deoxygenation rate k_1 is an important area of study. We have seen that k_1 is not only both time and temperature dependent, but also dependent upon the type of organic matter and the biological, chemical, and physical condition under which it is undergoing degradation.

In general we can remind the reader at the beginning of this chapter that normal domestic sewage was found to deoxygenate at a daily rate of the fraction of about 0.1 per day at 20°C under standardized laboratory examination conditions. Rates higher than this will result in a more rapid depletion of oxygen in the receiving stream and, conversely, rates lower than 0.1 per day will allow a slower demand for oxygen below the point of entry in a stream. To the downstream user the rate of deoxygenation becomes of prime importance, as it affects the resulting condition of the watercourse, as far as oxygen is concerned, at its point of use.

To date, there exist at least four generally accepted methods for computing the reaction rate of the BOD curve k_1: (1) the least-squares technique; (2) the slope method; (3) the moments method; and (4) the logarithmic method. These will be described and referenced briefly, with examples of each given for clarification. In addition, (5) our general laboratory procedure, and (6) Rhame's two-point method are also presented.

For an excellent treatise on the methods used for obtaining k, the reader is referred to a publication by Gaudy et al., (1967).

4-1. LEAST-SQUARES TECHNIQUE

Theriault (1931) first applied the least-squares method to a set of BOD data to obtain the most mathematically precise deoxygenation rate per day, k_1, and an ultimate value of BOD, L, where the rate of change in BOD with time becomes essentially zero. A description of the mathematical technique is in order here to show the reader how the two normal equations are developed which, when solved simultaneously, will yield values of K and L.

Given: Log Y (BOD) is roughly a linear $f(x)$ (time) $[f(x, t)]$; Have "n" values of y for corresponding values of x. When plotted, one obtains a scatter pattern.

Problem: Find the relation between y (BOD) and x (time) which will give the best possible fit, that is, find equation of line of best fit.

Solution: Use the method of least squares, which is based on the fact that the sum of the squares of the deviations from the mean is a minimum. The difference between a "y" value on the line of best fit and a corresponding observed value is termed a residual "R." The line of best fit is obtained when the sum of the squares of the residuals (ΣR^2) is a minimum; that is, a line must be drawn such that ΣR^2 = min.

What is now needed is the equation of this line.

The generalized equation of a straight line

(1) $y = mx + b$

where

m = slope
b = y axis intercept

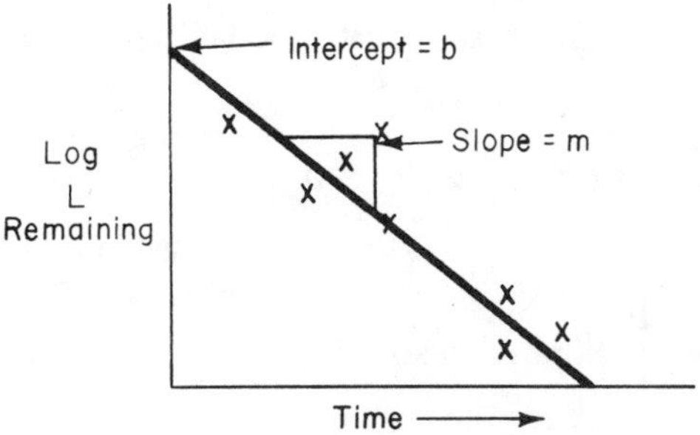

If values on the line of best fit are designated as y and corresponding observed values as y', the sum of the residuals will be

(2) $\sum R = \sum (y - y') = \sum (mx + b - y')$

and the sum of the squares of the residuals will be

(3) $\sum R^2 = \sum (y - y')^2$

or

(4) $\sum R^2 = \sum (mx + b - y')^2$

The problem now is to find values of "m" and "b" which will make $\Sigma R^2 =$ minimum. This condition will be satisfied if the first derivatives of Equation (4) with respect to m and b are set equal to 0.

(5) $$\frac{\partial \left(\sum R^2\right)}{\partial m} = 2 \sum R \frac{\partial R}{\partial m} - 0$$

(6) $$2 \sum R \frac{\partial R}{\partial m} = 2 \sum (mx + b - y)(x) = 0$$

or

(7) $\sum (mx + b - y')(x) - \sum (mx^2 + xb - xy')$

(8) $$\frac{\partial \left(\sum R^2\right)}{\partial b} = 2 \sum R \frac{\partial R}{\partial b} = 0$$

or

(9) $$2 \sum R \frac{\partial R}{\partial b} = 2 \sum (mx + b - y')(1) = 0$$

(10) $\sum (mx + b - y') = 0$

For "n" observations Equation (10) becomes

I. $\sum (mx + b - y') = m \sum x + nb - \sum y' = 0$ (First "normal" equation)

and Equation (7) becomes

II. $\sum (mx^2 + xb - xy') = m \sum x^2 + b \sum x - \sum xy' = 0$

(Second "normal" equation)

$$\left.\begin{array}{ll} \text{I.} & m \sum x + nb - \sum y' = 0 \\ \text{II.} & m \sum x^2 + b \sum x - \sum xy' = 0 \end{array}\right\} \text{Normal equations}$$

Solution of these two equations simultaneously for m and b gives the constants of the line of the best fit.

Theriault's Tables (1931) may be used to simplify calculations and remove the necessity of using log tables. He computed the function $(1 - 10^{-kt})$ for a range of k and t values. If y, L, and k are known, a function can be referred to in

Theriault's Tables which can be divided into y to obtain L, the ultimate BOD, or, from a knowledge of the fraction of the ultimate BOD exerted at any time, k_1 can be obtained from his Tables by using the fraction of days of incubation. The writer has included this table and others to assist the reader in these computations (see Appendix).

Procedure for Bimolecular (Second-Order) Reaction

Most workers have concluded that the first stage of oxidation of organic matter by microorganisms is one following a monomolecular reaction; that is, the rate of oxidation is directly proportionate to the amount of one reactant remaining to be oxidized. Orford and Ingram (1953) contended that this was a poor expression of a complex biological phenomenon. This is commonly referred to as a first-order reaction. I have found many instances of sewage-industrial waste mixtures which were characterized better by a second-order reaction; that is, by an oxidation rate which depends upon the concentration of two reactants remaining to be oxidized. As municipal wastes include more industrial complex chemicals, something other than monomolecular equations becomes more descriptive and accurate. In fact, Gaudy et al. (1967) caution the researcher "against indiscriminately fitting BOD data to such a curve (monomolecular) and we have purposely selected a BOD curve which does not follow first-order decreasing rate kinetics simply to emphasize the fact that the many methods for obtaining K and L do not apply to all cases." They have proven that some of the 16 methods used to compute k, based upon the monomolecular reaction rate, yielded results in considerable variation with the plot of the observed BOD data.

The two normal equations to be used when presuming a bimolecular reaction rate are derived as follows

$$(1)\quad \frac{-d(\text{reactants remaining})}{dt} = k[A]^2 = k(\text{reactants remaining})^2 \text{ or } k[A_{\text{react}}][B_{\text{reactant}}]$$

where (reactants remaining) $= L - y_t$ and y_t = amount BOD exerted at time t; L = ultimate, first stage BOD in ppm, then

$$(2)\quad \frac{d(L-y)}{dt} = k(L-y)^2$$

By dividing both sides of Step 2 by $(L - y)^2$ and multiplying by dt we obtain

$$(3)\quad \frac{d(L-y)}{(L-y)^2} = k\,dt$$

If we integrate both sides of Step 3 within the limits of $y = y$ and $y = 0$ and $t = t$ and $t = 0$ we obtain Step 4

$$\int_{y=0}^{y=y} \frac{d(L-y)}{(L-y)^2} = \int_{t=0}^{t=t} -k\,dt$$

(4) $$\frac{1}{L} - \frac{1}{(L-y)} = kt$$

Solving for y we obtain Step 5

(5) $$-y = \frac{k1^2 t}{-kLt + 1}$$

Now by dividing both the numerator and denominator of the right side by $k1^2$ we put Equation (5) in the form of a second-order reaction as Equation (6).

(6) $$y = \frac{t}{\dfrac{1}{kL^2} + \dfrac{1}{L}}$$

In a second-order reaction form of

$$y = \frac{t}{a + bt}$$

$$a = \frac{1}{kL^2} \quad \text{and} \quad b = -\frac{1}{L}$$

This generalization of Step 6 may be linearized by rearrangement as Step 7

(7) $$\frac{t}{y} = a + bt$$

where in plotting t/y vs. t a straight line should be obtained where a = the y intercept and is = $1/kL^2$ and b = the slope and is = $1/L$.

By least-squares treatment of Equation (7) we obtain the two equations, (8) and (9), which when solved simultaneously for a given set of observed BOD data will yield k and L values

(8) $$\sum\left(a + bt - \frac{t}{y}\right) = a \quad \text{or} \quad ta + b\sum t - \frac{\sum t}{y} = 0$$

(9) $$\sum\left(a + bt - \frac{t}{y}\right)t = 0 \quad \text{or} \quad a\sum t + b\sum t^2 - \frac{\sum t^2}{y} = 0$$

When b is obtained from the simultaneous solution of Equations (8) and (9), solve for L by the relationship $b = 1/L$. Then solve for k, by knowing L and a in the relationship $a = 1/kL^2$.

A tannery dehairing waste yielded the following BOD data in the laboratory, using the standard dilution bottle technique over a 10-day period:

Time (days)	*BOD exerted at 20°C (ppm)*
0	0
1	1447
2	3363
3	2416
4	3806
5	3052
6	4388
7	4155
8	4350
9	4109
10	4200

As the reader can see, these results are quite scattered but may by typical of some of the practical everyday BOD data obtained with industrial wastes in laboratory analyses using no special techniques or precautionary measures. The observed data (above) are plotted in Fig. 4-1 and a curve is fitted to the plotted data by eye. From this fitted curve the ultimate BOD value of 4,600 ppm is estimated and used to compute a reaction rate k. The data used for the least-squares computation are shown in Table 4-1. The k value of .113 and L value of 3,569 do not agree with the observed data, as can readily be seen when the calculated values are plotted (Fig. 4-1). Another trial of an estimated L value will be necessary to yield k and L values that more closely resemble the observed values.

Two normal equations are:

(1) $\sum \log (L - Y) = m \sum X + nb$

where n = no. of days of values, and

(2) $\sum X \log (L - Y) = m \sum X^2 + b \sum X$

Substituting in the above normal equations the summations above

(1) $33.06830 = 55m + 11b$

(2) $152.95548 = 385m + 55b$

Solving simultaneously by multiplying (1) by 7 we obtain

$$b = 3.569 = \log L \quad \therefore L = 3{,}750$$

$$-k = m = -0.113 \quad \therefore k = 0.113$$

In using the least-squares technique the user must decide whether the deoxygenation rate is monomolecular (first-order) or bimolecular (second-order). To do this he may use the least-squares technique to determine the equation best fit for each of the two reactions (first- and second-order). He ends up with two

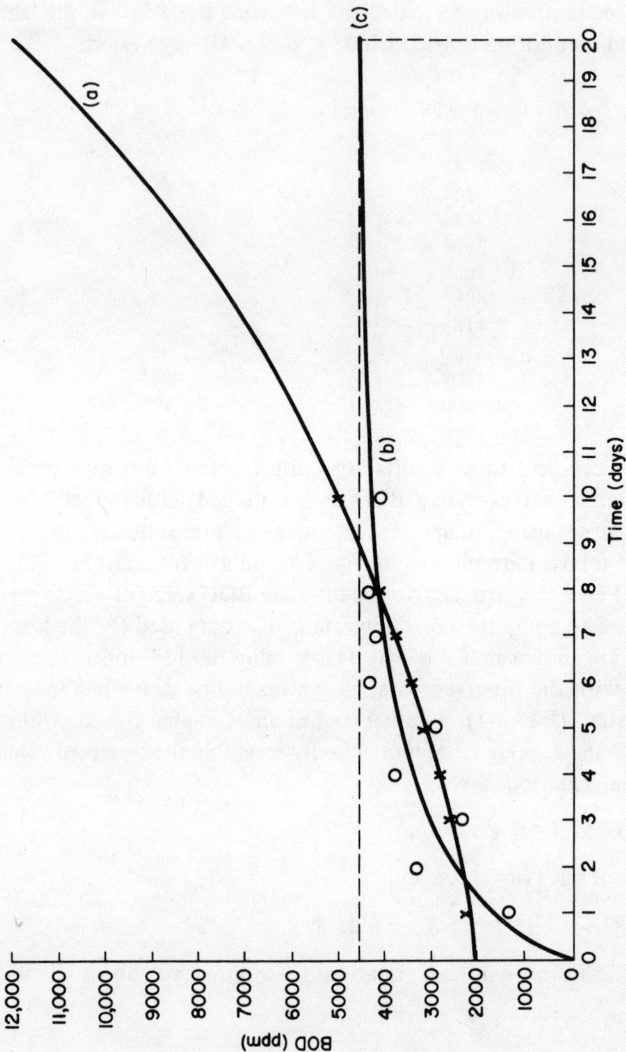

FIG. 4-1. BOD versus time. (o) = actual BOD values; (x) = least-squares values; (a) = theoretical curve; (b) = fitted curve; (c) = estimated values of 4,600 ppm.

TABLE 4-1. Least-squares Method of BOD Analysis for Computing K_1 and L

X Time in (days)	$*Y$ BOD exerted	$L - Y$ BOD remaining	Log $(L - Y)$	X log $(L - Y)$	X^2
0	0	4600	3.66276	0	0
1	1700	2900	3.46389	3.46389	1
2	2550	2050	3.31175	6.62350	4
3	3000	1600	3.20412	9.61236	9
4	3360	1240	3.09343	12.37368	16
5	3650	950	2.97772	14.88860	25
6	3870	730	2.86332	17.17992	36
7	4040	560	2.74819	19.23733	49
8	4160	440	2.64345	21.14760	64
9	4230	370	2.56820	23.11380	81
10	4260	340	2.53148	25.31480	100
ΣX=55			Σ=33.06830	Σ=152.95548	Σ=385

*Values of $L - Y$ were not taken from original data but were selected from curve fitted by eye (Fig. 4-1).

equations of best fit for the specific type reaction. In order to ascertain which of the two more closely fits the observed laboratory BOD test data, he must choose the one equation which yields the smaller ΣR^2, signifying a closer fit of the actual data. In other words, the "best fit" equation is the one for which the sum of the squares of the deviations from the equation curve and the observed data is smaller.

4-2. SLOPE OF METHOD FOR COMPUTING k_1

Thomas (1937) states that "such methods as the Reed and Theriault undoubtedly give the most consistent and, in general, the most accurate results. Unfortunately the computations involved are somewhat cumbersome. Moreover, the computations may have to be repeated because of the selection of the trial value."

The Thomas method assumes a linear relationship between the rate of change of BOD and the BOD value itself, and is useful when the interval BOD data are unequal. It is an attempt to simplify the least-squares technique and only applies when a monomolecular rate occurs.

$$\frac{dy}{dt} = k^1 (L - y)$$

This shows a linear relationship between dy/dt, the rate of change of BOD, and y, if we let $dy/dt = y^1$ and use the least-squares procedure for L and k. The least-square sum is

$$R = Lk - ky - y^1$$

which, in *a* form is

$$a + by - y^1 \text{ is a minimum}$$

where $a = L/k$ and $b = -k$; or

$$\sum R^2 = \sum (a + by - y^1)^2 \text{ is a minimum}$$

The two normal equations which can be written to solve for a and b when R is a minimum are:

$$na + b\sum y - \sum y^1 = 0$$

by multiplying by coefficient of a, and

$$a\sum y + b\sum y^2 - \sum yy^1 = 0$$

by multiplying by coefficient of b

In this case

$$\frac{dy}{dt} = y^1 = \frac{(y_2 - y_1)\left(\dfrac{t_3 - t_2}{t_2 - t_1}\right) + (y_3 - y_2)\left(\dfrac{t_2 - t_1}{t_3 - t_2}\right)}{(t_3 - t_1)}$$

which represents the slope of the curve at t_2 when the r's are equal intervals

$$\frac{dy}{dt} = y^1 = \frac{y_3 - y_1}{2\Delta t} \quad \text{or} \quad \frac{y_{n+1} - y_{n-1}}{t_{n+1} - t_{n-1}}$$

The BOD data shown on page 113 and plotted in Fig. 4-1 for a tannery unhairing waste were used in computing k_1 and L by the Thomas Slope Method. The calculations are shown in Table 4-2 below.

$$9a + 31086b - 3433 = 0 \rightarrow na + b\sum y - \sum Y' = 0$$

$$31086a + 115365924b - 8634707 = 0 \rightarrow a\sum y + b\sum y^2 - \sum Y'y = 0$$

$$b = -0.4031$$

$$a = 1774$$

$$-b = K' = 2.303k - 0.4031 \rightarrow -b = K' = 2.303k$$

$$\therefore k = \frac{.4031}{2.303} = \frac{0.174}{\text{day}}$$

$$L = \frac{a}{k'} = \frac{1774}{0.4031} = 4400$$

where

K' = reaction rate in common log form

k = reaction rate in natural log form

TABLE 4-2. Thomas Slope Method for Computing k_1 and L of Wastewater

$$*Y'_n = \frac{Y_{n+1} - y_{n-1}}{t_n + 1 - t_n - 1}$$

Time in days t	BOD observed y	y'_n	Y'_y	y^2
1	1,447	1,682	2,433,854	2,093,809
2	3,363	485	1,632,055	11,309,769
3	2,416	222	536,352	5,837,056
4	3,806	318	1,210,308	14,485,636
5	3,052	291	888,132	9,314,704
6	4,388	552	2,422,176	19,254,544
7	4,155	−19	−78,945	17,264,025
8	4,450	−23	−100,050	18,922,500
9	4,109	−75	−308,175	16,833,881
$n = 9$	$\Sigma y = 31{,}086$	$\Sigma Y'_n = 3{,}433$	$\Sigma Y'_y = 8{,}634{,}707$	$\Sigma y^2 = 115{,}365{,}924$

*The use of this method for tannery waste does not imply that the waste is degraded monomolecularly. On the contrary, there is some evidence that it may not decompose in this fashion. This may account for some of the variation obtained in k_1 and L values in this chapter, using different methods.

4-3. MOMENTS METHOD

Moore, Thomas, and Snow (1950) arrived at the constants by a simpler mathematical route—that of taking moments of the determined values about the vertical coordinate axis of the BOD versus the time curve.

The sum of a number of observed BOD values, Σy, divided by the sum of the products of the observations and their times, Σty, is a numerical quantity that depends only on the k value of the sewage observed; therefore, the value of this ratio can be determined in advance for a range of k values, and plotted as a curve of $\Sigma y / \Sigma ty$ vs. k. A similar curve can be computed that will yield the value of L. These curves apply only to a specified time sequence of observations which must be selected in advance. The authors prepared three time sequence curves (Moore et al., 1950), one of which we reproduce here as Fig. 4-2. The data obtained from the moments method of the tannery waste (Table 4-3) was used with Fig. 4-2 to find k_1 and L as follows

From Fig. 4-2: $k_1 = 0.204$

From Fig. 4-2: $\dfrac{\Sigma y}{L} = 5.38$ and $L = 4205 = \dfrac{(22{,}627)}{(5.38)}$

These curves were prepared by Thomas and Moore, using several types of wastes or similar wastes but conducted so as to yield varying reaction rates, such

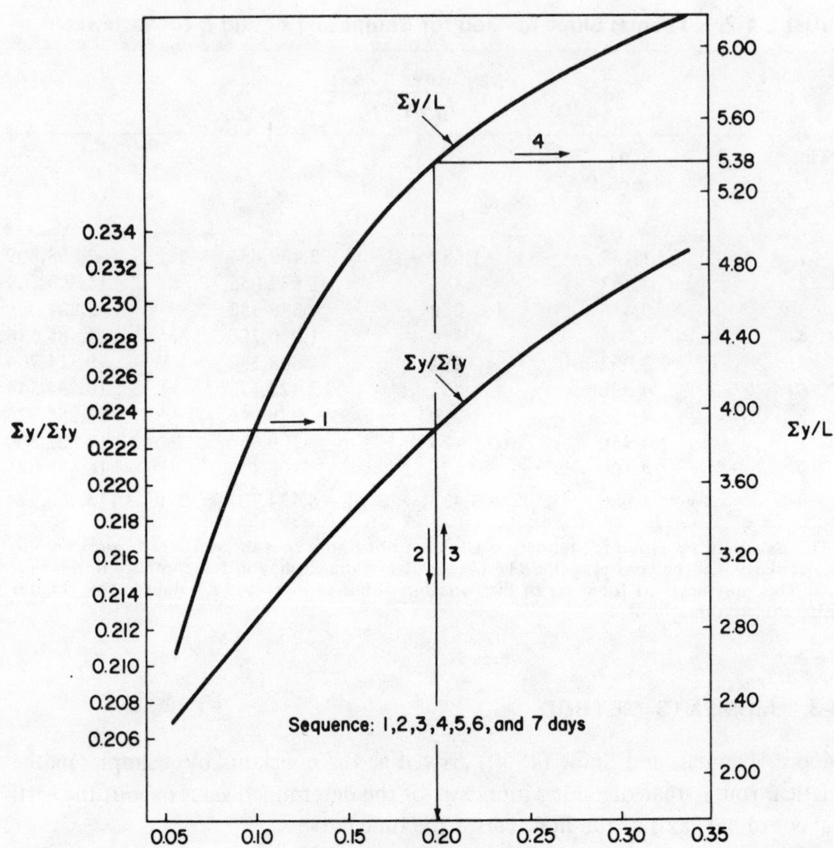

FIG. 4-2. Curve for K and L calculations. (*Thomas and Moore, 1950.*) The procedure for using this graph is to begin on the $\Sigma\, y/\Sigma\, ty$ ordinate at the left and follow the arrowed numbers from 1 to 4 to obtain k_1 of 0.204 and $\Sigma\, y/L$ of 5.38, which yields an L value of 4,205 when the 5.38 is divided into $\Sigma\, y$ of 22,627.

as by varying temperatures of reaction. They can only be used for the precise sequences of 1-, 2-, 3-, 4-, 5-, 6-, and 7-day BOD values (or an exact multiple of these days).

Thomas and Moore realized that the fact that BOD determinations must conform to the same time schedule as is used in setting up the curve constituted a decided disadvantage of the method. They felt, however, that the reduction in labor of computation goes far in compensating for this disadvantage. They also made the assumption that none of the sample data is more accurate than any other. In the case of the tannery waste BOD data, at least the adherence to the assumption is doubtful (see 2-day BOD value). The authors compared their re-

TABLE 4-3. BOD Data of Tannery Waste Used in Moments Method

Time in days *t	(BOD exerted) y	ty
1	1,447	1,447
2	3,363	6,726
3	2,416	7,248
4	3,806	15,224
5	3,052	15,260
6	4,388	26,328
7	4,155	29,085
	Σy=22,627	Σty=101,318

$$\frac{\Sigma y}{\Sigma ty} = 0.2233 = \frac{22,627}{101,318}$$

*Seven days only required in sequence to use prepared curves (Fig. 4-2).

sults with the Slope and Reed-Theriault methods (Table 4-4). The data demonstrate rather conclusively that the moment method yields valid results.

4-4. THE LOGARITHMIC METHOD

Orford and Ingram (1953) found that when the BOD is plotted against the logarithm of time, the resulting curve is approximately a straight line

$$y_t = m \log t + b$$

However, they state that the general equation of the straight line above is unsuitable, because m and b vary both with the magnitude and the rate of reaction.

They transform, mathematically, therefore, the above general equation by dividing each side by the 5-day BOD intercept of the line to give

$$y_t = s(M \log t + B)$$

where

s = 5-day BOD intercept of the line

$M = \frac{m}{s}$ = BOD rate parameter

$B = \frac{b}{s}$ = BOD rate parameter

TABLE 4-4. Values of *k* and *L* for Domestic Sewage as Determined by Analysis of Data by Various Methods. [*After Moore, et al., 1950.*]

Experiment No.	Reed-Theriault Method			Slope Method			Moment Method		
	k	L(ppm)	ΣR^2	k	L(ppm)	ΣR^2	k	L(ppm)	ΣR^2
3	0.158	197	248	0.146	204	275	0.165	195	265
5	0.205	171	191	0.192	174	215	0.213	170	194
6	0.192	140	132	0.181	145	186	0.204	141	174
7	0.224	205	250	0.210	212	328	0.231	205	260
10	0.199	199	383	0.177	206	439	0.208	198	410
13	0.194	173	120	0.186	176	120	0.199	172	132
14a	0.262	158	87	0.284	157	139	0.259	159	85
14b	0.255	150	51	0.275	149	95	0.253	150	66
14c	0.191	128	574	0.192	128	574	0.198	129	589
17	0.186	225	300	0.197	222	318	0.187	226	320
Arith. mean	0.207	175	–	0.204	177	–	0.212	174	–
Stand. dev.	0.030	30	–	0.041	31	–	0.027	29	–

Experimental Material	Reed-Theriault Method		Slope Method		Moment Method	
	k	L(ppm)[1]	k	L(ppm)[1]	k	L(ppm)[1]
1. Sulfite pulp waste, 0.02%	0.152	4.89	0.151	4.99	0.151	4.90
2. Sulfite pulp waste, 0.02%	0.130	4.97	0.132	5.01	0.128	4.99
3. Sulfite pulp waste, 0.02%	0.135	5.07	0.137	5.14	0.125	5.16
4. Penicillin waste, 0.045%	0.132	5.54	0.148	5.42	0.121	5.72
5. Penicillin waste, 0.03%	0.118	3.99	0.146	3.67	0.109	4.11

[1] L values for the industrial wastes refer to the diluted material.

They found, by experiment, that for domestic sewage oxidation at 20°C, the straight line through the points, when extrapolated to the log t axis, intercepts the log t at .333 days. Also, the above equation has the following form

$$y_t = s_{20}(0.85 \log t + 0.41)$$

when 0.85 and 0.41 are the BOD rate parameters for domestic sewage (see Fig. 4-3a). Since the rate of oxidation can be expressed as a time relationship, the equation

$$y_t = s(m \log t + B)$$

can be made usable by making time the rate parameter of the equation and comparing the oxidation time of any BOD curve with the oxidation time of normal domestic sewage. The simplified equation becomes

$$y_t = s(.85 \log at + 0.41)$$

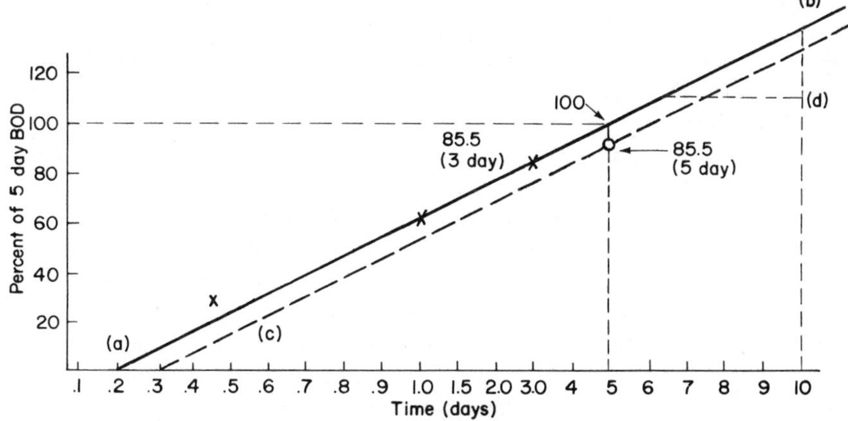

FIG. 4-3a. Example of computation of logarithmic constants, a and s, by graphical means. (ab = plot of observed data; cd = standard curve ‖ to observed data and passing through log t = .333.) In practice it is not necessary to draw the line ‖ by mechanical means. An observed time value is multiplied by the factor a to convert the equation to the standardized form. For example, 6 days times 1.67 = 10 days. (*Orford and Ingram, 1953.*)

where

a = x axis intercept of domestic (normal) sewage BOD curve divided by the x axis intercept of the observed BOD curve
  = 0.333/x axis intercept
  = $\frac{0.333}{0.333}$ = 1.0 for standardized domestic sewage BOD curve
s = BOD intercept of the line at $5/a$ days
  = 5-day BOD value which would have been obtained at standardized domestic sewage oxidation rates (a = 1.0)

$$a = \frac{.333}{0.20} = 1.67 \quad \text{The value of } s = \frac{5 \text{ day}}{1.67} = 3 \text{ day}$$

3-day value of 85.5 now becomes s value. The equation now becomes

$$y_t = 85.5(0.85 \log 1.67 + 0.41)$$

Therefore oxidation rate by Logarithmic Formula is

$$\frac{dy}{dt} = \frac{0.85s}{2.303t}$$

Oxidation rate by Monomolecular Formula is

$$\frac{dy}{dt} = k(L - y_t)$$

In Orford's method, the oxidation rate is directly proportional to s, the BOD strength parameter, and inversely proportional to time. Thus, a is not an oxidation rate parameter, but is the inverse ratio of the time required to exert a given amount of BOD to the time required to exert the same BOD for standardized domestic sewage (average of many purely domestic sewages). Since velocity is a ratio of the amount of the time, a should be considered an oxidation velocity or speed ratio, and not an oxidation rate.

Orford and Ingram (1953) claimed that a new, simpler mathematical equation to express biological oxidation was developed which gives as good a fit to observed BOD data as does the monomolecular equation, and eliminates many of the difficulties of the latter, including instability of the constants and complicated calculations. In the "logarithmic BOD equation" the oxidation rate at time t is proportional to the strength parameter, s, and inversely proportional to the time. In the monomolecular equation the oxidation rate is proportional to the amount of material remaining to be oxidized $(L - y_t)$. Thus, the logarithmic equation is a fundamentally different mathematical expression, and is not a reformation of the monomolecular equation. The logarithmic equation plots as a straight line on semilogarithmic graph paper when y_t values are plotted on the coordinate axis, and time values plotted on the logarithmic axis.

Because observed BOD data follow the straight-line logarithmic curve closely, it is possible to draw a straight line through the plotted data and determine the constants of the equation without resorting to complicated mathematical analyses to determine the line of best fit. When observed BOD data are plotted on semilogarithmic paper, the different oxidation stages plot as straight lines with distinctly different slopes; thus, it is possible to identify the various special stages (such as nitrification, initial lag period, and final oxidation stages) by observing the time at which the breaks occur in the straight lines of best fit for each stage.

The value of a doubles for each 10°C rise in temperature between 10°C and 30°C, for materials other than sewage. The writers give corresponding values of a, s, and k, and L at temperatures from 9°C to 30°C (see Table 4-5).

Orford and Ingram (1953) found that for several industrial wastes the logarithmic equation follows the observed points as well as it does for sewage, if not better. Specialized wastes when oxidized alone oxidized more slowly than sew-

TABLE 4-5. Logarithmic Constants, *a* and *s*, at Various Temperatures

Temperature	a	s	k	L
9°C	0.25	4.2	.0508	4.58
20°C	0.50	4.93	.0993	5.52
30°C	1.00	4.83	.1501	6.24

TABLE 4-6. Values of Logarithmic Constants for Certain Industrial Wastes. *[After Orford and Ingram, 1953.]*

Waste	a	s	k	L
Corn Products	0.50	417	.114	411
Paperboard	0.83	421	.140	516
Sugar	0.67	590	.114	658
Sugar & Sewage	1.67	565	.275	686

age, whereas the oxidation of sugar waste with a large proportion of sewage apparently accelerated.

Logarithmic and monomolecular equation constants for the BOD of various industrial wastes are given in Table 4-6 (Orford and Ingram, 1953). A comparison between the constants in the monomolecular and logarithmic formulas is given in Table 4-7 (Orford and Ingram, 1953).

$$\text{Log}\,(L - y_t) = -kt \qquad \text{Monomolecular Law}$$

$$y_t = s(0.85 \log at + .41) \qquad \text{Logarithmic Equation}$$

The value of k decreases and the value of L increases with increasing time of observation in observed BOD curves.

The values of k and L were computed by the moments method in which a was given unity. After k and L were determined, the k and L values were determined for any other time ratio by merely multiplying k by the time ratio a.

For practical use of the relationships between the constants of the two equations, the 10-day values of the constants were chosen as good parameters to express the entire curve; thus, $k = .150a$ and $L = 1.25s$ (see Table 4-7).

This method overcomes two of the difficulties in the standard BOD test:

1. The relatively long 5-day interval is shortened.
2. The standard 5-day BOD test measures BOD without evaluating the effect of differences in oxidation rate.

TABLE 4-7. Comparative Values of Constants

Time	k	L
3 days	$0.218a$	$1.04s$
5 days	$0.197a$	$1.10s$
7 days	$0.180a$	$1.16s$
10 days	$0.150a$	$1.25s$
14 days	$0.130a$	$1.34s$

Orford and Ingram recommend a 1 and 2- or 1 and 3-day test to compute a and s.

When using the tannery waste data and applying the Orford and Ingram method we obtain the following results:

Using the Fitted Curve Fig. 4-1

Time (days)	BOD (ppm)	% of 5-Day BOD
1	1,500	41.6
2	2,500	69.5
3	3,000	83.2
4	3,500	97.2
5	3,600	100.0

$$a = \frac{.333}{X \text{ axis intercept}} = \frac{.333}{.36} = 0.925$$

$$s = \text{BOD intercept at } \frac{5}{a} \text{ days} = \frac{5}{.925} = 5.4 \text{ days}$$

$$s = 1.10$$

$$\therefore k = .150(.925) = 0.13875 \quad \text{and}$$

$$L = 1.25s = 1.25(1.10) = 1.38(\text{5-day BOD}) = 1.38(3{,}600) = 4995$$

(See Fig. 4-3b.)

4-5. GENERAL LABORATORY METHOD

In the routine analyses in the author's laboratory, as well as elsewhere (Gaudy et al., 1967), the practice has been to determine (by approximation or graphical estimation) an ultimate first-stage BOD value from data taken over the first 10 days. This is usually a relatively easy prediction, as shown in Fig. 4-4.

Once the ultimate first-stage BOD value has been estimated, we can apply the monomolecular law, which states that the rate of oxidation is proportional to the organic matter remaining. Thus, a plot of the BOD remaining on a log scale versus the time of oxidation plotted arithmetically will yield a straight line sloping downward and away from the junction point. The slope of this line is defined as the reaction rate k_1. The intercept of the log axis will be the predicted value of L, the ultimate first-stage BOD. The least-squares technique may be applied to this data to obtain the best straight line. However, when this is done, the

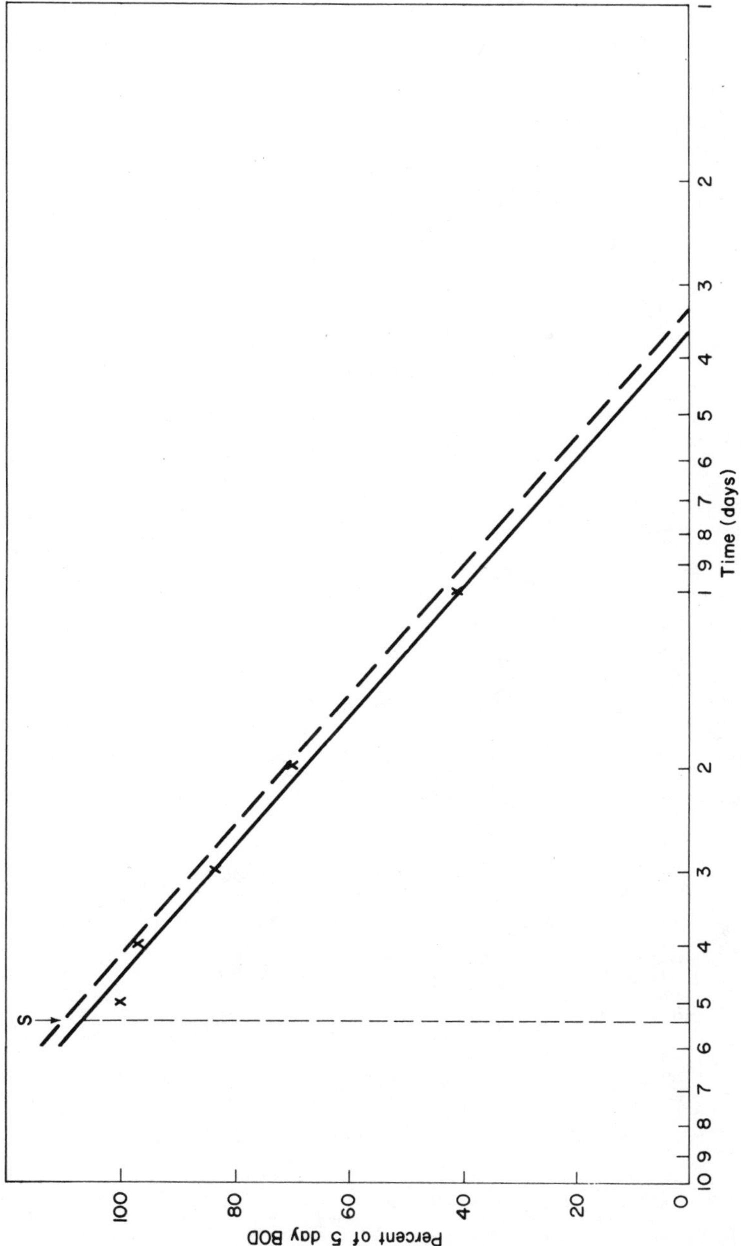

FIG. 4-3b. Logarithmic plot of tannery waste data.

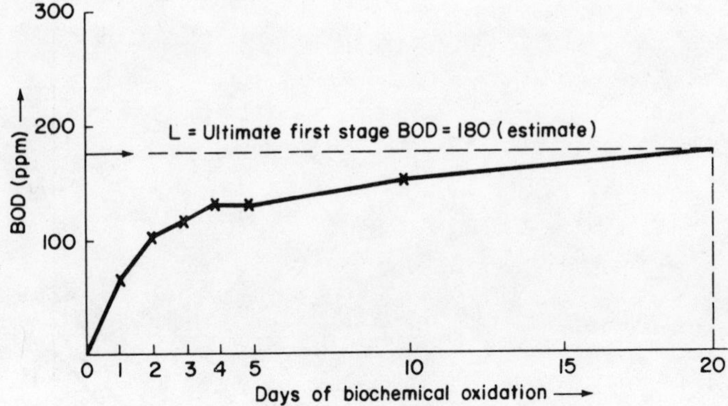

FIG. 4-4. Estimation of ultimate BOD from 10-day BOD data.

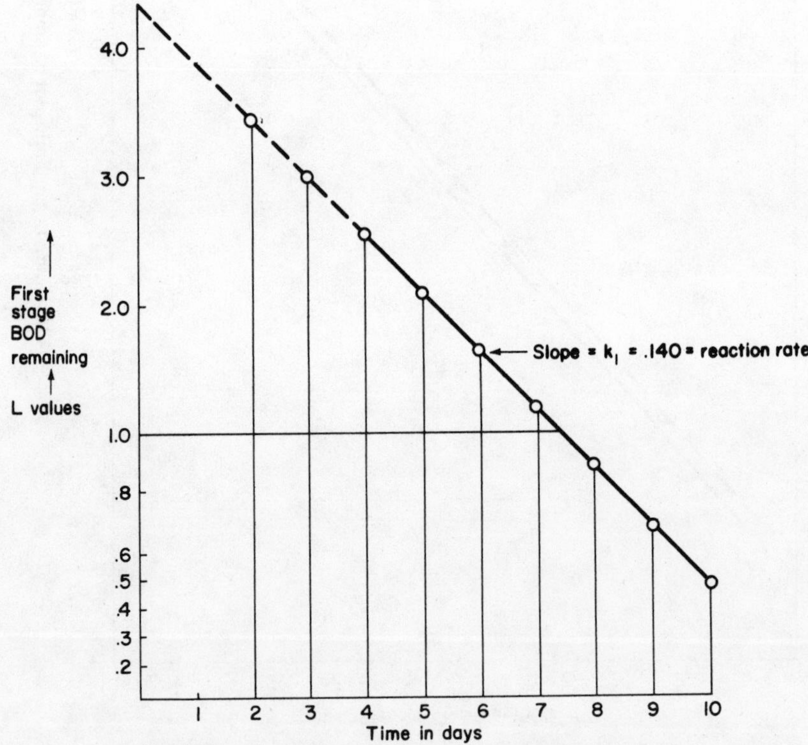

FIG. 4-5. Graphical solution to deoxygenation rate K_1.

actual L value obtained will usually be somewhat different from the original estimation. Because of this, general practice dictates fitting the line by eye, using the estimated L value as one point on the line (see Fig. 4-5).

The major advantage of this method is its ease of computation once a well-defined straight line has been ascertained, as shown in Fig. 4-5. Since L had to be estimated initially in order to plot Fig. 4-5, the value of k_1 obtained is only an approximate one. The accuracy of k_1 is directly related to the ability of the analyst to estimate a precise L value from the original BOD data (Fig. 4-4).

In any event, the k_1 value obtained is usually close enough to the actual value to allow comparison of laboratory rates of deoxygenation of other wastewaters.

General Laboratory Method Applied to the Tannery Waste Data

In Fig. 4-1 the projected, estimated L value is 4,600 ppm. The individual BOD values exerted over the 10 days with tannery wastes are shown in the second column in Table 4-1. Columns 3 and 4 of this table also give the BOD and log of the BOD remaining at each time interval. A plot of the log of the BOD remaining versus time of reaction is shown in Fig. 4-6. An ultimate, first-stage BOD of 3,850 is obtained when, eliminating the 9th and 10th days from the data (which do not appear to represent first-stage reaction anyway), k_1 is shown to be 0.117.

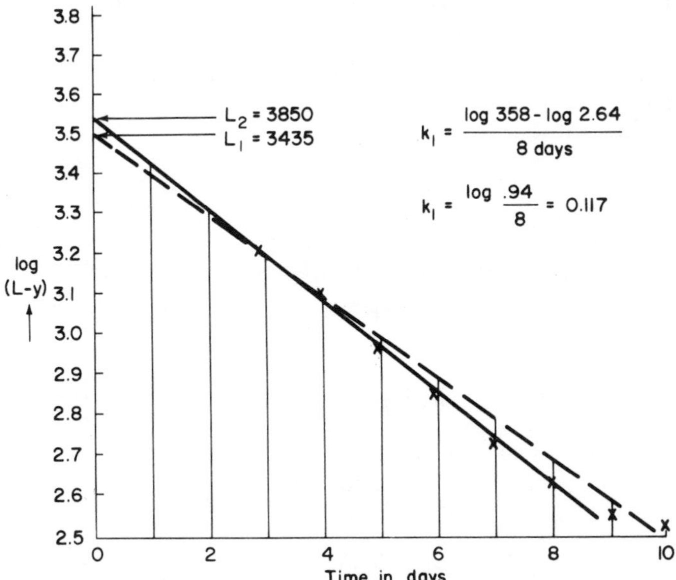

FIG. 4-6. Log of BOD remaining versus time of reaction.

4-6. RHAME's TWO-POINT METHOD

Rhame (1956) developed a relationship between the BOD values at two times, the second time being exactly double that of the first. The two expressions he uses were developed from empirical data and are as follows:

$$L = \frac{X^2}{(2X - Z)}$$

and

$$k_1 = \frac{1}{T - t} \log \frac{X}{Z - X}$$

where

X = BOD at t days
Z = BOD at $2t$ or T days, since $T = 2t$

Rhame recommends using t values as high as possible. The author found that, with most data, t of 5 days and T of 10 days produced results comparable to those obtained with other methods. Ultimate BOD, L, and deoxygenation rate, k_1, for the tannery waste are computed, using Rhame's method as shown in Table 4-8.

This method is limited by the fact that it assumes only one straight-line relationship between two preselected points on the curve to hold true over the en-

TABLE 4-8. Rhame's Method as Applied to Tannery Waste

$$L = \frac{X^2}{(2X - Z)}$$

X = Dilution BOD at 5 days (t) = 3,052 ppm
Z = Dilution BOD at 10 days (T) = 4,200 ppm

$$L = \frac{93147404}{1904} = 4{,}892$$

$$K_1 = \frac{1}{T - t} \log \frac{X}{Z - X} = 0.0849$$

For 4 and 8 days

$$L = \frac{14{,}500{,}000}{7{,}612 - 4{,}350} = 4{,}450$$

$$K_1 = \frac{1}{4} \log \frac{3{,}806}{544} = 0.210$$

NOTE—Readers will note the variation in L and K_1 when using 4 and 8 days as the two points rather than 5 and 10 days.

tire curve. It is, at best, a rough approximation. It has the great advantage of being simple to compute. If only one type of waste is being studied and the same two points are selected repeatedly, an accurate comparison of reaction rates can be obtained.

Lee (1951) at Queens University in Ontario, Canada concluded, with a rapid modification of the Moore, Thomas, Snow (1950) method that k, at 20°C was 0.1. He found that most of the higher values were a result of failing to recognize the effect of the initial demand for oxygen.

A summary of the previous major methods of determining k values is given by our author in Table 4-9 on the following page.

4-7. ULTIMATE OXYGEN DEMAND (UOD)

When settled sewage is treated by a biological process, carbon is oxidized to CO_2, but some evidence exists that there is little or no oxidation of NH_3. Thus the five-day BOD of the effluent from such a process is cimparatively low, because *nitrogen is not oxidized during the five-day period* and because the *liquid contains little organic carbon, but the ultimate demand is comparatively high since it includes the oxygen equivalent of the nitrogen present in the effluent.*

At treatment plants where the biological process produces a nitrified effluent, or an effluent that starts to nitrify during the incubation period of the BOD test, the ratio of the ultimate oxygen demand to the five-day BOD may not be as low as for settled sewage. Southgate and Gameson (1956) present data to show the effect of primary and secondary treatment on the ratios of ultimate to five-day BOD (see Table 4-10).

TABLE 4-10. Ratios of Carbonaceous Oxygen Demand to Five-Day BOD and of Organic Carbon to Total Unoxidized Nitrogen in Two Sewage Effluents. [*After Southgate and Gameson, 1956*)

Type treatment	Avg. 5 day BOD (ppm)	Ratio carbonaceous ultimate demand; 5 day BOD	Ratio C:N
Primary settled effluents			
1	250	2.7	3.3
2	210	3.1	2.7
Activated sludge effluents			
1	104	4.0	2.0
2	22	11.0	0.85

NOTE–The level of oxygenation will depend *both* on the *quantity of oxidizable matter* discharged and on the *nature of the stream.*

TABLE 4-9. Deoxygenation Rates Summary

Method	Advantages	Disadvantages	When Used
Least-Squares	Best fit of data mathematically	Cumbersome; apt to be misleading	Math accuracy is desired
Slope	Less cumbersone, and useful when data of unequal daily intervals are available	Still depends on much computation. Useful only when reaction is truly monomolecular.	When various intervals of days are used
Moment	Simpler, and hence, less labor	Requires standard curve for comparison which must utilize same time sequence. Can be used only with the precise time sequence.	When standard curves of proper sequence are available
Logarithmic	Compares directly the rate with domestic sewage, and is done in less than five days required	Difficult to comprehend	When comparing rates of various wastes, and when time is valuable
General Lab	Uses generally available lab data; can be utilized graphically, and thus easier visualized than the other mathematical methods.	Only an estimation	When various day BOD data is obtained, and when visual comparison of reaction rates is desired
Rhame	Easy to compute Easy to get only two BOD values	Assumes a straight-line relationship	When lab space and analysis are at a premium and when extreme accuracy is not desired

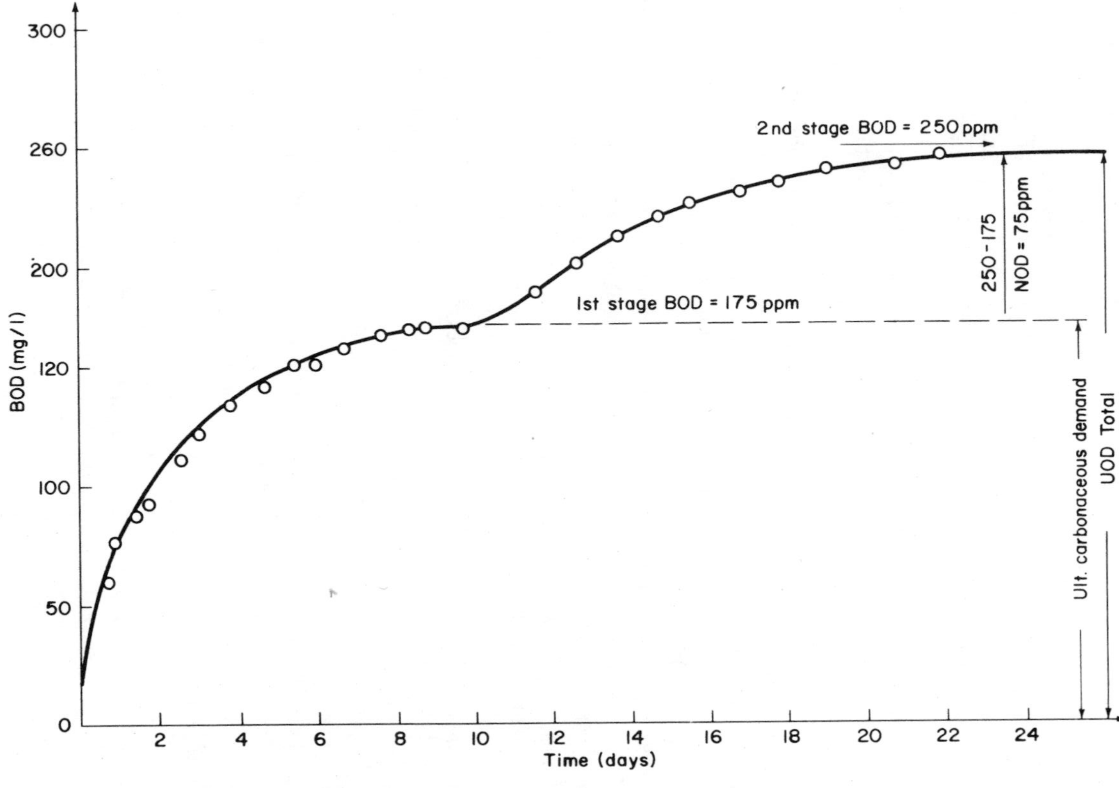

FIG. 4-7. BOD versus time for Brookside sewage, 4-1-65 to 4-23-65, at 20°C.

The ultimate oxygen demand is increasing in significance in stream sanitation, since the competition for water resources in receiving streams increases steadily. In the type of treatment process being used today the carbonaceous BOD is greatly reduced, which presumably allows the nitrogenous cycle to take over. When the time of flow from water outlet to the sea is lengthened by either continued reuse or artificial impoundment, the UOD becomes greater. These practices make it imperative to consider the nitrogenous (second-stage) oxygen demand as well as the carbonaceous (first-stage) demand in predicting oxygen depletion. A typical 22-day oxygen demand curve was prepared from a domestic sewage of a subdivision (see Fig. 4-7.) Many regulatory agencies, including the New York State Department of Environmental Conservation, put limits on the quantity of NOD in addition to the ultimate first-stage BOD which may be discharged into a given stream (Mt. Pleasant and Schlickenrieder, 1971). However, it remains to be proven more conclusively that the NOD is exerted as soon as the carbonaceous BOD is removed or reduced.

There are many other ramifications of a decision to limit the summation of NH_3—N and organic nitrogen in a stream. One obvious effect is that of substituting nitrites and nitrates in lieu of reduced and unoxidized forms of nitrogen. Nitrates have already been shown to be dangerous to health in drinking water, and recently (Brody, 1971) nitrites have also been revealed to cause brain damage in animals. Nitrites also apparently cause cancer in laboratory animals when combined with amines to form nitrosoamines. In addition, nitrates are readily available to algae for growth—more so than ammonia or organic nitrogen forms. Simply converting nitrogen from the reduced to the oxidized form may alleviate the oxygen deficiency problem in streams, but may cause other problems of even greater magnitude.

In Table 4-11, the deoxygenation rates of k_1 and ultimate first-stage BOD values L, are summarized for each of the methods described. One should select a method that best suits one's time schedule, accuracy requirement, laboratory testing facilities, and specific overall objective.

TABLE 4-11. Values for *k* and *L* for Tannery Waste as Determined by Various Methods

Method	k	L(ppm)
Least-squares	0.113/day	4,600 (By eye)
Slope method	0.174/day	4,400
Moments method	0.204/day	4,205
Logarithmic	0.1385/day	5,200
General lab method	0.117/day	3,850
Rhame's 2-point	0.210/day	4,450

QUESTIONS

1. Upon what basic principle is the least-squares technique for determining k_1 based? What are its advantages and shortcomings?
2. What are the mechanics of finding k_1 and L with this method?
3. What is meant by monomolecular rates and bimolecular rates of oxidation?
4. How would we determine at which rate the wastewater decomposes? Why would we be interested?
5. What is the definition of the slope method of determining k_1? What are its advantages and shortcomings?
6. What is the foundation or the basic premise of the moments method of determining k_1? What are its advantages and disadvantages?
7. What is the basis of the logarithmic method of determining k_1?
8. What is the general laboratory method of determining k_1?
9. What is the major assumption which must be made to validate the use of Rhame's two-point method?
10. What is meant by NOD? Discuss its implication in the BOD test.
11. As a stream resource scientist, would you recommend a limit to NOD discharge? Upon what factors would your decision depend?

REFERENCES

Brody, J. E.: Brain Change Tied to Food Additive, *New York Times*, October 16, 1971, p. 33M.

Gaudy, A. F., et al.: *Methods for Evaluating the First Order Constants k and L for BOD Exertion*, M-1 Center for Water Research in Engineering, Oklahoma State University, Stillwater, Oklahoma, May 1967.

Lee, J. O.: Simplified Method for Analysis of BOD Data, *Sewage and Industrial Wastes*, **23**:164, May 1951.

Moore, E. W., H. A. Thomas, and W. B. Snow: Simplified Method for the Analysis of BOD Data, *Sewage and Industrial Wastes*, **22**(10):1343, 1950.

Mt. Pleasant, R. C., and W. Schlickenrieder: Implications of Nitrogenous BOD in Treatment Plant Design, *J. San. Eng. Div. (ASCE)*, SA5, October 1971, p. 709.

Orford, H. E., and W. T. Ingram: Deoxygenation of Sewage, *Sewage and Industrial Wastes*, **25**(4):424, 1953; *Sewage and Industrial Wastes*, **25**(5):566, 1953.

Rhame, G.: A Two-Point Method for Estimation of First Stage BOD, *Sewage and Industrial Wastes*, **28**(9):1087, 1956.

Southgate, B. A., and A. L. H. Gameson: Recent Developments in the Control of Stream Pollution, *Surveyor* (London), **115**(3344):349–354, 1956.

Theriault, E. J.: The Oxygen Demand of Polluted Wastes, Public Health Bulletin 173, United States Public Health Service, Washington, D.C., 1927. *Chem. Abstracts*, 254644, 1931.

Thomas, H. A.: The Slope Method of Evaluating the Constants of the First Stage BOD Curve, *Sewage Works Journal*, **9**(3):425, 1937.

5
STREAM REAERATION

Nature fortunately provides a mechanism for counteracting the effects of deoxygenation, which we studied in Chapter 4. That mechanism is known as reaeration, a means whereby oxygen as well as other gaseous components of air is renewed in flowing streamwater. The fact that the oxygen supply is renewed in water courses is no guarantee, however, that sufficient amounts will therefore be available at the proper location for a specific water use. Reaeration is also a rate phenomenon. The parameters that control the net effect of reaeration are vastly different, however, from those that affect deoxygenation.

In this chapter I attempt to define the factors that affect reaeration, and to present some mathematical procedures for predicting actual rates of reaeration.

The first studies of significance were those of Adeney from 1914–1919 (Adeney and Becher, 1919), which showed that the rate of aeration is proportional to the oxygen saturation deficit. The greater the deficit the greater the rate of solution of oxygen.

5-1. STREETER-PHELPS OHIO RIVER STUDIES

Streeter and Phelps (1925) applied this finding of Adeney and Becher to stream reaeration and developed the two following formulas:

$$\frac{dD}{dt} = K_1 L - K_2 D$$

where

D = oxygen deficit
L = ultimate carbonaceous oxygen demand

or the differentiated equation was solved to yield

$$D = \frac{k_1 L_A}{k_2 - k_1} \left[10^{-k_1 t} - 10^{-k_2 t}\right] + D_A \cdot 10^{-k_2 t}$$

This equation has been used in practically all important studies of stream assimilative capacity up to now. It has the best theoretical mathematical rationale and validity. It is important to observe that although the value of the reaeration rate k_2 can be computed from this formula if the other information is known, *it cannot be used to predict reaeration rates with reliable results.* Until recently, no other formulations were available for predicting k_2; this equation has been used to predict k_2. The value of k_2 obtained may be far too low, owing to *organic pollution collected on the streambed.* In addition to this source of error in computing k_2, the measurements of *time of water travel, DO, ultimate carbonaceous stage BOD,* and the *deoxygenation coefficient* k_1 must be made. Errors in any one of these are reflected in the computed value of k_2.

In the early work of Streeter (1926) it was proposed that k_2 depended upon many variables, such as *velocity, depth, slope,* and *channel irregularity*

$$\because K_2 = \frac{CV^n}{H^2}$$

where

K_2 = reaeration coefficient, per day
V = mean velocity, ft/sec
H = mean depth of water *above extreme low water,* ft
C,n = constants for a particular river stretch, the values of which depend in part on the channel slope and irregularity (roughness)

Very little consistency in K_2 values was obtained by using this formulation, mainly because of the *existence of pollution* and effects of algae and other aquatic organisms.

5-2. O'CONNOR AND DOBBINS COMPUTATION OF K_2

O'Connor and Dobbins (1956) developed two formulas for the prediction of the value of K_2.

For those *streams showing a pronounced vertical velocity gradient,* a formula for nonisotropic[1] turbulence was determined as

$$k_2 = \frac{480 D_L^{1/2} S^{1/4}}{H^{5/4}}$$

For comparatively deep channels, where it was possible that the turbulence might approach an isotropic condition,[1] their production value was obtained as

[1]Turbulence existing in a stream having a pronounced velocity gradient. Nonisotropic conditions assumed to exist when Chezy's C is < 14–20 when $U = C\sqrt{HS}$. When impossible to compute C, he assumes nonisotropic turbulence when depth is < 5 feet. Isotropic in this case refers to a fairly similar velocity from one layer of depth to the next.

$$k_2 = \frac{127(D_L U)^{1/2}}{H^{3/2}}$$

where

$'D_L$ = coefficient of molecular diffusion, liquid film ft^2/ day
S = slope of river channel, ft/ft
H = average depth of flow, ft
U = mean velocity of flow, ft/sec
k_2 = reaeration, per day

$$'D_L = 2.037(1.037)^{T-20}\left(10^{-5} \times \frac{cm^2}{sec}\right)$$

$\frac{ft^2}{day} \times 10^{-6}$	Temperature
$D_L = 1464$	10°C
$D_L = 1704$	15°C
$D_L = 1944$	20°C
$D_L = 2208$	25°C
$D_L = 2544$	30°C

O'Connor and Dobbins found good agreement between their formulas and Streeter-Phelps', but since both researchers' formulations are *theoretical,* some doubt as to their validity remains.

A serious difficulty with both formulas is that the computed reaeration rate for a particular reach is dependent, to a large extent, upon the type and concentration of pollution in solution at the time.

Pollution in most cases was found to suppress reaeration rates (Kehr, 1938; Downing and Truesdale, 1955; Gameson, Truesdale, and Varley, 1956). It is presumed, however, that this effect was measured at similar dissolved oxygen deficits. Kehr, (1938) for example, found that 6 ppm of soap reduced reaeration 51 percent at 20°C. The reduction was directly proportional to the flow. O'Connor and Dobbins state that turbulence may be assumed to be nonisotropic when Chezy's C ($V = C\sqrt{RS}$) is less than 14-20, and isotropic when C is greater than this range. Their basic assumption is that *the rate of surface renewal equals the velocity gradient at the surface* (R = hydraulic radius and S = Slope; V = velocity).

Churchill, however, concludes that equal velocity gradients do not necessarily indicate equal states of turbulence. In fact, the states of turbulence may be greatly different. Consequently, it seems extremely unlikely that stream reaeration rates are controlled by the velocity gradient.

5-3. ISAACS' MODIFICATION OF K_2 DETERMINATION

Isaacs (1967), using an experimental idealized simulated stream, found K_2 to be consistent with the general relationship Streeter and Phelps proposed (1926). He found that K_2 was proportional to the average stream velocity, and inversely proportional to the average stream depth raised to the $\frac{3}{2}$ power. Isaacs' equation was more definitive and took the precise form of

$$k_2 = 0.06339 \times \frac{D_m^{1/2}}{\partial^{1/6} g^{1/6}} \times \frac{V}{H^{3/2}}$$

where

D_m = the molecular diffusivity of oxygen into water
∂ = the kinematic viscosity of water
V = the average stream velocity, ft/sec
H = the mean depth of flow, ft
g = the gravitational constant, ft/sec^2

However, Isaacs used a constant surface roughness in this investigation. No one has been able to devise a means of measuring the surface roughness of the bottom of a natural stream. Since the roughness of the bottom will affect the velocity profile of a stream, a mean velocity is needed for use in the above equation.

Isaacs et al. (1969) found by experimentation with channels of different shapes and roughnesses that

$$k_2(20°\text{C}) = C \frac{V}{H^{3/2}}$$

and that C varied with roughness as follows:

Roughness	Value of constant C	No. of observations	Correlation coefficient
Circular	2.7587	16	0.984
Square	2.9281	16	0.972
Triangular	2.8229	16	0.987
Composite data	2,8330	48	0.981

They recommended the following equation:

$$k_2(20°\text{C}) = 2.833 \frac{V}{H^{3/2}}$$

where

k_2 = day^{-1}
V = mean stream velocity, ft/sec
H = mean depth, ft

The constant 2.833 represents an average of all roughnesses examined by Isaacs. He concluded (Isaacs et al., 1969) that it is possible to predict K_2 within a reasonable degree of accuracy in channels of variable surface roughness, from a knowledge of the *mean* velocity of flow and the *mean* depth of flow. The effects of channel roughness are evidently included in the constant when mean values of depth and velocity are obtained and used.

5-4. CHURCHILL METHOD FOR PREDICTION OF STREAM REAERATION RATES

In 1957, Churchill, Elmore, and Buckingham (1962) studied reaches of the lower Clinch, Holston, and French Broad Rivers, and in 1958 and 1959 the South Holston, Watauga, Nottely, and Hiwassee Rivers. Discharges varied from 1.145 cfs to 14,300 cfs (measured by staff gages at both ends of reaches); DO values were obtained at cross sections for 24-hour periods. It was found that both vertical and horizontal velocity gradients existed. For most of the reaches, 16–19 separate hourly measurements of the reaeration rate were made under the same conditions of flow. They used submerged floats to determine sampling time, and measured k_2 as follows:

$$\frac{dD}{dt} = KD$$

$$k_2 = \frac{\log_{10} D_2 - \log_{10} D_1}{t_2 - t_1}$$

Therefore, by measuring DO deficit at two points and determining the time of flow between them, they could calculate the k_2, which is independent of the absolute amount of DO present. Therefore, k_2 was determined by field measurements.

By selecting reaches that varied in type but were constant within each one in terms of the physical variable that affects the value of k_2, it was possible to observe the quantitative effect of each such "independent" variable on k_2.

The reaches selected were free from significant organic pollution (average 5-day BOD in all cases was 0.81 mg/l). DO measurements were made by amperometric titration with a rotating platinum electrode as the indicator and a saturated calomel electrode as the reference electrode (standard deviation ±.007 ppm DO). The presence of organic matter affects the overall reaeration, since oxygen level is lowered by biodegradation; however, the physical phenomenon is presumed the same with or without organic matter.

The first step Churchill carried out was to list all physical variables involved in the reaeration problem. They are grouped so that each group is dimensionless.

	Variable		
Reaeration coefficient	k_2	1	T^{-1}
Velocity	V	2	LT^{-1}
Mean depth	R	3	L
Energy slope	S	–	None
Resistance coefficient	f	–	None
Density	p	4	ML^{-3} or FT^2L^{-4}
Dynamic viscosity	μ	5	$ML^{-1}T^{-1}$ or FTL^{-2}
Surface tension	σ	6	MT^{-2} or FL^{-1}
Molecular diffusion, liquid film	D_L	7	L^2T^{-1}
Diffusion coefficient (vertical)	ϵ	8	L^2T^{-1}

According to the classical Buckingham II theorem of dimensionless analysis, there should exist three fewer dimensionless groups than there are variables (S or f may be used, but not both). Eight variables, and therefore five dimensionless groups, are possible.

Dimensionless groups of interest here are

(1) $\dfrac{k_2R}{V}$

(2) $\dfrac{\mu}{D_L}$

In addition, the slope S and the resistance coefficient f are each dimensionless.

(3) $\dfrac{V\rho R}{\mu}$ = Reynolds Number

(4) $\dfrac{\sigma}{\mu V}$

(5) $\dfrac{\mu}{\epsilon}$

For computation of k_2 from experimental data, the five dimensionless groups may be arranged in the following formula:

$$k_2 = a\frac{V}{R}(f)^{b_1}\left(\frac{\mu}{\rho D_L}\right)^{b_2}\left(\frac{V\rho R}{\mu}\right)^{b_3}\left(\frac{\sigma}{\mu V}\right)^{b_4}\left(\frac{\mu}{\rho\epsilon}\right)^{b_5}$$

where a and b are constants to be evaluated from the data by multiple regression procedures.

Since the term k_2R/V is dimensionless, the equation for k_2 can be, as far as dimensions are concerned, as simple as

$$k_2 = a\frac{V}{R} \quad \text{but improved by} \quad k_2 = a\frac{V^{b_1}}{R^{b_2}}$$

and the other dimensionless terms can be added one at a time, or all together, if such additional terms improve the accuracy of prediction to a statistically significant extent.

Numerical values for density, dynamic viscosity, and surface tension of water were taken from standard reference tables. Other hydraulic properties of the river were measured for each of the 30 experiments.

Churchill et al. (1962) developed 19 equations and decided to use the one which was the simplest, and essentially as accurate as any, as follows:

$$k_2(20^\circ\text{C}) = 5.026\,\frac{V^{0.969}}{R^{1.673}}$$

The plots of this equation are shown in Fig. 5-1.

However, even after the true value of k_2 is obtained, it probably will not agree with the computed value in the stream because of:

1. Existence of organic matter, which depresses the sag curve.
2. Algae.
3. Sludge deposits.
4. Existence of pollution, such as detergents and oil, which can affect the physiochemical properties, and thus alter reaeration. No one, so far, has been able to define these effects precisely in any mathematical way so as to arrive at a closer value of k_2.

5-5. REAERATION COEFFICIENTS DEVELOPED BY OTHERS

Some other researchers who have contributed to our knowledge of reaeration prediction in flowing waters since 1960 include Zak (1960), Krenkel and Orlob, (1962), and Owens, Edwards, and Gibbs (1964). Although it is not possible to present in this book all the detailed studies made by each worker, the following formulations may be helpful to the reader. Zak provides a graph plotting k_2 versus the turbulent liquid diffusion coefficient D_T when the river flows are either fast or slow (Fig. 5-2).

In Zak's computation of D_T, the turbulent diffusion coefficient in m^2/sec, g is the gravitational constant in m/sec^2, n is the roughness coefficient, C is Chezy's coefficient, h is the mean depth, V the mean velocity in m^2/sec, and k_2 the reaeration rate in days^{-1}.

Krenkel and Orlob (1962) derived

$$k_2(20^\circ\text{C}) = (4.302 \times 10^{-5})\,D_L^{1.15} \times h^{-1.915}$$

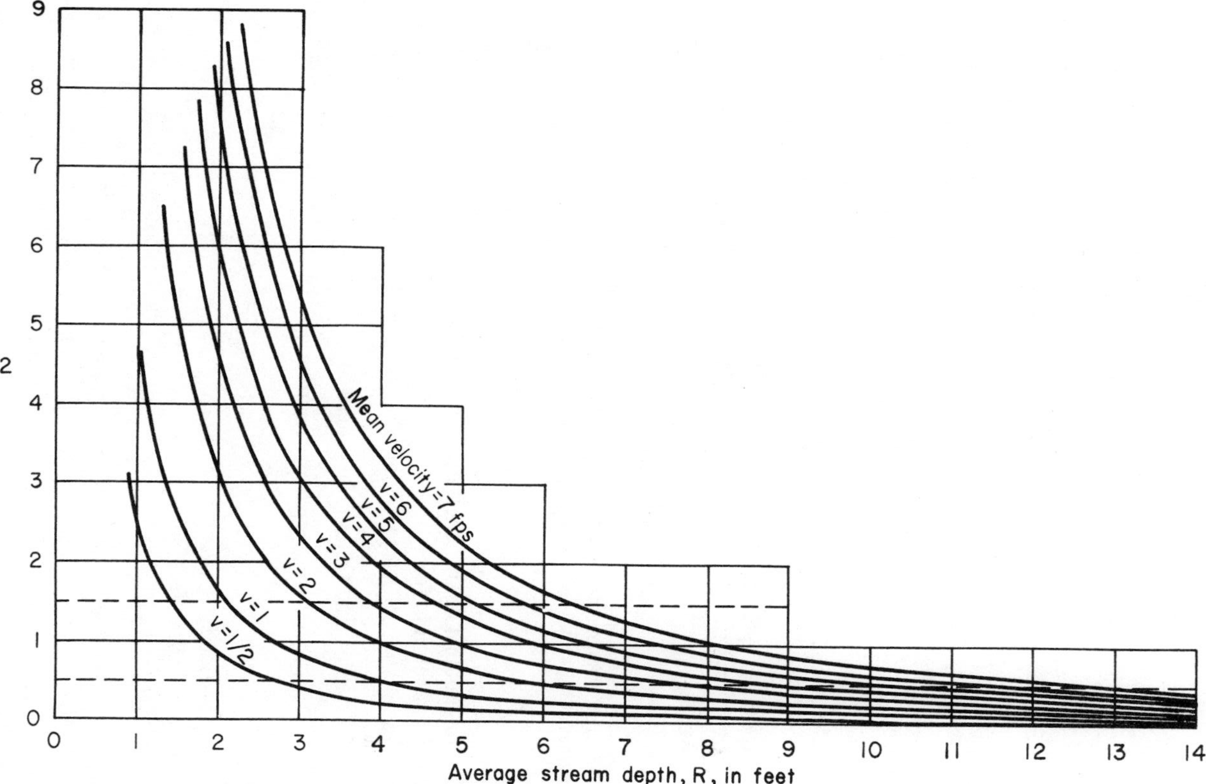

FIG. 5-1. Solution to equation $k_2 = 5.026\ V^{0.969}/R^{1.673}$. *(After Churchill et al., 1962. By permission of publishers.)*

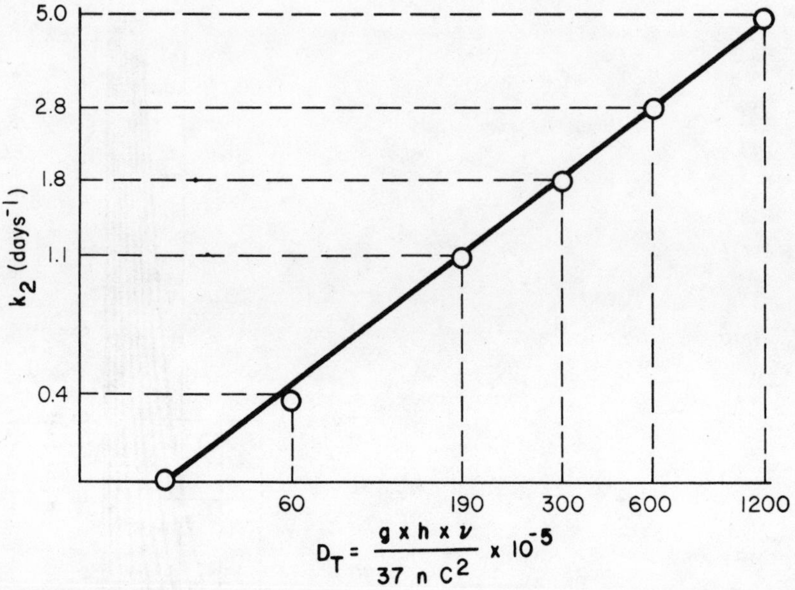

FIG. 5-2. Zak's diagram for k_2 determination. (*After Zak, 1960. By permission of publishers.*) Oirginal in Russian.

where

D_L = longitudinal mixing coefficient in ft^2/min
h = mean depth in feet
k_2 = reaeration coefficient in min^{-1}

Owens, Edwards, and Gibbs (1964) derived

$$k_2(20°C) = 9.4\ V^{0.67} h^{-1.85}$$

where

V = mean velocity in ft/sec
h = mean depth in feet
k_2 = reaeration coefficient in $days^{-1}$

for streams with velocity variation range from 0.1 to 5.0 ft/sec and depths from 0.4 to 11.0 feet.

A research study by Negulescu and Rojanski (1969) derived the following equation:

$$k_2 = 0.0153 D_L \left(\frac{V}{h}\right)^{1.63}$$

where

D_L is computed from $D_L = 310\,(V/h)^{-0.78}$ in cm^2/sec
h = depth in meters where the value is less than 0.5 m
V = velocity in m/sec

The Gas Tracer Technique was originally developed and used by Tsivoglou (1965-68) to measure true k_2. It followed the previously used and so-called Disturbed Equilibrium Technique (1955-59 and 1961) developed by Edwards et al., Gameson et al., and Gameson and Truesdale at the Water Pollution Research Laboratory (WPRL) in England. The latter technique consisted of measuring DO values at both upstream and downstream ends of given stream reaches. They obtained two levels of DO by putting Na_2SO_3 and a cobalt catalyst directly into the stream, and depended upon photosynthesis, reaeration coefficients, respiration, average velocity, and saturation concentration remaining constant in the reach. This latter technique was severely hampered by the necessity of using only very small streams, sampling at night when little or no photosynthesis existed, and the questionable validity of the assumption that respiration is independent of the dissolved oxygen level. The Tracer Technique, however, does not require the measurement and evaluation of any added sources of DO or significant depletions of the same. Tsivoglou's basis for using the Gas Tracer Technique was that the ratio of the rate coefficient for the absorption of a tracer gas from water to the rate coefficient for the absorption of oxygen by the same water is independent of streamflow conditions, including that of temperature. Rathbun (1977) found that the Gas Tracer Technique was far preferable to the Dissolved Oxygen Balance Technique first used by Streeter and Phelps, or the Disturbed Equilibrium Technique used by the WPRL. The basic Gas Tracer Technique (Tsivoglou, 1965, 1968) uses radioactive Krypton as the tracer gas; tritium is used to correct for dispersion and dilution; and Rhodomine-WT fluorescent dye is used to determine when to sample for the radioactive tracer. The United States Geological Survey modified the technique by using a low molecular weight hydrocarbon gas to avoid objections of adding radioactivity to water courses. Dye concentrations are measured fluorometrically. Bennett and Rathbun (1971) present a more detailed evaluation of the Gas Tracer Technique as well as the other techniques along with their respective errors. Rathbun (1977) concluded that no one equation for computing stream reaeration was best for all streams that he studied. He found that the predicted coefficients appeared to be much larger than could be accounted for by errors in measurement. Thus, some variables still may be needed for inclusion in the reaeration equations.

The following table, Table 5-1, summarizes the major equations and pros and cons of each, as well as valuable suggestions of when to use each.

TABLE 5-1. Reaeration Rates

Method	Advantages	Disadvantages	When Used
1. Streeter–Phelps $K_2 = K_1 - \frac{L}{D} - \frac{\Delta D}{230 \Delta t}$	Accurately describes actual results; beneficial when V approaches zero.	Must have K_1; must have t; not readily usable for predicting K_2 under critical conditions; math is difficult.	When flow time is readily measurable and k_1 is obtained.
2. O'Connor and Dobbins $K_2 = \frac{480 D_L^{1/2} S^{1/4}}{H^{5/4}}$ or $k_2 = \frac{127 (D_L U)^{1/2}}{H^{3/2}}$	Doesn't require flow time or k_1.	Is questionable since it may change from point to point; difficult to use with accuracy when heavy organic pollution exists.	When stream is quite uniform and flow time is not available.
3. Isaacs Modification $k_2 = \frac{2.833 V}{H^{3/2}}$	Easier to use than O'Connor's.	Is average of all possible stream roughness.	Quick and quite accurate general value.
4. Churchill's $k_2 = \frac{5.026 V^{0.969}}{R^{1.673}}$	More accurate than Isaacs' and proven in the field (TVA).	Is not too dependable where pollutants, especially BOD, are present.	In clean streams.

5-6. EFFECT OF TEMPERATURE ON k_2 AND OXYGEN LEVELS

The influence of temperature seems to be exerted primarily by affecting the velocity of movement of oxygen molecules in the water and in the air at the surface of the water. According to Churchill et al. (1962), this phenomenon of increased velocity of movement of oxygen molecules results in a faster saturation of the exposed water film, and probably saturation to a greater depth per equal time. Higher water temperatures increase the rate of molecular diffusion of gaseous O_2 in the surface film of water, and thus increase the rate of stream reaeration. However, the oxygen solubility also decreases with an increase of temperature, and thus the oxygen deficit–the major reaeration driving force–also decreases.

Churchill et al. (1962) found that when k_2 is plotted against temperature,

$$k_2 = ae^{bt}$$

where

a = intercept constant
b = rate constant
t = temperature, °C

and that

$$k_2(t°\text{C}) = 0.4364\, e^{0.0238\,t}$$

or

$$k_2(t°\text{C}) = k_2(20°\text{C}) \times 1.0238^{(t-20)}$$

Further, he states that "the reaeration rate increases with water temperature at the *geometric rate of 2.41%* per degree centigrade." Therefore, k_2, as well as other oxygen relationships, varies greatly with temperature and hence time of day and time of year. The stream analyst must take this into consideration when deciding upon its effect on oxygen sag–and, since k_1 increases at about twice the rate at which k_2 increases with an increase in temperature, Fair's f value, k_2/k_1, will decrease as the temperature rises.

From the foregoing material it is clear that reareation rates are quite variable, depending upon the particular stream characteristics and temperature. This variation in k_2 can cause a considerable difference in the amount of dissolved oxygen present in a given body of water. For example, one group of researchers (Gameson, Truesdale, and Varley, 1956) gives the following oxygen plots for four different waters (Fig. 5-3).

Some rationale for the emphasis upon reaeration in streams is obtained when we make a comparison of the concentration of oxygen present in water (mg/liter) and in air (mg/liter) on a *weight basis.*

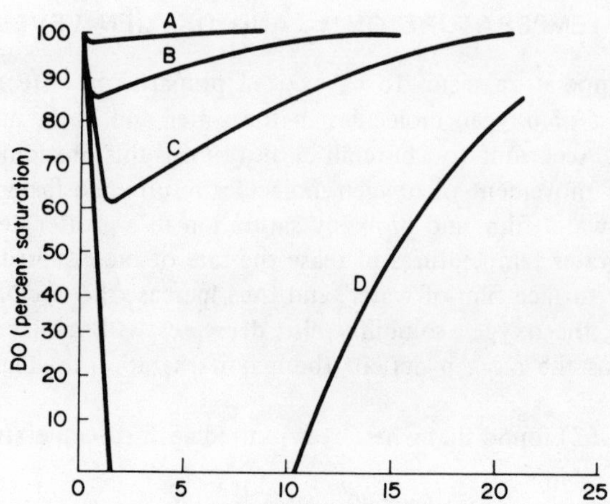

Calculated effect of a given pollution on rivers of different character.

Depth—3 feet — Ultimate BOD at start—50 ppm

A – Lakeland Beck — f = 200 cm/hr
B – Tributary of River Lea — f = 30 cm/hr
C – Effluent Channel — f = 10 cm/hr
D – Nearly Stagnant Water — f = 1 cm/hr

where f = surface renewal rate

FIG. 5-3. Variation in oxygen content from a shallow turbulent stream to a stagnant body of water.

Air. 1 mole air occupies *22.4* liters at STP; 1 mole air weights *29* grams. Therefore the unit weight of air = (29 grams)/(22.4 liters) = 1.3 grams/liter at STP; weight of oxygen = 1,300 mg/l X 20.9 percent O_2 (in air) = 273 mg/l.

Therefore, O_2 in water at STP = 8–9 mg/l, whereas O_2 in air at STP = ~273 mg/l.

Thus, the concentration of oxygen by weight is 30–34 times greater in air than in water.

Your author suggests using either Churchill's, O'Connor's, or Isaacs' formulations for computing k_2 in absence of oxygen deficits greater than 25 percent of saturation. The Streeter-Phelps modification must be used when oxygen levels are low, since reaeration is greatly increased by extreme deficits.

QUESTIONS

1. Upon what general mathematical relationship does the rate of change of oxygen deficit with time depend?
2. Upon what differentiated form?

3. Describe what data or information must be known or available in (2) in order to compute the oxygen deficit at a specific point in a stream.
4. Can you compute k_2 from the sag curve equation above (2)? If yes, what are the sources of error?
5. What is a more realistic and practical method first developed by Streeter in 1926 for describing the general computation of k_2?
6. What are C and n in the above equation and upon what do they depend?
7. Upon what are the formulations of O'Connor and Dobbins based?
8. Once again, what is the serious defect with both the O'Connor and the Streeter equations?
9. Does the velocity gradient at the surface of a stream necessarily dictate the rate of surface renewal of oxygen?
10. How does Isaacs' modification of K_2 determination affect its value, and what is the major drawback of his work? Does k_2 depend upon the roughness of the channel? If so, why?
11. What general composite formulation does Isaacs give for k_2 determination?
12. How did Churchill, in his 1957 studies, actually measure the true k_2?
13. How did Churchill find the correlation of k_2 to the proper dimensionless group?
14. What equation did Churchill finally arrive at, and why?
15. Why will the true k_2 value not correlate directly with the computed value?
16. Have any other formulations on k_2 determination been proposed? Do they really offer any new insight into its real value?
17. Does temperature affect k_2? How?
18. How many more times is oxygen more soluble in air than in water?

REFERENCES

Adeney, and Becher: The Determination of the Rate of Solution of Atmospheric Nitrogen and Oxygen by Water, *Phil. Mag.*, 38:1919.

Bennett, J. P., and R. E. Rathbun: Reaeration in Open-Channel Flow, *U.S. Geological Survey Open-File Report*, Bay St. Louis, Miss., April 1971.

Churchill, M. A., H. L. Elmore, and R. A. Buckhingham: The Prediction of Stream Reaeration Rates, *J. San. Eng. Div.* (ASCE), SA 4, July 1962, paper 3199.

Downing and Truesdale: *Journal of Applied Chemicstry*, 5:570, 1955.

Edwards, R. W., M. Owens, and J. W. Gibbs: Estimates of Surface Aeration in Two Streams. *Journal of the Institute of Water Engineers*, London, England, Vol. 15, No. 5, August 1961, pp. 395–405.

Gameson, A. L. H., G. A. Truesdale, and A. L. Downing: Re-aeration Studies in a Lakeland Beck, *Journal of the Institution of Water Engineers*, London, England, Vol. 9, No. 7, November 1955, pp. 571–594.

Gameson, A. L. H., and G. A. Truesdale: Some Oxygen Studies in Streams, *Journal of the Institution of Water Engineers*, London, England, Vol. 13, No. 2, March 1959, pp. 175–187.

Gameson, Truesdale, and Varley: *Water and Sanitary Engineer*, 6:52, 1956.

Isaacs, W. P.: Atmospheric Oxygenation and Biological Deoxygenation in an Idealized

Stream Flow Model. Ph.D. dissertation, Oklahoma State University, Stillwater, Oklahoma, May 1967.

Isaacs, W. P., et al.: Proceedings of the 24th Industrial Waste Conference, *Purdue Univ. Eng. Est. Ser. No. 135,* May 6-8, 1969, p. 1464.

Kehr, R. W.: Measures of Natural Oxidation in Polluted Streams IV. Effect of Sewage on Atmospheric Reaeration Rates Under Stream Flow Conditions, *Sewage Works J.,* **10:** 228, March 1938.

Krenkel, P. A., and G. T. Orlob: Turbulent Diffusion and the Reaeration Coefficient, *J. San. Eng. Div.* (ASCE), **3** (88):SA2, p. 53, March, 1962.

Negulescu, M., and V. Rojanski: Recent Research to Determine Reaeration Coefficient, *Water Research,* Pergamon Press, London, **3:**(0)189, 1969.

O'Connor, D., and W. Dobbins: The Mechanism of Reaeration in Natural Streams, *J. San. Eng. Div.* (ASCE) SA6, 1115-1-1115-30, December 1956.

Owens, M., R. W. Edwards and J. W. Gibbs: Some Reaeration Studies in Streams, *Int. J. Air Water Poll.,* **8:**(0)469, 1964.

Rathbun, R. E.: Reaeration Coefficients of Streams–State of the Art, *Journ. Hydraulics Division* (ASCE) HY4, p. 409, April 1977.

Streeter, Harold: *The Rate of Atmospheric Reaeration of Sewage Polluted Streams,* Reprint 1063, Public Health Reports, 1926.

Streeter, H., and E. Phelps: A Study of the Purification of the Ohio River, U.S. Publ. Health Service Bull. No. 146, Washington, D.C., 1925.

Tsivoglou, E. C., et al., Tracer Technique of Atmospheric Reaeration I–Laboratory Studies, *Journ. of Water Pollution Control,* **37**, 10, October 1965, pp. 1343-1362.

Tsivoglou, E. C., et al., Tracer Measurements of Stream Reaeration II–Field Studies, *Journ. of Water Pollution Control,* **40**(2):285-305, Pt. 1, February 1968.

Zak, G. L.: *Samoociscenie Vodolmox.* Ministerstvo Kommunalnogo Hozeaistva, Moskva, 1960.

6
STREAM OXYGEN SAG ANALYSES

A curve can be constructed from the combined effects of deoxygenation and reaeration which will plot the course of dissolved oxygen along a stretch in a river (see Fig. 6-1) for a single point discharge of biodegradable organic matter. This curve is, therefore, very important in stream sanitation work.

The curve of deoxygenation will be the reciprocal curve of the BOD reaction which is the rate of withdrawal of oxygen, which starts at a maximum and diminishes continuously toward zero.

Reaeration starts at zero rate, since we assumed the water was saturated to start with. Since the reaeration rate is always proportional to the saturation deficit, the rate increases as the deficit increases.

As deoxygenation and reaeration proceed, a minimum DO point results which is called the *critical point*, after which the reaeration becomes dominant and the dissolved oxygen starts to rise. This critical DO and the time to reach this point are both very important for the oxygen quality and use of the stream.

6-1. ANALYSES OF SAG CURVE BY STREETER-PHELPS FORMULATION

This equation, as presented in 5-1, states that the rate of increase in the deficit is proportional to the algebraic sum of the two terms

$$\frac{dD}{dt} = K_1 L - K_2 D$$

The first ($K_1 L$) represents the deoxygenation reaction. It increases the deficit proportional to the residual BOD (L) and the reaction (deoxygenation) rate K_1.

The second ($K_2 D$) represents the reaeration reaction. It decreases the deficit proportional to the deficit existing (D) and the reaeration rate K_2. Besides being

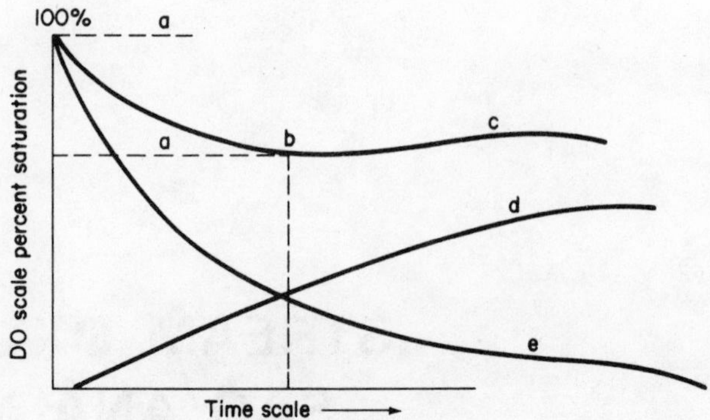

FIG. 6-1. Deoxygenation, reaeration, and oxygen sag curve. (a) = maximum deficit; (b) = critical point; (c) = sag curve (summation of reaeration and deoxygenation); (d) = reaeration (cumulative); (e) = deoxygenation (cumulative).

dependent upon temperature, as is K_1, it also depends upon the depth, velocity, and physical characteristics of the stream.

The integrated form of the above equation is

Sag Equation
$$D_t = \frac{k_1 L_a}{k_2 - k_1}(10^{-k_1 t^*} - 10^{-k_2 t^*}) + D_a \cdot 10^{-k_2 t^*}$$

where L_a and D_a are the initial ultimate BOD and oxygen deficits in the stream, respectively; k_1 and k_2 are the deoxygenation and reaeration rates expressed as common logs; and D_t then is the deficit at time t (days).

Therefore, if we know D_a, L_a and the k values, we are able to compute the deficit at any time downstream. *The sag curve is the locus of the deficit values plotted as a continuous curve with time.*

Since at the critical point in the sag curve the reaeration rate is equal to the deoxygenation rate, and at this precise moment the change in deficit is zero, we can say that

$$\frac{dD}{dt} = k_1 L - k_2 D_c = 0$$

$$-k_2 D_c = -k_1 L$$

$$D_c = \frac{k_1 L}{k_2}$$

---

*Values for these quantities for various values of $K_1 t$ and $K_2 t$ are presented in Appendix C, page 419.

but

$$L = L_a \times e^{k_1 t}$$

$$D_c = \frac{k_1}{k_2} L_a \times e^{-k_1 t_c}$$

in common logs

$$D_c = \frac{k_1}{k_2} L_a \times 10^{-k_1 t_c}$$

or

$$\log D_c = \log \frac{k_1}{k_2} L_a - k_1 t_c$$

where D_c is the oxygen deficit at the critical oxygen sag point. G. M. Fair (1939) introduced his formulation of relating the value k_2/k_1 to an "f" ratio. A range of values of f is given in Table 6-1. Presumably one should use personal judgment in selecting the appropriate f value within the range for each stream classification given.

Fair found that there is a decrease in the value of f of about 3 percent for each 1° rise in temperature. Although k_2 increases with temperature, k_1 also increases, and the solubility of oxygen in water decreases, resulting in less deficit. Therefore, the k_2 rate becomes lowered due to the decreased deficit.

By differentiating the sag equation and solving for t_c, the time to the critical sag point, we obtain

$$t_c = \frac{1}{k_2 - k_1} \log \frac{k_2}{k_1} \left(1 - \frac{D_a(k_2 - k_1)}{L_a k_1}\right)$$

TABLE 6-1. Range of "f" Values for Various Waters.
[*After Fair, 1939. By permission of publisher.*]

	Nature of receiving water	Value of f 20°C
(a)	Small ponds and backwaters	0.5–1.0
(b)	Sluggish streams and large lakes or impounding reservoirs	1.0–1.5
(c)	Large streams of low velocity	1.5–2.0
(d)	Large streams of normal velocity	2.0–3.0
(e)	Swift streams	3.0–5.0
(f)	Rapids and waterfalls	above 5.0

Obviously with Fair's f this is simpler and can be expressed without really calculating k_2, by substituting for k_2/k_1 the value of f from the above table, as

$$t_c = \frac{1}{k_1(f-1)} \log \left((f)\left[1-(f-1)\frac{D_a}{L_a}\right]\right)$$

We observe the *initial* or *upstream DO* and *BOD* in the *river* as well as the *temperature*, and either observe the BOD at a station downstream at a measured distance in time of flow to calculate the k_1 value or assume a definite f value (saving the k_2 calculation) in order to predict mathematically the *oxygen deficit at the critical time downstream.*

In a similar manner we may use Fair's f value in place of k_2/k_1 to simplify and to approximate the computation of D_c from the equation at the top of the page.

$$\log D_c = \log \frac{L_a}{f} - k_1 t_c$$

As an example of the use of the sag equation, we wish to compute the oxygen deficit at Station 5, knowing that stream constants are

$$k_1 = 1.8 \qquad k_2 = 5.72$$

and

$$D_{\text{sta.4}} = 1{,}494 \frac{\text{pounds } O_2}{\text{day}}$$

$$L_{\text{sta.4}} = 5{,}565 \frac{\text{pounds BOD ultimate}}{\text{day}}$$

and the time of flow between stations is 0.146 days; then

$$D_5 = \frac{k_1}{k_2 - k_1} L_4(10^{-k_1 t} - 10^{-k_2 t}) + D_4 \times 10^{-k_2 t}$$

$$= \frac{1.8}{5.72 - 1.80} (5565)(10^{-1.8(.146)} - 10^{-5.72(.146)}) + 1494 \times 10^{-5.72(.146)}$$

$$= 2560(10^{-.263} - 10^{-.835}) + 1494 \times 10^{-.835}$$

$$= 2560(.5458 - .1462) + 1494(.1462)$$

$$= \frac{1{,}240\ \#}{\text{day}}$$

The D_5 value of 1,240 pounds per day can be converted to concentration in parts per million by dividing by the product of the streamflow at Station 5 in mgd, and 8.34, the pounds weight in a gallon of water, that is,

$$\frac{1{,}240\#/\text{day}}{\text{Qmg/day} \times 8.34\#/\text{gal}} = \text{ppm oxygen deficit}$$

By utilizing the findings of Streeter and Phelps (1925) from the Ohio River Studies, the stream analyst can calculate the deficit in oxygen existing at any location in a stream below a source of organic pollution. He should use the three basic equations already presented to arrive at his conclusions as follows:

(1) $K_1 = \frac{1}{\Delta t} \log \frac{L_A}{L_B}$ = deoxygenation rate (see Chapters 3 and 4)

(2) $K_2 = K_1 \frac{\bar{L}}{\bar{D}} - \frac{\Delta D}{2.3 \Delta t \bar{D}}$ = reaeration rate (see Chapter 5-1)

(3) $D_t = \frac{K_1 L_A}{K_2 - K_1} (10^{-K_1 t} - 10^{-K_2 t}) + D_A 10^{-K_2 t}$ = dissolved oxygen deficit downstream at time t (as derived in this chapter)

When using Streeter-Phelps formulations above, the following definitions hold:

K_1 = deoxygenation rate, per day
Δt = time of travel in days
L_A = ultimate upstream first-stage BOD in ppm or pounds
L_B = ultimate downstream first-stage BOD in ppm or pounds
K_2 = reaeration rate, per day
$\bar{L}$ = average ultimate first-stage oxygen demand in reach (stream section between points A and B) in ppm or pounds
$\bar{D}$ = average oxygen deficit in reach in ppm or pounds
ΔD = change in oxygen deficit from upstream to downstream sampling points in ppm or pounds
t = time of streamflow from upstream to downstream points of sampling in days
D_t = dissolved oxygen deficit in ppm or pounds at time t downstream

6-2. STREAM ANALYSIS BY STATISTICAL METHODS

Churchill and Buckingham (1956) propose a method of analysis based on the multiple correlation of all the principal factors producing and controlling the extent of the oxygen sag below a source of pollution. *Times of* water *travel* determinations are not necessary. Long-term BOD determinations are not required, since the BOD constants k_1 and k_2 and ultimate carbonaceous BOD L are not used in the analysis.

A number of factors operate concurrently in a stream below a source of pollution to produce an oxygen deficiency in the river. The basic procedure correlates these factors statistically, all at the same time, with the measured decrease, or drop, in oxygen concentration from *above pollution* to the *low point of the oxygen sag.* The factors used must be readily measurable in quantitative units. When pollution is discharged into a reasonably well aerated stream, the factors subject to direct quantitative observation which affect the extent of the oxygen sag are *BOD*, *stream temperature*, and *stream discharge.*

Churchill and Buckingham believe that these controlling factors can be correlated by relating the observed values of the so-called independent variables with corresponding observed values of the dependent variable, *DO drop*, resulting in an equation that can be used with confidence to predict the extent of the DO drop for various assumed values of the independent variables.

In general, Churchill (1956) has found, by making 24 stream surveys and taking 24 stream samples at appropriate points, that a good correlation exists between BOD, DO, temperature, and streamflow. In other words, he found that the dissolved oxygen sag occurring in a stream depends upon only three variables: BOD, temperature, and flow. By using the least-squares method, the line of regression can be computed so that for any desired BOD loading the dissolved oxygen sag can be predicted. This method eliminates the often questionable and always cumbersome procedure for determining times of flow between stations and resulting stream reaction rates (k_1, k_2, and k_3).

The author and his students (Simmons, Nemerow, and Armstrong, 1957) found that good correlation exists in the Churchill and Buckingham method if each sample is observed under maximum and minimum conditions of each of the three stream variables of streamflow, BOD, and temperature. Only six samples were required to produce practical results. Additional samples may add some small degree of refinement to the results, but the refinement probably would not offset the effort and cost expended in planning, collecting, and analyzing the samples and calculating the results.

This method yields, after a great deal of arithmetic calculation (which even when performed with the aid of a calculator is cumbersome and error prone), an equation that can be useful in further stream reaction predictions. However, little insight is gained into the reasons for the stream's reaction. It is intended to "give only the final score rather than the scoring plays." With the advent of electronic data processing, the problem should be lessened significantly.

The simplest type of multiple correlation is expressed by the linear equation

$$Y = a + b_1 X_1 + b_2 X_2 + b_3 X_3$$

in which

Y = the dependent variable, DO drop in ppm
X_1 = 5-day BOD in ppm
X_2 = water temperature in °C.

X_3 = stream discharge factor in ft^3/sec
a, b_1, b_2, b_3 = constants derived from the data actually used

Development of Normal Equations for Determining Line of Best Fit by Least-Squares Procedure

The least-squares line for a given series may be obtained by the use of a set of "normal equations." These can be derived mathematically, but for our purpose can be obtained by multiplying the "type" equation

$$(Y = a + b_1 X_1 + b_2 X_2 + b_3 X_3)$$

by the coefficients of each unknown (a, b_1, b_2, b_3). The coefficient of the first unknown, b_1, is X_1. Multiplying the type equation by X_1 we have

$$X_1 Y = b_1 X_1^2 + b_2 X_1 X_2 + b_3 X_1 X_3$$

The formula must be summed up for all points

(1) $\sum X_1 Y = b_1 \sum X_1^2 + b_2 \sum X_1 X_2 + b_3 \sum X_1 X_3$

This is the form of the first normal equation. The other two are found in exactly the same manner.

(2) $\sum X_2 Y = b_1 \sum X_1 X_2 + b_2 \sum X_2^2 + b_3 \sum X_2 X_3$

(3) $\sum X_3 Y = b_1 \sum X_1 X_3 + b_2 \sum X_2 X_3 + b_3 \sum X_3^2$

A practical example of solving these three equations simultaneously to ascertain the constants and resulting dissolved oxygen drop is shown in a study of the Neuse River, North Carolina (Simmons, Nemerow, and Armstrong, 1957), in Tables 6-2 and 6-3 and the computations in Table 6-4. The sampling stations are shown in Fig. 6-2. It should be noted that Stations 2 and 4 are located on tributaries of the Neuse River and carry portions of the sewage effluents from Raleigh.

Another correlation procedure must be used to compute the allowable BOD loading at the source of pollution. The BOD equation can be derived from the same least-squares method by correlating the upstream BOD loads with the temperature, discharge, and resulting BOD at the sag point in the stream.

As with all methods of stream analysis, ideally one should sample the stream under all critical conditions of the above four variables. The closer one comes to this ideal situation, the more accurate will be the stream equation derived by this method from the analytical data. To conclude that the equation is useless if critical values of each variable do not occur during the stream survey is not valid. No worker has yet shown how much actual discrepancy will occur in deriving the stream equation under less than ideal conditions.

TABLE 6-2. Basic Data—Neuse River, N.C. [*After Simmons, Nemerow, and Armstrong, 1957. By permission of publishers.*]

Date (1956)	Station 1				Station 2				Station 3				Station 4				Station 5				Station 6			
	Temp. (°C)	DO (ppm)	BOD (ppm)	Disch. (cfs)	Temp. (°C)	DO (ppm)	BOD (ppm)	Disch. (cfs)	Temp. (°C)	DO (ppm)	BOD (ppm)	Disch. (cfs)	Temp. (°C)	DO (ppm)	BOD (ppm)	Disch. (cfs)	Temp. (°C)	DO (ppm)	BOD (ppm)	Disch. (cfs)	Temp. (°C)	DO (ppm)	BOD (ppm)	Disch. (cfs)
June 26	25.8	4.7	2.6	140	25.8	0	18	30	26.8	2.4	7.4	170	25.2	0	38	14	26.8	2.4	6.0	220	27.0	4.1	3.1	215
June 30	29.0	8.6	6.0	125	27.5	0	31	15	29.0	6.3	6.8	140	28.5	0	43	10	29.0	4.9	7.4	183	29.0	10.3	6.1	200
July 4	28.0	5.3	1.4	132	26.0	0.6	13	33	27.0	2.6	5.0	165	24.8	0.8	11	48	26.5	0.1	5.6	285	27.5	2.6	4.0	220
July 22	25.5	5.6	1.2	480	26.0	3.8	6.8	84	26.0	6.0	2.3	570	26.0	1.4	69	30	26.5	5.3	1.4	630	27.0	4.5	1.9	780
Aug. 5	25.0	5.4	1.0	460	23.5	2.2	16	62	25.5	5.2	2.6	525	24.0	0	41	14	25.5	5.0	3.0	580	26.0	3.6	2.4	525

TABLE 6-3. Multiple Linear Correlation of DO Drop, BOD, Temperature, and Discharge Using Three Samples. [*After Simmons, Nemerow, and Armstrong, 1957. By permission of publishers.*]

Item	Dissolved oxygen (ppm)			BOD at sag (ppm) X_1	Temp. (°C) X_2	1,000 cfs X_3	Y^2	YX_1	YX_2	YX_3	X_1^2	X_1X_2	X_1X_3	X_2^2	X_2X_3	X_3^2
	Sta. 1	at sag	Drop Y													
6/26/56	4.7	2.4	2.3	6.0	26.8	4.55	5.29	13.80	61.64	10.47	36.00	160.80	27.30	718.24	121.94	20.70
6/30/56	8.6	4.9	3.7	7.4	29.0	5.46	13.69	27.38	107.30	20.20	54.76	214.60	40.40	841.00	158.34	29.81
7/4/56	5.3	0.1	5.2	5.6	26.5	3.51	27.04	29.12	137.80	18.25	31.36	148.40	19.66	702.25	93.02	12.32
Sums			11.2	19.0	82.3	13.52	46.02	70.30	306.74	48.92	122.12	523.80	87.36	2,261.49	373.30	62.83
Means			$\bar{Y}$	$\bar{X}_1$	$\bar{X}_2$	$\bar{X}_3$	–	–	–	–	–	–	–	–	–	–
			3.73	6.33	27.43	4.51										
Correction items*							$n\bar{Y}_2$	$n\bar{Y}\bar{X}_1$	$n\bar{Y}\bar{X}_2$	$n\bar{Y}\bar{X}_3$	$n\bar{X}_1^2$	$n\bar{X}_1\bar{X}_2$	$n\bar{X}_1\bar{X}_3$	$n\bar{X}_2^2$	$n\bar{X}_2\bar{X}_3$	$n\bar{X}_3^2$
							41.74	70.83	306.94	50.47	120.21	520.89	85.64	2,257.21	371.12	61.02
Corrected sums							4.28	-0.53	-0.20	-1.55	1.91	2.91	1.72	4.28	2.18	1.81

*n = number of observations.

TABLE 6-4. Computation for Coefficients in the Multiple Linear Correlations. *[After Simmons, Nemerow, and Armstrong, 1957. By permission of publishers.]*

The following three "normal" equations are based on the principles of the least squares.

$$b_1 \sum X_1^2 + b_2 \sum X_1X_2 + b_3 \sum X_1X_2 = \sum X_1 Y \quad \text{(1a)}$$

$$b_1 \sum X_1X_2 + b_2 \sum X_2^2 + b_3 \sum X_2X_3 = \sum X_2 Y \quad \text{(1b)}$$

$$b_1 \sum X_1X_3 + b_2 \sum X_2X_3 + b_3 \sum X_3^2 = \sum X_3 Y \quad \text{(1c)}$$

Computation for coefficients for three samples

Substituting values from Table II in Eq. (1a)		$1.91b_1 + 2.91b_2 + 1.72b_3 = -0.53$	(2)
Dividing Eq. (2) by −1.91 then		$-b_1 - 1.523560b_2 - 0.900524b_3 = +0.277487$	(3)
Multiplying Eq. (2) by −1.523560		$-2.91b_1 - 4.433560b_2 - 2.620523b_3 = +0.807487$	(4)
Substituting values from Table II in Eq. (1*b*)		$2.91b_1 + 4.28b_2 + 2.18b_3 = -0.20$	(5)
	$(\sum_2)$	$-0.153606b_2 - 0.440523b_3 = 0.607487$	(6)
Dividing $\sum_2$ by −0.153560		$-b_2 - 2.868735b_3 = 3.956024$	(7)
Multiplying Eq. (2) by −0.900524		$-1.72b_1 - 2.62b_2 - 1.548901b_3 = +0.477278$	(8)
Now, multiplying $\sum_2$ by −2.868735		$+0.44b_2 + 1.263744b_3 = 1.742719$	(9)
Substituting values from Table II in Eq. (1*c*)		$1.72b_1 + 2.18b_2 + 1.81b_3 = -1.55$	(10)
Adding Eqs. (8, 9, and 10)	$(\sum_3)$	$1.524843b_3 = -2.815441$	(11)
Solving for b_3		$b_3 = -1.846381$	
Substituting value of b_3 in Eq. (7)		$-b_2 - 2.868735\ (-1.846381) = 3.956024$	
Solving for b_2		$b_2 = 1.340754$	
Substituting values of b_3 and b_2 in Eq. (3)		$-b_1 - 1.523560(1.340754) - 0.900524(-1.846381) = +0.277487$	
Solving for b_1		$b_1 = -0.657496$	
Check by substituting in Eq. (10)		$1.72(-0.657496) + 2.18(1.340754) + 1.81(-1.846381) = -1.55$	
		$1.540999 = 1.55$	

TABLE 6-4. (*Continued*)

then

$$b_1 = -0.6575$$
$$b_2 = 1.3408$$
$$b_3 = -1.8464$$

$$R^2 = \frac{b_1 \sum YX_1 + b_2 \sum YX_2 + b_3 \sum YX_3}{Y^2} = \text{correlation coefficient}$$

$$= \frac{(-.6575)(-0.53) + (1.3408)(-0.20) + (1.8464)(-1.55)}{4.28}$$

$$= \frac{0.3485 - 0.2682 + 2.8619}{4.28} = \frac{2.9422}{4.28} = 0.687430$$

$$R = \sqrt{.687430} = 0.83 = \text{correlation coefficient}$$

$$a = \bar{Y} - b_1\bar{X}_1 - b_2X_2 - b_3\bar{X}_3$$

$$= 3.73 - (-0.6575)(6.33) - (1.3408)(27.43) - (-1.8464)(4.51)$$

$$= 3.73 + 4.1620 - 36.7781 + 8.3273$$

$$= -36.7781 + 16.2193 = -20.5588$$

Sample calculation

$$Y = a + b_1X_1 + b_2X_2 + b_3X_3 = -20.5588 + (-.6575)\,X_1 + 1.3408(X_2) - 1.8464(X_3)$$
$$= -20.5588 + (-0.6575)(6.0) + (1.3408)(26.8) + (-1.8464)(4.55)$$
$$= -20.5588 - 3.9450 + 35.9334 - 8.4011$$
$$= 3.0285$$

6-3. EFFECTS OF SEDIMENTATION, CHANNEL SCOUR, AND PHOTOSYNTHESIS ON STREAM OXYGEN SAG COMPUTATIONS

Stream analysts have decried the poor correlation that exists between observed dissolved oxygen and computed values when using the Streeter-Phelps formulations. The use of the multiple linear correlation method has neither completely nor satisfactorily eliminated this lack of correlation. Most stream analysts agree in fixing the cause of poor correlation upon the so-called unsteady state of receiving streams. Streams are dynamic and viable bodies of water. As such, they change in their reactions to pollution from time to time, at the same location as well as from location to location on their paths. At least three major phenomena cause this unsteady-state condition to exist: (1) sedimentation, (2) scouring, and (3) photosynthesis. Despite the fact that the basic principles of all three are well understood, workers have been unable to predict accurately when and to what extent they will occur. Some consideration of the effects of each is in order

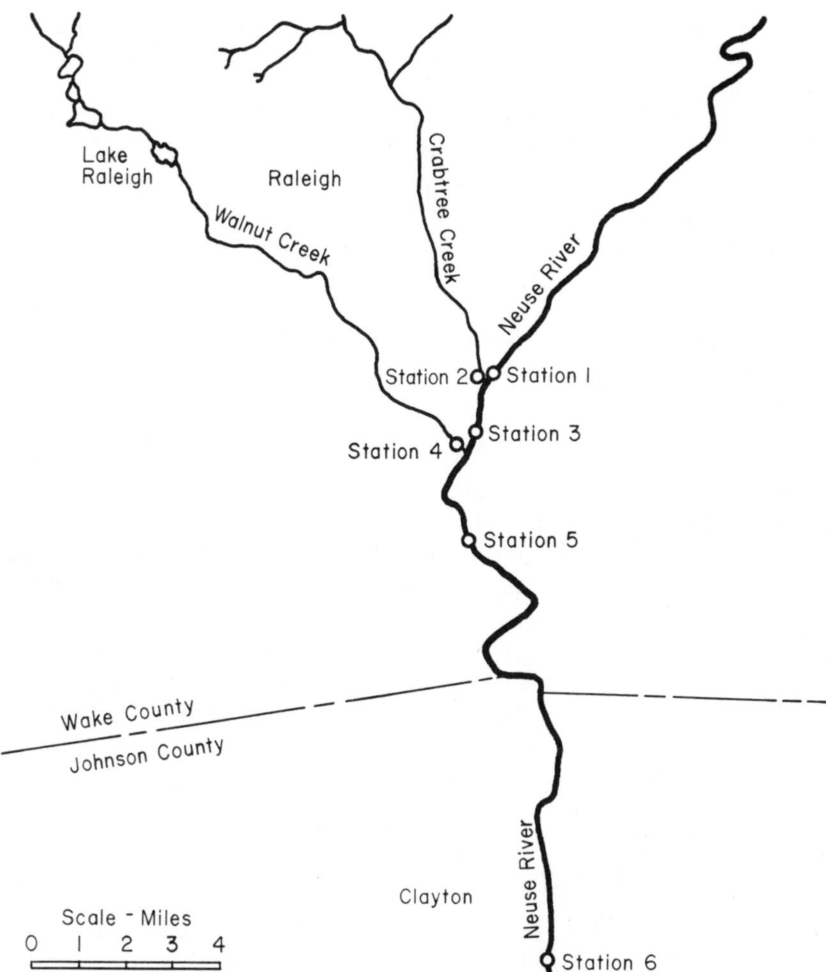

FIG. 6-2. Sampling stations on the Neuse River in North Carolina. (*From Simmons, Nemerow, and Armstrong, 1957. By permission of publisher.*)

here. Slime growths can also create an unsteady state. However, most investigators consider this effect as one enhancing the deoxygenation rate k_1.

Organic Sedimentation

We have already discussed the oxygen demand of benthal deposits in Chapters 1-4. We are concerned here with the effects of settling and the subsequent benthal decomposition on oxygen sag in a receiving stream. Some oxygen-demanding matter is usually associated with all suspended organic matter in streams. When this type of organic matter settles out of the main flow in a stream and deposits on the bottom, the deoxygenation rate k_1 is calculated to be greater than it really is, because the oxygen demand of this material actually disappears from the flowing stream. In reality, however, this oxygen demand is delayed for some future time when it will decompose benthally. In addition, the demand will be localized in a stretch of the stream directly above the settled organic matter. The net effect of deposited sludge then is to remove a portion of oxygen demand from the streamflow and thus raise oxygen levels downstream. Likewise, the oxygen level will decrease more rapidly in the stream flowing over the deposited sludge at some later time when it decomposes. This is pictured graphically in Fig. 6-3.

Scouring

When unsteady and turbulent flow exists in a stream such that the deposited sludge is dislodged and disrupted from the bottom of pools, oxygen levels in the flowing water change. The constant, slow, and localized oxygen demand is once again transferred to the flowing stream. A slight lowering of oxygen may be noticed in downstream flowing water that carries this scoured sludge. If and when the scoured deposits resettle, a sudden localized lowering of the oxygen level will recur at the new downstream location. Scouring, then, results in a slight lowering of oxygen in the stream and a rather drastic new drop in oxygen

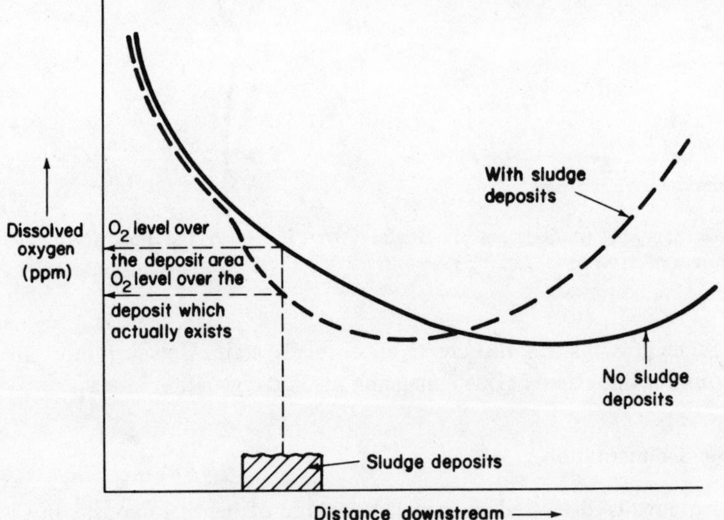

FIG. 6-3. Organic sedimentation effect in receiving stream.

at a new downstream location where the scoured solids are redeposited. The anaerobic deposit may be stirred up and become aerobic, thus exerting an increased oxygen demand.

Photosynthesis

The phenomenon of growth of microscopic green plants directly affects the oxygen levels in streams. Growing plants use energy in the form of sunlight, carbon dioxide, and minerals from air and water, to produce oxygen biomass. Oxygen levels will be higher in daylight (often supersaturated) and lower after sundown than under conditions of no photosynthesis in streams. These effects are more fully described in Chapter 2 and are depicted graphically in Fig. 2-5.

O'Connor (1967) gives a mathematical expression to determine oxygen release due to photosynthesis

$$P_{(t)} = \frac{P_m \sin t}{p\pi} \qquad \text{where } 0 \leqslant t \leqslant P$$

P_m is the maximum rate over period p; $P_{(t)}$ is the rate of oxygen release due to photosynthesis.

6-4. THOMAS' GRAPHICAL ANALYSIS FOR SHORTENED SAG CURVE ANALYSIS

Thomas (1948) gives as the most general form of the oxygen-sag equation

$$D = \frac{k_1 L_A}{k_2 - (k_1 + k_3)} \left(10^{-(k_1 + k_3)t} - 10^{-k_2 t}\right) + D_2 10^{-k_2 t}$$

where k_3 = the constant of proportionality reflecting the composition of the waste and receiving water and the quiescence of the stream at the point under consideration. In regions of considerable turbulence, k_3 at times may be negative, indicating channel scour, which reverses the effect of deposition. A positive k_3 indicates stream sedimentation of suspended organic matter.

The rate of removal of BOD due to decomposition is proportional to the BOD remaining and is = $2.3k_3L$, according to Thomas.

All three reaction rate constants are affected by temperature, as shown in Table 6-5.

According to Thomas (1948), the ranges of values of the stream pollution constants for most American streams handling nontoxic wastes during warm months are

k_1 per day	.06–.36
k_2 per day	.06–.96
k_3 per day	–.36–+.36

TABLE 6-5. Temperature Factors for Oxygen Balance Constants. [*After Thomas, 1948. By permission of publisher.*]

Temp. °C	16	18	20	22	24	26
k_1	0.832	0.912	1.000	1.096	1.202	1.317
k_2	0.939	0.969	1.000	1.032	1.065	1.099
k_3	0.752	0.867	1.000	1.153	1.331	1.535
L	0.920	0.960	1.000	1.040	1.080	1.120

Your author has found numerous streams possessing "apparent" k_1 and k_2 values much higher than those given by Thomas. Not all of these differences can be attributed to noncomputed k_3 values.

Thomas gives an example that illustrates the drastic effect that these three stream constants have on the sag curve:

A town discharges *1.0 mgd* of settled sewage of *100 ppm* ultimate first-stage BOD and *0 ppm* DO into a stream with an average flow of *11 mgd*, *4 ppm* of ultimate BOD, and *9.2 ppm* of DO. The stream and sewage temperatures are *68°F*. The combination of these two fluids results in a low of *12 mgd*, *12 ppm ultimate BOD*, and *8.43 ppm* DO. Since at *68°F* the saturation of DO is *9.2 ppm*, the initial oxygen deficit is $D_A = 9.2 - 8.43 = 0.77$ *ppm*. The resulting oxygen deficits, obtained by using the Streeter-Phelps equations for various values of k_1, k_2, and k_3, are compiled in Table 6-6.

Thomas, one of the few researchers to consider the effect of sedimentation and channel scour on k_3, found that it may be positive (as in Case III) during the greater part of the time; then it may quickly become negative during a flood, or during a period of stagnation in warm weather that causes fragments of the bottom deposits to be carried to the surface by gases of anaerobic decomposition.

According to Thomas, Case IV indicates the marked effect of a small negative value of k_3 upon the DO, namely, a substantial increase in magnitude of the DO deficit.

The constant k_3 may also take into account the effect of irregular, noncontinuous pollution, dilution, and other disturbing influences within the reach. If such irregularities occur, the time-average value of k_3 will not necessarily be zero. In some streams, deposition, channel scour, and noncontinuous pollution and dilution are not major factors, and it is permissible to assume k_3 to be zero. This assumption has usually been made in previous applications of the oxygen balance formulation.

Thomas (1948) also states that the Streeter-Phelps equation is *unwieldy* and in most practical applications can be solved *only by tedious trial-and-error procedures.*

This disadvantage may be overcome by the use of nomograms, one particu-

TABLE 6-6. Oxygen Deficits with Various Values of Stream Pollution Capacity (ppm). *[After Thomas, 1948. By permission of publisher.]*

Time (days)	Case I $k_1 = .12$ $k_2 = .12$ $k_3 = .00$	Case II $k_1 = .12$ $k_2 = .24$ $k_3 = .00$	Case III Sediment $k_1 = .12$ $k_2 = .12$ $k_3 = .06$	Case IV Scour $k_1 = .12$ $k_2 = .12$ $k_3 = -.06$
0	0.8	0.8	0.8	0.8
$\frac{1}{2}$	2.1	1.9	2.0	2.1
1	3.2	2.6	2.8	3.2
2	4.2	3.2	3.6	4.7
3	4.7	3.1	3.7	5.5
4	4.6	2.7	3.4	5.9
5	4.4	2.3	3.0	6.0
6	4.0	1.9	2.6	5.9
8	3.0	1.2	1.8	5.3
10	2.1	0.7	1.1	4.5

larly useful type of which is presented in Fig. 6-4 for computing the oxygen sag with less time and effort than by using the standard Streeter-Phelps formulations.

Assuming the data of the example given in Table 6-6 for Case I,

$$\frac{k_2}{k_1} = 1.0$$

$$D_A = 0.77 \text{ ppm}$$

$$L_A = 12 \text{ ppm}$$

Using a straightedge, form a straight line called an *isopleth*, connecting point $\frac{0.77}{12} = .064$ [1] on the D_A/L_A scale at the left with, say, the point representing the first day, $k_2 t = 0.12 \times 1 = .12$ [2] at the intersection of the line 0.12 on the $k_2 t$ scale on the $k_2/k_1 = 1.0$ [2] curve. Next read 0.265 [3] on the D/LA scale at the intersection of the isopleth. Finally, the value of the deficit at the end of the first day may be computed

$$\frac{D}{L_A} = .265$$

$$\therefore \underset{\substack{\uparrow \\ (1 \text{ day})}}{D} = 0.265\underset{\uparrow \to L_A}{(12)} = 3.2 \text{ ppm}$$

Of particular interest is the value of the maximum or critical deficit, D_c, which may be determined by increasing the inclination of the isopleth until it

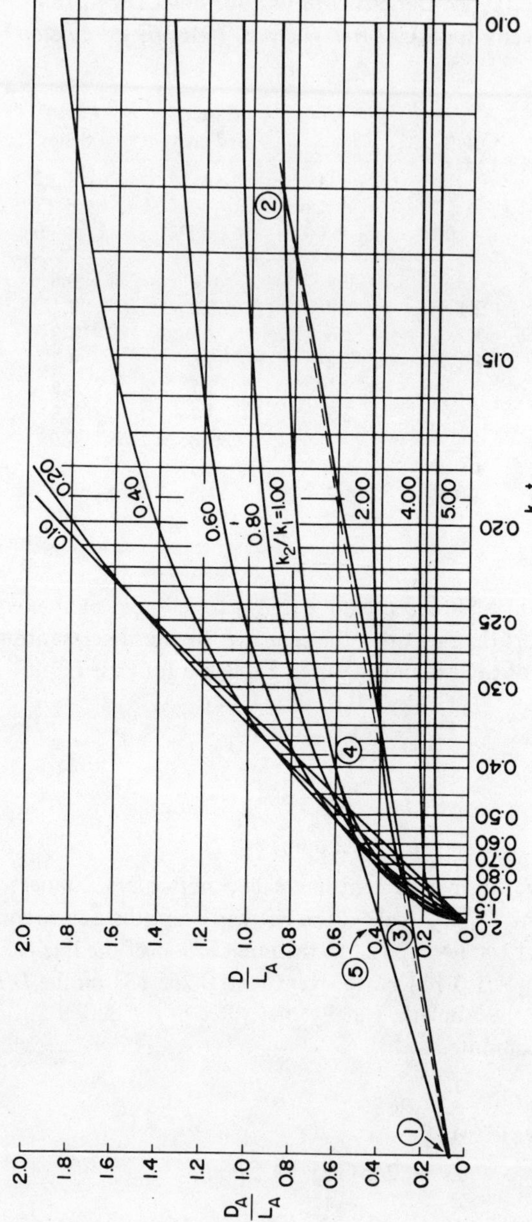

FIG. 6-4. Nomogram for the dissolved oxygen sag. Oxygen deficits downstream from a point of pollution may be determined from initial BOD and DO and stream self-purification constants. (*After Thomas, 1948. By permission of publisher.*)

becomes tangent to the $k_2/k_1 = 1$ curve, **[4]**. At this inclination the value of D/L_A reads 0.39, **[5]**, so that the critical deficit is 0.39(12) = 4.7 ppm.

Maximum Permissible BOD Load

Assuming that no deposition occurs, the maximum BOD load, L_A, that may be introduced so that the oxygen concentration will remain above a specified value downstream, may be approximated with the following formula (Thomas, 1948).

$$\text{Log}\, L_A = \log D_c + \left[1 + \frac{k_1}{k_2 - k_1}\left(1 - \frac{D_A}{D_c}\right)^{.418}\right] \log \frac{k_2}{k_1}$$

With the data of Case I, our problem is to find the L_A that will permit the DO of the stream to remain above 2.0 ppm.

The allowable deficit $D_c = 9.2 - 2.0 = 7.2$

$$D_c = \text{saturation} - \text{required oxygen level}$$

then

$$\text{Log}\, L_A = \log 7.2 + \left[1 + \frac{.12}{.12 - .12}\left(1 - \frac{.77}{7.2}\right)^{.418}\right] \log \frac{.12}{.12}$$

when $k_1 = k_2$ equation becomes $\text{Log}\, L_A = \log D_c + .434(1 - D_A/D_c)^{.418} = 1.27$

$L_A = 18.7$ ppm, the maximum permissible BOD load

Thomas calculates k_1 and k_2 and assumes k_3 to be included in the value of k_1 as follows

$$k_1 + k_3 = \frac{1}{\Delta t} \log \frac{L_A}{L_B}$$

$$k_2 = k_1 \frac{\bar{L}}{\bar{D}} - \frac{\Delta D}{2.3 \Delta t \bar{D}}$$

Thomas' method has proven acceptable when sags occur as a result of one particular waste loading into a receiving stream that remains fairly constant in physical characteristics. Since the number of streams where this occurs is diminishing with time, this method is becoming limited.

6-5. HULL'S MODIFICATION OF SAG CURVE ANALYSIS

Cecil Hale Hull (1960) developed a *simplified technique for determination of theoretical and effective coefficients in polluted streams.* He derives a formula for '*f*' from the basic Streeter-Phelps equation. The technique proposed for use of this formula appears to offer many possibilities for studying the oxygen balance of water in both laboratory and field investigations.

The value of the technique is in its simplicity. It requires only three measurements (temperature, DO, and BOD). The need to determine k_1, k_2, or the time of flow t is eliminated.

The difficulty in applying the Streeter-Phelps formula has always been that the so-called "constants" k_1 and k_2 are not really constants. They vary with *temperature*, *turbulence*, *waste loading*, *streamflow*, *and weather*, as well as other, more obscure factors. Despite the variations, they are still being used. Actual determination of these coefficients has been costly and time-consuming, and the final results are of questionable reliability.

Hull proposes a simplification of one of Fair's equations as the basis for an easy method of determining the self-purification coefficient.

Fair derived the following formulas for the maximum deficit:

$$D_c = \frac{L_a}{f} 10^{-k_1 t_c} \tag{1}$$

$$D_c = \frac{1}{f}(L_a - X_c) \quad \text{or} \quad D_c = \frac{L_c}{f} \tag{2}$$

where

$f = k_2/k_1$

D_c and t_c = the maximum or critical deficit at the lowest point of the sag corresponding to a time of flow t_c from the reference point

X_c = first-stage BOD exerted during time-of-flow to critical point

L_a = first-stage BOD at reference point a

These equations afford a simple means of determining the self-purification coefficient at the critical sag point, wherever it occurs.

By equating (1) and (2)

$$\frac{L_a}{f} 10^{-k_1 t_c} = \frac{1}{f}(L_a - X_c) = D_c \tag{3}$$

Therefore

$$f \cdot D_c = L_c$$

or

$$D_c = \frac{L_c}{f} \tag{4}$$

and

$$f = \frac{L_c}{D_c} \tag{5}$$

Hull points out that f in (1) and (2) represents an average or effective value integrated over the entire stretch of the river from the initial point a to the critical point c.

Also, the substitution of L_c for $L_a \cdot 10^{-k_1 t_c}$ and $L_a - X_c$ implies that no BOD has been added to or removed from the stream in the intervening stretch, except by the BOD reaction.

River stretches that conform to restrictions are rare; however. Hull derives this same relationship (5) in another manner with a somewhat different interpretation of the significance of f.

At the critical point the slope of the sag curve (dD/dt) is zero. Streeter's original equation reads

$$\frac{dD}{dt} = K_1 L - K_2 D \tag{6}$$

By assigning subscripts c to each term, Equation (6) can be rewritten

$$\frac{dD}{dt} = K_{1_c} L_c - K_{2_c} D_c = 0 \tag{7}$$

Equation (7) states merely that at the point where the DO concentration is not changing, the deoxygenation rate is exactly balanced by the reoxygenation rate.

Equation (7) can be rewritten for that precise situation as

$$\frac{K_{2c}}{K_{1c}} = \frac{L_c}{D_c} \text{ or common logs } \frac{k_{2c}}{k_{1c}} = \frac{L_c}{D_c}$$

or, modifying Fair's f by adding subscript c

$$f_c = \frac{k_{2c}}{k_{1c}} = \frac{L_c}{D_c} \tag{8}$$

which is exactly the same as that derived in Equation (4) in another way

$$L_c = f_c D_c \tag{9}$$

Equation (8) provides a means of determining f at a specific location, the bottom of the sag, under a given set of conditions. In addition, self-purification coefficients can be determined by Equation (8) for a variety of streamflows and other controlling factors. In this way, it should not take very long to accumulate a series of values of f, showing the purification capacity of the stream under various conditions.

Once the critical sag point is located, it requires only the determination of the temperature, DO, and the ultimate first-stage BOD, all at the critical sag point. No time of flow or k_1 or k_2 measurements are required.

General Procedure Used in Hull's Modification

1. A DO profile survey run by a rapid boat trip will locate the sag point.
2. Sample at sag or as close as possible for temperature, DO, and BOD (ultimate).
3. Move upstream far enough to ensure a slightly higher DO than found at sag point. Sample same as at sag.
4. Move downstream to a good reference point, such as a bridge access with a uniform channel, below the sag point, and sample same as at sag.
5. Plot the data obtained on the same graph and draw smooth profiles bracketing the sag point, as shown in Fig. 6-5.
6. On the DO profile, accurately locate the sag point, and from it, draw a line normal to the sag curve through the temperature and BOD profiles.
7. Take the values of DO, temperature, and BOD from the curves at their intersection with the normal line.
8. Convert the DO to the deficit D_c for the temperature at the sag point.
9. Using the values of L_c and D_c, solve for f_c by means of Equation (9).

Hull states that in a true Streeter-Phelps system involving only atmospheric reaeration and first-stage BOD, the observed oxygen sag curve would coincide with the theoretical oxygen sag curve, and

$$h_c = f_c = \frac{L_c}{D_c} \tag{10}$$

But where algae, sludge deposits, or COD occur, the observed will not agree with the theoretical sag analysis, and

$$h_c \neq f_c$$

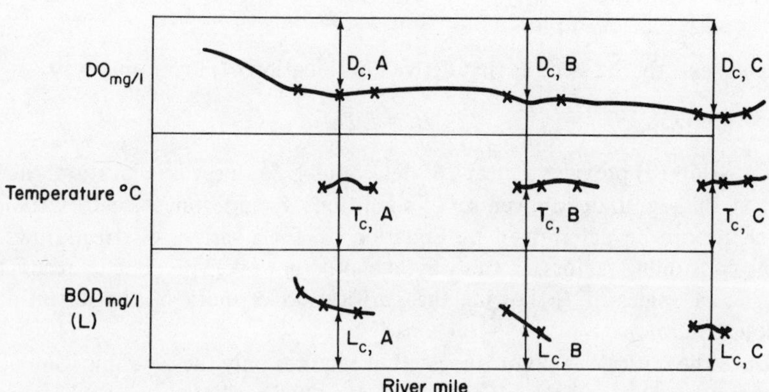

FIG. 6-5. DO saturation values. (*After Hull, 1960. By permission of publishers.*)

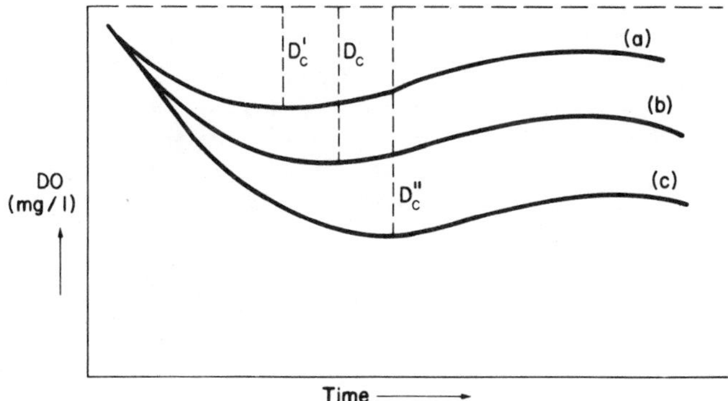

FIG. 6-6. DO saturation value. Curve (a) = algae; curve (b) = theoretical Streeter-Phelps; curve (c) = benthal and COD.

but

$$h'_c = C_3 f'_c = \frac{L'_c}{D'_c} \tag{11}$$

where C_3 equals the ratio of the h'_c to f'_c, and L'_c and D'_c are observed values.

Hull defines h'_c as the *effective* or *apparent* self-purification coefficient at the observed critical sag point, and f'_c as the theoretical self-purification coefficient at the observed critical sag point. Hull depicts the variation in actual sag curves from the theoretical in Fig. 6-6.

The new symbol, h_c, serves to caution the users of the Streeter-Phelps oxygen-sag formula, as modified by Fair, that they are really dealing with an empirical equation which only approximately describes the sag curve in most natural streams.

h cannot be assumed to be a constant over a wide range of the various streamflows, temperatures, and sunlight.

For most purposes, however, the value of the coefficient h_c, determined from the observed critical sag point of a stream, would serve as a readily obtained estimate of the average value of h throughout the deoxygenation zone.

The h value so obtained could be substituted for f in Equation (1) to predict the critical oxygen deficit for any given BOD loading.

Example[1]–Hypothetical stream for four consecutive reaches with different self-purification characteristics (Fig. 6-7).

The DO deficit and BOD concentration upstream are assumed to be 1 and 10 ppm, respectively. The critical point values were taken from the graph and used

[1]Stream constants in each reach assumed as in Table 6-7.

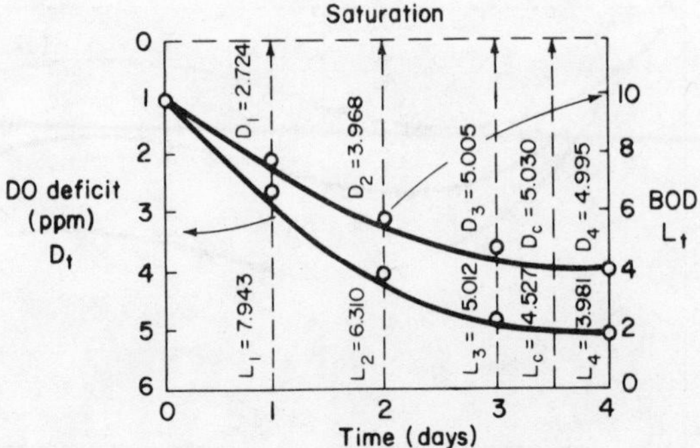

FIG. 6-7. Sag curve plot. *(After Hull, 1960. By permission of publisher.)*

to calculate

$$f_c = \frac{L_c}{D_c} = \frac{4.53}{5.03} = 0.90 \quad \text{by Hull's Modification}$$

The computed value of f_c (.90) agrees with the value originally assumed for Reach 4 (Table 6-7).

The critical point values of L and D can also be computed. The *critical time* can be determined from Fair's relationship

$$k_1 t_c = \frac{1}{f-1} \log \left[f - f(f-1) \frac{D_a}{L_a} \right] \tag{12}$$

TABLE 6-7. Computed DO Deficits and BOD for a Hypothetical Stream with Assumed Purification Characteristics, Initial BOD, and Initial DO Deficit

Reach	(1)	(2)	(3)	(4)	(5)
k_1	0.10	0.10	0.10	0.10	
k_2	.075	.05	.025	.09	
f	.75	.50	.25	.90	
t	0.0	1.0	2.0	3.0	4.0
D	1.0	2.724	3.968	5.005	4.995
L	10.0	7.943	6.310	5.012	3.981
t_c				3.442	
L_c				4.527	
D_c				5.030	

Knowing the critical time, it is possible to determine the value of L_c from the following equation (Streeter-Phelps first-order):

$$L_c = L_a \cdot 10^{-k_1 t_c} \tag{13}$$

The *critical deficit* is given by Fair's Equation (1)

$$D_c = \frac{L_a}{f} \cdot 10^{-k_1 t_c} \tag{1}$$

Values computed by this method where the sag curve is relatively flat are quite good:

$$f_c \text{ (approx)} = \frac{5.012}{5.005} = 1.00 \text{ for } t = 3 \text{ days}$$

$$f_c \text{ (approx)} = \frac{3.981}{4.995} = 0.80 \text{ for } t = 4 \text{ days}$$

$$f_c \text{ (computed)} = \frac{4.527}{5.030} = 0.900$$

∵ error in self-purification coefficient is about ±0.1.

As long as the critical sag point stays in Reach 4, the value of f_c will remain unchanged, *despite* the change in initial deficit and BOD loading. This is true because the ratio of L_c/D_c stays constant and linear between L_c and D_c (shown in Fig. 6-7a). The linear relationship for the constant self-purification coefficient depends only on *physical*, *chemical*, and *biological* factors that control reoxygenation and deoxygenation at that point. Therefore, upstream conditions influence the location of critical sag point, and the value of f_c is not a function of upstream conditions.

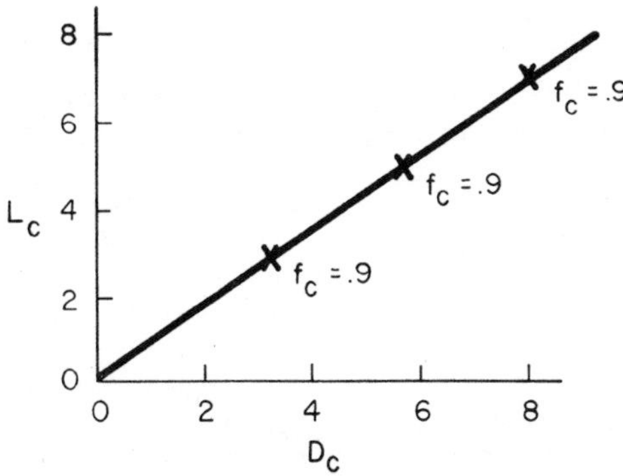

FIG. 6-7a. Critical BOD versus critical oxygen deficit.

6-6. McBRIDE'S NOMOGRAPHS TO FACILITATE USE OF STREETER-PHELPS EQUATIONS

McBride (1982) acknowledges the validity of the Streeter-Phelps equations for predicting stream reactions to pollution loads. However, he also believes that the nomograph approach that he recommends will provide reasonable predictions for most rivers when rapid and approximate answers are needed—usually in an emergency situation. In developing this model he assumes that: (1) the BOD is exerted in first-order kinetics; (2) the rate of deoxygenation and the rate of removal of ultimate BOD are equal; (3) the rate of reaeration is proportional to the river dissolved oxygen deficit; and (4) there are no tributary inflows in the reach being modeled.

Use of Nomographs

All uses of these nomographs require the following data:

D_o = initial river DOD, g/m^3
k_L = BOD decay coefficient evaluated for the standard laboratory BOD test, days^{-1}
k_1 = river deoxygenation coefficient, days^{-1}
k_2 = river reaeration coefficient, days^{-1}

The coefficients k_L, k_1, and k_2 are all defined to the base e. Use of the coefficients k_L and k_1 is in recognition of the possibility that the rate of river deoxygenation may exceed the rate of BOD decay in the laboratory BOD test, that is, $k_1 \geqslant k_L$.

First the value of k_L is entered onto Fig. 6-7b to obtain the value of α. Then the k_1 and k_2 coefficients are grouped into the dimensionless "self-purification constant" f calculated from

$$f = \frac{k_2}{k_1} \tag{1}$$

Oxygen Sag Prediction—Location. The location of the maximum downstream river DOD, x^* (km), can be obtained if α, f, D_o, L_o, and u are given,

where

L_o = initial river BOD$_5$, g/m^3
u = river velocity, m/s

The steps taken are as follows:

- The initial deficit-load ratio (R_o) is calculated from

$$R_o = \frac{D_o}{\alpha L_o} \tag{2}$$

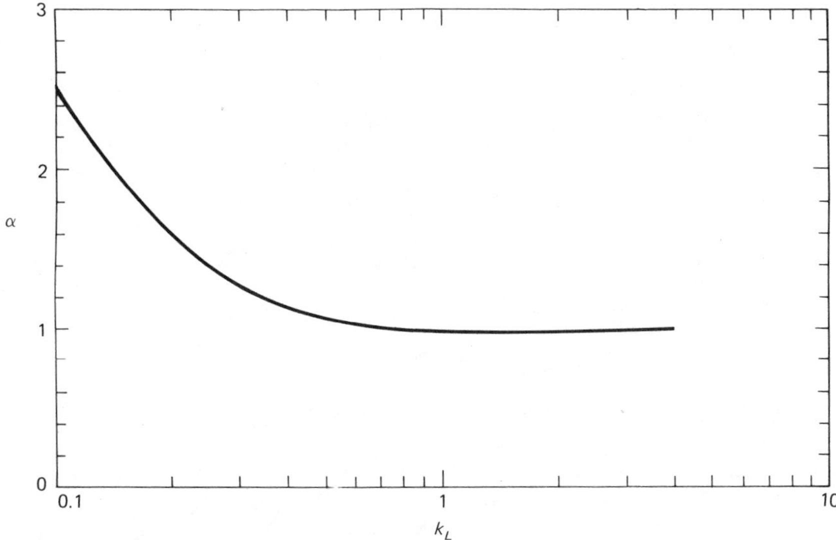

FIG. 6-7b. Graphs of α versus k_L.

Note—α = ratio of $BOD_{ultimate}$: BOD_5
k_L = BOD decay coefficient (base e) evaluated for the standard laboratory BOD test, $days^{-1}$.

- Using these values of f and R_o, the value of the dimensionless time-of-travel, τ^*, from the initial point to the critical point is read from the nomograph in Fig. 6-7c.
- The location of the maximum downstream river DOD (for constant river velocity) is calculated from

$$x^* = 86.4 \frac{u\tau^*}{k_1} \tag{3}$$

The numerical factor of 86.4 arises because of the mixture of units in the equation (x^* in km, u in m/s, k_1 in $days^{-1}$). The above form is retained because these units are conventional for the variables in the equation.

Oxygen Sag Prediction—Magnitude. The magnitude of the maximum downstream river DOD, D^* (g/m^3), can be obtained if α, f, D_o and L_o are given. The steps taken are as follows:

- The initial deficit-load ratio, R_o, is calculated as above.
- Using the calculated values of f and R_o, the value of the critical downstream deficit-load ratio, R^*, is read from the nomograph in Fig. 6-7d.

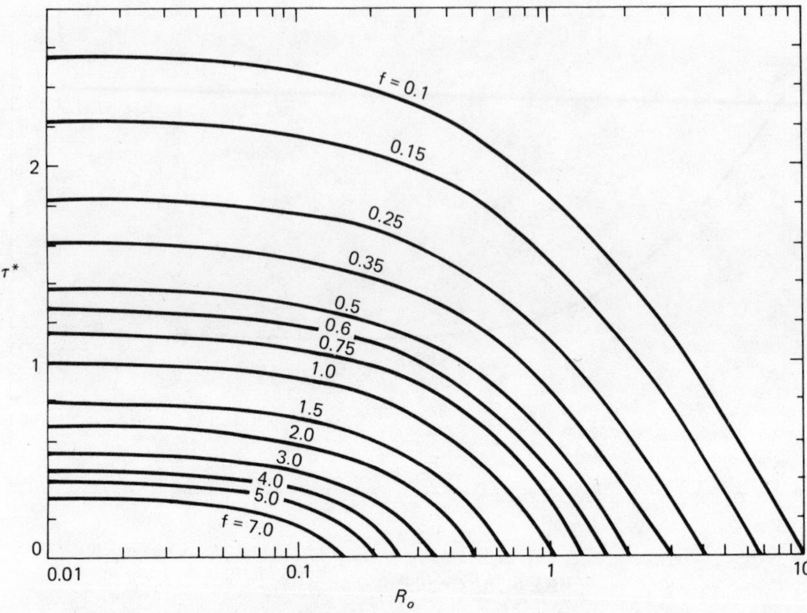

FIG. 6-7c. Nomograph of τ^* versus R_o and f.

Note–$R_o = \dfrac{\text{initial river DOD}}{\text{initial river BOD}_{\text{ultimate}}}$

$\tau^* = k_1 t^*$

where

k_1 = river deoxygenation coefficient (base e), days^{-1}

and

t^* = river time-of-travel from initial point to critical point, days.

- The magnitude of the maximum downstream river DOD is calculated from

$$D^* = \alpha L_o R^* \tag{4}$$

Assimilative Capacity Prediction. The river assimilative capacity, L_o (g/m^3), can be obtained if α, f, D_o and D^* are given. The steps taken are as follows:

- The dimensionless initial river DOD is calculated from

$$d_o = \frac{D_o}{D^*} \tag{5}$$

- Using the calculated values of f and d_o the value of the dimensionless initial BOD$_{\text{ultimate}}$, b_o, is read from the nomograph in Fig. 6-7e.

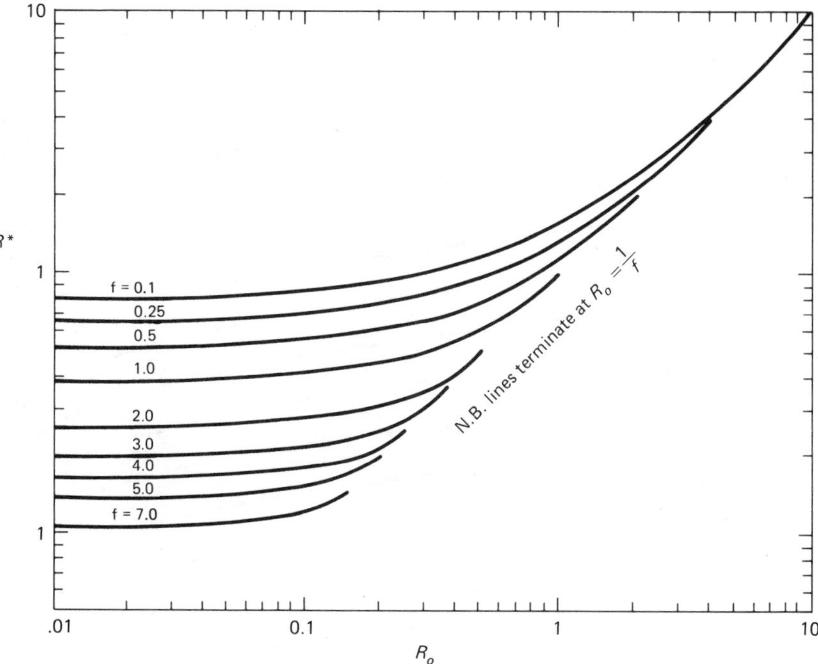

FIG. 6-7d. Nomograph of R^* versus R_o and f.

$$\text{Note}-R^* = \frac{\text{maximum downstream DOD}}{\text{initial river BOD}_{\text{ultimate}}}$$

$$R_o = \frac{\text{initial river DOD}}{\text{initial river BOD}_{\text{ultimate}}}$$

$$f = \frac{\text{river reaeration coefficient}}{\text{river deoxygenation coefficient}}.$$

- The river assimilative capacity is calculated from

$$L_o = \frac{b_o D^*}{\alpha} \tag{6}$$

Example Use of Nomographs

Consider a problem with the following data from McBride (1982):

initial river DOD, $D_o = 1.5\ \text{g/m}^3$

BOD test decay coefficient, $k_L = 0.4\ \text{days}^{-1}$

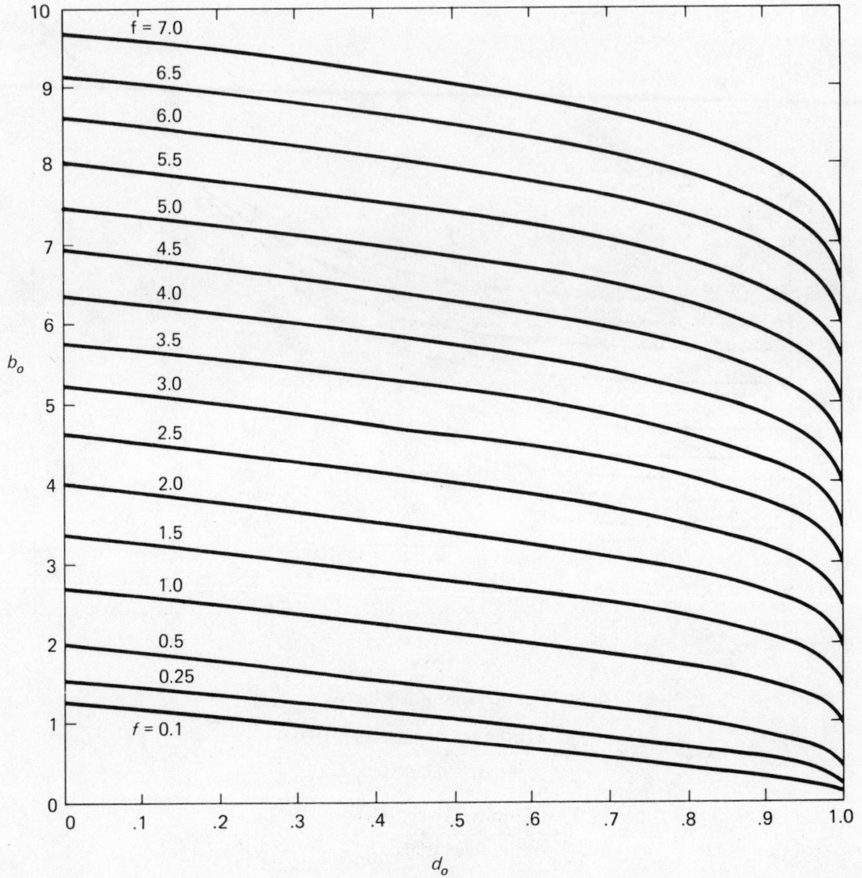

FIG. 6-7e. Nomograph of b_o versus d_o and f.

$$\text{Note}-b_o = \frac{\text{initial river BOD}_{\text{ultimate}}}{\text{maximum permissible downstream DOD}}$$

$$d_o = \frac{\text{initial river DOD}}{\text{maximum permissible downstream DOD}}$$

$$f = \frac{\text{river reaeration coefficient}}{\text{river deoxygenation coefficient}}.$$

$$\text{river deoxygenation coefficient,} \quad k_1 = 0.6 \text{ days}^{-1}$$

$$\text{river reaeration coefficient,} \quad k_2 = 0.75 \text{ days}^{-1}$$

Therefore, $f = 1.25$ from Equation (1) and $\alpha \simeq 1.15$ from Fig. 6-7b.

Oxygen Sag Prediction—Location and Magnitude. Suppose that the constant river velocity is $u = 0.5$ m/s and the initial river BOD_5 is $L_o = 9.2$ g/m^3. Then $R_o \simeq 0.14$ from Equation (2), and so using Fig. 6-7c we have $\tau^* \simeq 0.75$. From Equation (3) the location of the maximum downstream river DOD is

$$x^* \simeq 86.4 \times 0.5 \times 0.75/0.6 = 54 \text{ km}$$

Also, using Fig. 6-7d and the above values of f and R_o, we find $R^* \simeq 0.38$. Then, from Equation (4) the magnitude of the maximum downstream river DOD is

$$D^* \simeq 1.15 \times 9.2 \times 0.38 \simeq 4.0 \text{ g/m}^3$$

Assimilative Capacity Prediction. In this case, the initial river BOD_5, L_o, is the unknown and we shall set the maximum permissible downstream DOD, D^*, to 4.0 g/m^3. Using Equation (5) we obtain $d_o = 0.375$. This value of d_o and the above value of f entered into Fig. 6-7e give $b_o \simeq 2.65$. Then using Equation (6) the river assimilative capacity is given by

$$L_o \simeq 2.65 \times 4/1.15 \simeq 9.2 \text{ g/m}^3$$

which recovers the value used for oxygen sag prediction, as would be expected.

Discussion

The nomographs presented herein have the great advantage that no detailed computations are required. Their use involves three simple steps: the formation of dimensionless ratios (α, f, and R_o or d_o); the reading of a nomograph; and the calculation of the required answer by a simple arithmetic formula. This feature makes them particularly useful in emergency situations where rapid calculations of the effects of discharges are required. Of course data on the values of k_L, k_1, k_2, D_o, and L_o or D^* must be on hand to enable the dimensionless ratios to be calculated.

Four points of caution must be made about use of these nomographs (according to McBride (1982):

- The prediction of the magnitude of the oxygen sag can result in a value of river DOD that exceeds the saturation river dissolved oxygen, that is, a negative dissolved oxygen is predicted. For example, if in the above example $L_o = 45$ g/m^3, then $R_o \simeq 0.029$ and so $R^* \simeq 0.35$. Equation (4) then predicts that $D^* \simeq 18.1$ g/m^3, which is absurd. This problem is attributable to the second assumption of the Streeter-Phelps model, in that the rate of exertion of BOD cannot be independent of the dissolved oxygen concentration at low levels of dissolved oxygen. Clearly however, if a D^* value greater than dissolved oxygen saturation is predicted, then severe river pollution and destruction of aquatic life will occur.
- The effects of longitudinal dispersion have been ignored. This may result in

overconservative predictions being made, since the effect of dispersion is to reduce the peak values of river BOD, particularly for rapidly varying waste discharges. However, for waste discharges of a longer period (1 day, as in the case of the local milk dumpings) peak reduction attributable to dispersion will be small.

- The effects of inflows downstream of the initial point have been ignored, in accordance with the fourth assumption of the Streeter-Phelps model. This is strictly valid only if the BOD and DO of the inflow are similar to the river BOD and DO at the confluence.
- In predicting river assimilative capacity using Fig. 6-7e, it has been assumed that the initial river DOD, D_o, is the same as the DOD just upstream of the initial point. It is, therefore, unaffected by inflows at that point. For large inflows with a DOD different from D_o, this may not be strictly true, and some trial-and-error procedure may be necessary.

This development of solution nomographs is not extendable to cases where the Streeter-Phelps model is modified to include other processes, because of the increase in the number of parameters required to describe the model. In this case, some other method of solving the model equations must be used.

6-7. THE VELZ METHOD

Clarence Velz, (1970) whom I consider the real "father" of the use of mathematics–and especially statistical analysis–in stream surveys, developed what he terms "the rational accounting system." Velz has been quoted extensively in likening oxygen resources in a flowing stream to assets in a checking account in one's local bank. His accounting system depends upon completion of information needed in two forms: Form A and Form B. We will review these forms in this section in order to understand how Velz's method more closely follows the true oxygen level in various reaches of a stream than most (if not all) other computed procedures.

Velz's major complaint with the oxygen sag equation developed from Streeter and Phelps' formulations is concerned with measuring k_2, the reaeration rate. He feels that "lacking a rational basis for determining k_2, there has developed a tendency to misuse it in the sag equation as though it were constant over long reaches of river, whereas from the fundamental formulation it is known to be variable from reach to reach." He also deplores any "short-cut" methods, some of which this author has also observed being used by many engineering firms. For example, he states that "also there is a tendency to use the sag equation without determining the time of passage (flow time) from reach to reach on the basis of measured physical characteristics of the channel. This produces a dissolved oxygen profile in relation to the time of the course" (1970, p. 206). He also believes that the physical methods of determining k_2, such as those proposed by O'Connor and Dobbins (1956) and Churchill (1962) are erroneous, since they both depend upon the measurement of turbulence, which is as difficult to measure directly as reaeration itself.

Statistical methods such as Churchill and Buckingham's (1962) multiple regression technique are also questioned by Velz, who recognized the limitation to conditions actually observed in the initial samplings only. In other words, projection of initial data to critical conditions is dangerous, even when good statistical correlation of the initial data is shown.

Velz also points out that in using the sag equation where the reaeration rate varies from reach to reach with velocity and depth, the direct solution for the critical time and oxygen deficits cannot be applied. He recognizes the theoretical nature of the sag curve formula, because rivers of uniform channels are seldom, if ever, actually found. True values of L_c and D_c, then, can only be obtained by applying the sag curve formula over each succeeding and varying reach. Thus, the simplicity of the original sag curve formulation is lost. However, Velz recalls that Streeter and Phelps initially proposed that the channel depth, time of passage, and the necessity of computing k_2, reach by reach, were usually overriding factors that were eliminated by using the integrated oxygen sag equation.

Velz then suggests that the rational method "is less burdensome and more flexible in taking full advantage of detailed characteristics of the waste loads and of the channel" (1970, p. 218). Computation of the dissolved oxygen profile by the rational method is based on measurements of stream depth, width, and volume (usually at one-mile intervals with the channel low-water profile adjusted to actual sampled flow regime) from which time of passage is computed. Although the reader is urged to study Velz's rational method in detail as presented in his excellent text, the writer includes here illustrations of the method as recorded and computed in Form A and Form B accounting procedures.

The first phase is an orderly accounting of the liability side of the ledger, deoxygenation. Form A is used for this purpose, with Velz dividing the liability into two phases: (1) specific character of BOD, and (2) amortization rate in the receiving stream.

Velz suggests including, as stations in his rational accounting system, each successive effluent outlet, junctions of major tributaries, dams, hydro or thermal power plants, breaking points in channel configuration and effective depth, and critical stream velocity.

The second phase is the accounting of the asset side of the ledger, and striking a net balance between assets and liabilities to provide the dissolved oxygen profile along the course of the river channel. Form B, which does not necessarily include all stations used in Form A, is what Velz recommends, for convenience. The division of the stream channel into a number of reaches for reaeration calcu-

FORM A DEOXYGENATION (1) Specific Character

Stream ______________________

Runoff ______________ CFS at Reference Gauge

Drought Probability

Special Conditions ______________________

BOD Loading:[2] Residual at mile ______ from upstream sources and landwash

WASTE LOADS:			BOD POPULATION EQUIVALENT (PE)–DISCHARGED TO STREAM			
River Mile (1)	Source (2)	Type of Treatment (3)	Immediate Demand (4)	Colloidal and Dissolved Fraction (5)	Settleable Solids Fraction (6)	Total (7)

[2] Refers to pounds of BOD in stream per day divided by 0.17 to equate it to that discharged by an equivalent number of persons.

FORM A DEOXYGENATION (2) AMORTIZATION

				BOD (POPULATION EQUIVALENTS)				
STATION (River Mile)	MEAN WATER TEMPERATURE °C	K_1 at MEAN TEMPERATURE	TIME OF PASSAGE FROM INITIAL STATION (days)	DISCHARGED AT STATION	RESIDUAL FROM UPSTREAM STATIONS	TOTAL AT STATION	SATISFIED IN STREAM BETWEEN STATIONS	TOTAL SATISFIED ABOVE STATION
(1)	(2)	(3)	(4)	(5)	(6)	(7)	(8)	(9)

Form A. Reproduced with permission. From *Applied Stream Sanitation*, by Clarence Velz.

Station River Mile (1)	Runoff at Station (CES) (2)	Mean Temperature (3)	Effective Depth (ft) (4)	Frequency of Turnover (Min/Max) (5)	River Volume (mg) (6)	Total Demand Satisfied Above Station (P1) Form A (Col 9) (7)	DISSOLVED OXYGEN (POP. EQUIVALENTS)						Net Oxygen Balance at Station ICol. (11) plus Col (13) (14)	Percentage of Saturation at Station [Col (14) divided by Col (10) (15)
							Added by Runoff between Stations (8)	Total Added by Runoff above Station (9)	Total Runoff at Station at Saturation (10)	Net at Station I Col (9) minus Col (7) (11)	Added by Reaeration Between Stations (12)	Total Added by Reaeration above Stations (13)		
	Obtained from a plot of flows at various station vs. tributary drainage areas		Obtained usually from plots of occupied channel volume versus channel surface area	A function of depth obtained from a plot of Mix Interval (Min) vs. depth	Obtained from the summation curve of occupied channel volume	Obtained directly from Form A (Col. 9) or from a graphical plot of Col. 9 and stations along channel	Oxygen assets of runoff obtained from saturation values for prevailing temperature generally about 85% of saturation			Partial balance between assets and liabilities with the exception of reaeration				

lation and oxygen balance depends upon how well-defined a profile is wanted, and other factors, such as time of flow, channel variations, junction of tributaries, source of immediate demand, and location of benthal activity in the channel. Naturally, one major objective should be to locate clearly and to quantify the critical sag points. All reaches should possess reasonably constant physical characteristics.

In order to assist the reader in completing the forms used in the accounting systems, the writer has included in the forms, in the appropriate columns, key statements indicating from where or how data is computed. In Velz's text, detailed practical illustrative examples are given, and the student is urged once again to refer to this if further details are necessary.

Computations for Form B, Columns 1 through 11, are rather straightforward, as indicated in the columns. However, beginning with Column 12, reaeration must be computed. Reaeration is related to the oxygen deficit, and since each reach is selected so as to possess a straight-line relationship of oxygen from the beginning to the end of the stretch, the average deficit of the two values must be obtained and used. This is the most difficult part of the Velz method, and is accomplished by a series of trial-and-error values for the oxygen concentration of the downstream end of the reach. Velz believes that with practice, the stream analyst can accomplish this with about three approximations. He gives an example in which the entire channel oxygen profile of 10 separate reaches is determined by this procedure with a maximum of only two approximations. The first reach only is reproduced here with some editorial explanations for ease of understanding.

The stretch (reach) consists of 1.85 miles (mile point 87.0 to mile point 85.15) and contains 172.8 mg; is 6.06 feet deep; has a mean temperature of 21°C; a mix interval of 17.4 minutes (see Fig. 6-8 as given by Velz [1970]); and a dissolved oxygen per mix at 100% deficit = 0.430% (see Fig. 6-9 as given by Velz).

Since mile 87.0 is an upstream clean stream location, Velz assumes it to be 85 percent saturated at that location. The problem is to find the percent saturation at mile 85.15, the lower end of the reach. Velz makes an initial assumption that this level is only 80%; thus the average saturation deficit in the reach is

$$\frac{(100 - 85) + (100 - 80)}{2} = \frac{15 + 20}{2} = 17.5\%$$

Now, to prove this assumption, his reaeration compilations must be used to result in a downstream oxygen saturation of 80%.

Step 1

$$\underset{(\text{Avg. \% deficit})}{0.175} \times \underset{(\text{\% DO per mix})}{0.430} \times \frac{1{,}440}{17.4}\ \begin{matrix}(\text{Min/day})\\(\text{Min/mix})\\(\text{mixes/day})\end{matrix} = 6.23\% \text{ per day}$$

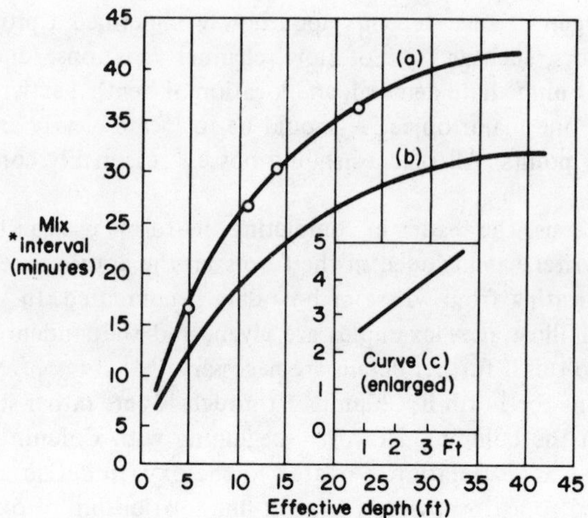

FIG. 6-8. Relation between effective depth and mix interval. (a) = usual freshwater streams; (b) = tidal estuaries; (c) = shallow, relatively high-velocity streams. Mix interval is apparently related to the time required for a given layer to be exposed again to surface aeration. (Reproduced with permission. From *Applied Stream Sanitation* by Clarence Velz. © 1970 by John Wiley & Sons, Inc.)

Step 2

$$\frac{4{,}900^{*}\ (\text{cfs})}{1.547\ \ (\text{cfs/mgd})} \times \frac{8.99 \text{ ppm DO}}{\text{at saturation } 21^{\circ}\text{C}} \times \frac{8.34\ (\#/\text{mg})}{.24\ \ (\text{PE}/\#)}$$

$$= 989{,}380 \text{ PE at saturation}$$

Step 3

$$0.85 \times 989{,}380 = 840{,}980 \text{ PE at 85\% saturation at mile 87.0}$$

Step 4

$$8.99 \text{ ppm DO} \times 8.34/.24 = 312 \text{ PE/mg at saturation}$$

Step 5

$$0.0623\ \frac{(\%\ \text{DO})}{(\text{per day})} \times 312\ \frac{(\text{PE})}{(\text{mg})} \times 172.8\ (\text{mg}) = 3{,}360 \text{ PE reaeration in reach}$$

Step 6

798,180	+	3,360	=	801,540 PE
(Net oxygen resources at Station mile 87.0 from Col. 11, Form B)		(Reaeration in reach)		(Total oxygen resources at mile 85.15 including reaeration)

---

*Upstream (mile 87) runoff.

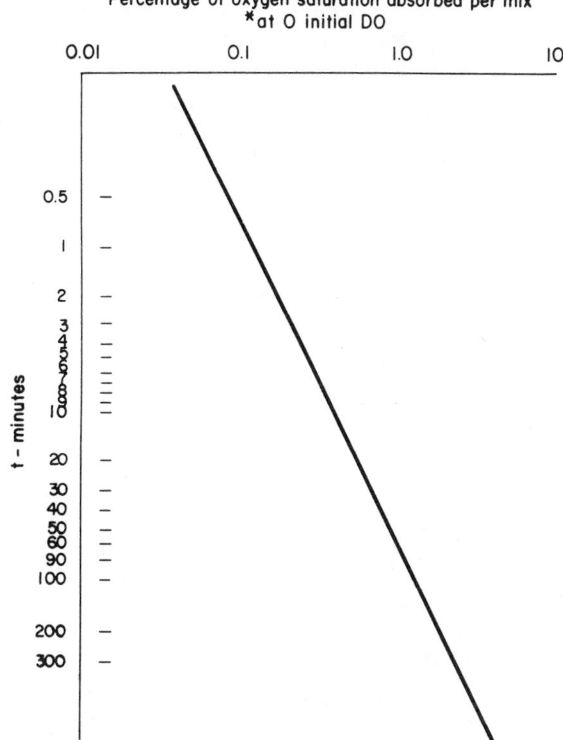

FIG. 6-9. Standard reaeration curve. (Reproduced with permission. From *Applied Stream Sanitation* by Clarence Velz. © 1970 by John Wiley & Sons, Inc.)

Step 7

$$\frac{801{,}540}{989{,}380} = 81\% \text{ saturation}$$

[80% ≠ 81%, therefore our assumption of 80% is too low and we must make a second higher one.]

Our second assumption (again from Velz's example) is 81%. Therefore

$$\frac{(100 - 85) + (100 - 81)}{2} = \frac{15 + 19}{2} = 17\% \text{ average deficit}$$

The ratio of the second assumption of average deficit to the first one times the previous computed value of reaeration in the reach should yield a new assumption as follows:

$$17/17.5 \times 3{,}360 = 3{,}260 \text{ (new assumed reaeration in stretch)}$$

TABLE 6-8. Oxygen Sag Computations

Method of computing oxygen sag	Advantages	Disadvantages	When used
1. Streeter-Phelps	good agreement with practice under certain conditions	1. requires values of K_1 and K_2 2. is not dependable when multiple inputs of organic pollution exist 3. requires reevaluation of K_1 and K_2 under many conditions–especially critical conditions	1. when single source of pollution exists and flow time is relatively easy to obtain 2. when the critical conditions are sampled or easy to project to
2. Churchill	1. no flow time nor reaction rates are required 2. very useful when a variety of pollution inputs exist 3. stream equation can be used under most conditions observed during collection of data 4. no need for evaluating scour, photosynthesis, benthal conditions	1. requires numerous samples 2. some difficulty in deriving the equation from data	1. when multiple inputs exist and no values of K_1 and K_2 are known 2. when time is not a limiting factor
3. Thomas's nomograph	overcomes unwieldy Streeter-Phelps equation	still requires use of K_1, K_2, and t.	1. when time is very important 2. when values of K_2, t, and K_2/K_1 are included in nomograph
4. Hull's modification	no need to compute exact values of K_2 or K_1 in order to compute D_c	must locate exactly the point of critical sag	when K_1 and K_2 are not known
5. Velz method	1. rational method is less burdensome and is more flexible–based on measurements every mile, and makes an asset-liability balance 2. doesn't require t, or reaction rates	1. many stations and much information required. 2. All reaches are assumed to possess constant characteristics–physical; chemical; biological	when massive accurate data is available and reaction rates are either variable or nonexistent

plus the former net oxygen resources at Station 87.0 = 798,180 equals a new total oxygen resources at Station 85.15 = 801,440. The percent saturation equals

$$\frac{801{,}440}{989{,}380} \times 100 = 81\%$$

[Since 81% = 81%, Velz's second assumed oxygen level downstream is an accurate one.]

Velz repeats this procedure for the next uniform reach, using the new upstream oxygen saturation value of 81%. In this way, one is able to derive the entire sag curve. It must be pointed out that although we do not depend upon theoretical formulations (such as recommended by O'Connor and Dobbins, or Churchill) for computing the reaeration, we rely upon the validity of Figs. 6-8 and 6-9 to define the physical mixing and oxygen transfer occurring.

Some deviations from actuality may result from uisng these curves and assuming then to be valid, especially if one incorporates too long or too variable reaches along the channel. If that occurs, one would only be approaching the presumptions made by the other formulations–that the entire channel is characterized by one set of constant physical parameters.

A summary of the five major methods of computing oxygen sag in a critical receiving water stretch is shown in Table 6-8.

6-8. ASSESSMENT OF POLLUTION CAPACITY OF STREAM

In order to solve any scientific problem, one must not only understand the cause, but also ascertain the effect of any given degree of treatment prescribed. This principle is also true for stream pollution abatement–the object of all wastewater treatment. The most common and best understood problem of stream pollution is that of oxygen deficiency. The cause is excessive discharges of degradable organic matter. The cure is to oxidize some, most, or all of this organic matter prior to discharge into the watercourse. One difficulty involved is that of deciding how much of the organic matter should be oxidized in order to protect the best usage of the stream. Too little oxidation treatment before discharge results in potential stream pollution from oxygen consumption. Too much oxidation treatment can place an unnecessary heavy financial burden upon the producer of products and services. This section is intended to produce a sequence of mathematical and other scientific procedures that will yield the acceptable and realistic percentage of BOD reduction required to maintain a given oxygen water quality level.

General Procedures for Assessment

Preliminary step—Inplant steps are used first to reduce pollution to a minimum. The degree of final wastewater treatment depends primarily upon the *condition* and *best usage* of the receiving stream; therefore, determine both.

Best usage—Generally defined by existing laws, but, if not, then it should be *observed* and *evaluated* by stream analyst.

Standard—After the best use has been ascertained, the standards for that water quality must be determined by the analyst. In the United States these can also be obtained by consulting published standards, called water quality criteria. If these standards are not enacted into law or agreed upon by all bodies concerned, the stream analyst must review the literature and select the most appropriate value for limits for each major contaminant involved.

Now the condition of the receiving stream needs to be studied. The condition is really *an overall evaluation* similar to that of *your own personal condition.* You have a body temperature, pulse, cardiac character, lung capacity, blood count, weight, eyesight, hearing, etc. In addition, other outside situations can affect your mental health or psychological outlook on life. One could conclude that, if the level of the physical and psychological factors is high enough, the person is in "good condition" and vice versa. The same will eventually be necessary for proper evaluation of the condition of a stream. (I have proposed one that relates the existing level of a contaminant to the allowable limit, and averages the sum of these values to determine the stream condition for a specific use. The reader is referred to Chapter 7-4 for the development of this system of water quality evaluation.) However, the art of stream sanitation is neither ready to move in this direction immediately nor is it totally in agreement with any such complex system (as I propose). Therefore, for the present we generally see to it that a particular waste contains nothing seriously detrimental to the stream, such as toxic metals, acids, excessive radiation, inflammables, etc., and then make a determination of the amount of the organic matter that *can be discharged safely into a particular watercourse.*

Safely implies a statistical analysis and subjective decision after an objective array has been made.

Safely means that a study of *critical* periods must be made.

For that very reason a doctor will have you climb steps before a cardiogram—not eat before a blood count, etc.

The critical period in a stream is when the *temperature* is the highest, the *flow* is the lowest, and the *pollution* load of organic matter is the greatest.

Only a *very lucky gambler* or a *persistent stream analyst* (with much time and money at his disposal) can study the stream patient under the optimum combination of these critical conditions. Therefore, the stream analyst is faced with yet another compromise. He must sample when he can and accept the existing conditions and then project the existing situation to the critical condition. His job is usually more difficult, therefore, than that of the medical doctor. The great debates that occur today in stream analysis are those of projecting present conditions to those deemed critical. In many cases *we must presume* (which really means, make an educated guess) that the stream reaction to waste action will remain constant, or at least that the net results of the reaction of all constants will be the same as under the measured conditions. Some stream analysts make adjustments in the individual stream reaction constants for critical conditions. However, agreement among analysts is lacking and they differ not only from technical points of view, but also from the position from which they view the final results.

But finally, decisions must be made, and I have made the first decision and assumptions. These can be described best in the following 9 steps.

Specific Procedures

Step 1. Determine the stream deoxygenation constant from

$$K_1 = \frac{1}{\Delta t}\left(\log \frac{L_A}{L_B}\right)$$

when A is the location just downstream of the point source of pollution and properly mixed with the stream; B is some distance downstream of a point source of an organic pollutant; time, t, is the time of streamflow between A and B; and L is the ultimate BOD at each location (determined from analysis of stream samples).

Step 2. Determine critical conditions–usually the minimum flow to occur for a continuous period of seven days once in 10 years. Also select average temperature that occurs during this seven-day period. In addition, make a study to determine whether the organic wasteload is *at least* as likely to occur during this period as during any other period of the year.

Step 3. Correct K_1 for critical temperature and wasteloads of organic matter

$$(K_1 = K_1 \times 1.047^{(T-20)})$$

or about 4.7 percent change per °C.

Step 4. Determine the stream reaeration constant from

$$\text{A:} \quad K_2 = K_1 \frac{\bar{L}}{\bar{D}} - \frac{\Delta D}{2.3 \Delta t \bar{D}} \quad \text{(when possible)}$$

or from

$$\text{B:}\quad K_2 = C\,\frac{V^n}{H^m} \qquad \text{when} \quad C = 5.026,\; n = 1.673,\; m = 0.969$$

for relatively unpolluted streams as determined by Churchill.

$$\text{C:}\quad K_2 = 127\,\frac{(D_L U)^{0.5}}{H^{1.5}}$$

D_L depends mainly upon temperature and is 1944 at 20°C. Therefore, when D_L, ft^2/day, (the molecular diffusion at 20°C is known with assurance) U, the mean velocity, and H, the mean depth, are both measured and known for the entire stretch, K_2 can be computed with this formula

Step 5. Compare the cross-sectional area of the stream segment under critical and existing measured conditions. Adjust the K_2 value upward or downward, depending upon the relative turbulence and surface removal rates created under both conditions. Also correct K_2 for critical temperature, usually by

$$K_2(T°\text{C}) = K_2(20°\text{C}) \times 1.0241^{(T-20°\text{C})}$$

Step 6. Use either the *Streeter-Phelps Method* or the *Churchill Regression Method* for determining the dissolved oxygen deficit at the critical sag point downstream under existing conditions.

Streeter-Phelps $\quad D_c = \dfrac{K_1 L_A}{K_2 - K_1}\,[10^{-K_2 t} - 10^{-K_1 t}] + D_A \times 10^{-k_2 t}$

Churchill $\quad Y = a + b_1 X_1 + b_2 X_2 + b_3 X_3$

where

Y = DO sag from upstream to critical sag
X_1 = BOD at sag
X_2 = temperature at sag
X_3 = flow at sag

Step 7. Determine the relation of present organic loading to future expected maximum loading. An on-site analysis of existing loads must be made simultaneously with the stream study. Project, then, existing organic loading, L_A, to a maximum loading expected to occur during the critical streamflow period, and suggest organic loading that will be equal to or less than at least 90 percent of the seven-day periods.

Step 8. Insert this new value of organic loading in the stream sag equations, either as L_A in the Streeter-Phelps equation, or as X_1 for the Churchill equation, to solve for the *critical dissolved oxygen sag.*

Step 9. Determine the percent reduction of the new value of L_A or X_1 in

Step 8 which must be achieved in order to give an acceptable maximum oxygen sag in the river for its best usage.

QUESTIONS

1. What does the oxygen sag curve represent?
2. Define the critical sag point. How is it determined graphically?
3. Define all of the quantities in the sag equation and tell how they are computed.
4. What equation describes the solution for the critical dissolved oxygen deficit on a stream?
5. What is meant by Fair's f value and what is its range of values? What is so valuable about the use of f rather than its components?
6. How is the t_c found analytically? What is the derived equation?
7. How are the actual k_1 and k_2 values computed in a flowing stream by the Streeter-Phelps equation?
8. How is this information now used in computing the locus of the oxygen deficits downstream?
9. Discuss some of the advantages and disadvantages of using this equation to compute the oxygen deficit.
10. Upon what basic premise is the Churchill-Buckingham method based?
11. What are the three most important factors that can be correlated to the drop in oxygen concentration to yield a dependable stream equation?
12. What are the obvious advantages of this Churchill method over the Streeter-Phelps method? What are the disadvantages?
13. What is the form of the regression equation?
14. How are the three normal equations found in order to solve for the three unknowns?
15. How would you compute the allowable BOD loading at the source of pollution by the regression technique?
16. What factors are responsible for the poor correlation of observed oxygen deficit with calculated deficit?
17. How will the oxygen level in a flowing stream be affected by the sludge deposits
    (a) directly over the sludge bed?
    (b) a considerable distance downstream from the deposit?
18. What are the effects of scouring of sludge deposits on stream oxygen levels?
19. What are the effects of photosynthesis on O_2 levels in streams?
20. Why did Thomas find such wide divergence of k_1 and k_2 values in streams? How did he propose to overcome this and to simplify the Streeter-Phelps solution? How does he propose to find the allowable BOD load?
21. Upon what premise is the Hull method of computing stream constants based?
22. What is the major advantage of Hull's method?
23. What are Velz's major complaints with sag curve analysis?

24. What is Velz's major objection to the use of Churchill's procedures?
25. What is Velz's main basis for computing the DO profile?
26. What is the major drawback or possible source of error in Velz's rational accounting system?

REFERENCES

Churchill, M. A., and R. A. Buckingham: Statistical Method for Analysis of Stream Purification Capacity, *Sew. and Ind. Wastes*, **28**(4):517–537, 1956.

Churchill, M. A., H. L. Elmore, and R. A. Buckhingham: The Prediction of Stream Reaeration Rates, *J. San. Eng. Div.* (ASCE), SA4, July 1962, paper 3199.

Fair, G. M.: The Dissolved Oxygen Sag—An Analysis, *Sewage Works J.*, **11**:445, 1939.

Hull, C. H.: *Report No. III of the Low-Flow-Augmentation Project*, Johns Hopkins University, Baltimore, Md., April 1960.

McBride, G. B.: Nomograms for Rapid Solutions for the Streeter-Phelps Equations, *J. Water Pollution Control Fed.*, **54**(4):378, April 1982.

O'Connor, D.: The Temporal and Spatial Distribution of DO in Streams, *Water Resources Research*, **3**(1):65–79, 1967.

O'Connor, D., and W. Dobbins: The Mechanism of Reaeration in Natural Streams, *J. San. Eng. Div.* (ASCE) SA6, 1115-1–1115-30, December 1956.

Simmons, J. D., N. L. Nemerow, and T. F. Armstrong: Modified River Sampling for Computing Dissolved Oxygen Sag, *Sewage and Ind. Wastes*, **29**(8):936, 1957.

Streeter, Harold, and Earl Phelps: *Stream Sanitation*, John Wiley & Sons, Inc., New York, 1944.

Thomas, H. A.: Pollution Load Capacity of Streams, *Water and Sewage Works*, **95**(11):409, 1948.

Velz, Clarence J.: *Applied Stream Sanitation*, John Wiley & Sons, Inc., New York, 1970.

7
RECEIVING WATER QUALITY OBJECTIVES

Our goals in the matter of pollution abatement are not absolutely clear and certain. There are some who advocate maintaining and returning all receiving waters to a state of pristine purity. These are the idealistic, the preservationists, and those swayed by the emotional appeal of the first two types. During 1968 and 1969 this group seemed to gain in voice and apparent numbers–in 1972 both houses of Congress passed bills extremely restrictive of any discharge of contaminants into watercourses. This group is not to be confused with the traditional conservationists, who are dedicated to reasonable management of resources rather than complete nonuse. There are others who would protect receiving waters to serve only their best usage. This group permits the will of the local and/or the state population to decide the highest classification of a particular watercourse under question. Some streams would be used best primarily for drinking water, while others would be used optimally for fishing and other purposes. This thinking allows a maximum utilization of the receiving waters for the optimum benefit of the people. There exist other plans for maintaining water quality for specific purposes, but they are less prevalent and mostly represent the interests of one or another industrial or municipal consumer. It is plain, however, that most persons are united in the common goal of attempting to maintain the water quality at the highest level possible consistent with technical and economic limits.

7-1. RAW WATER CHARACTERISTICS

Uncontaminated river water will still not be pure when compared to such controls as rainwater or distilled water. It will contain dissolved gases and minerals as well as small amounts of organic matter. A typical sanitary analysis of "uncontaminated" river water may be:

pH	6.8
dissolved oxygen	9.0 ppm
temperature	13°C
color	5 ppm
turbidity	5 ppm
BOD (20°C, 5-day)	1.2 ppm
hardness	225 ppm as $CaCO_3$
total solids	500 ppm
chloride	10 ppm as Cl

Although this water may not be considered contaminated for most uses, it may be for some. In addition, its capacity to assimilate the additional pollutants must be considered on the basis of its present degree of impurity, its volume, and its buffering capacity. Certain amounts of contaminants, such as calcium, magnesium, iron, etc., are absolutely essential to aquatic life as well as to man. Pure or distilled water does not contain any of these essential "contaminants" that are needed for any type of life.

Rand (1969) gives more detailed chemical analyses of various types of natural waters (Table 7-1).

Cummins (1972) classifies fauna in generally unpolluted rivers according to their vertical and horizontal strata. The vertical animal life is *nekton* (all animals in the water flowing over the substrata) and *benthos* (bottom loving). The hori-

TABLE 7-1. Typical Water Analysis. [*After Rand, 1969.* * *By permission of publishers.*]

Chemical component	Expressed as	Natural water types				
		Rain	Surface soft	Ground soft	Surface hard	Ground hard
Calcium	$CaCO_3$, mg/l	16	30	29	80	142
Magnesium	$CaCO_3$ equiv., mg/l	3	16	32	40	59
Sodium and Potassium	Na, mg/l	6	9	26	19	20
Bicarbonate	$CaCO_3$, mg/l	12	42	60	106	143
Chloride	Cl, mg/l	5	7	9	23	23
Sulfate	SO_4, mg/l	10	12	17	38	59
Nitrate	N, mg/l	0.1	1.5	–	0.4	0.06
Iron	Fe, mg/l	0.0	1.1	1.8	0.0	0.18
Silica	SiO_2, mg/l	0	30	41	18	12
Carbon dioxide	$CaCO_3$, mg/l	4	4	59	4	14
pH	–	6.8	6.9	6.6	7.8	7.4

*Computed from original data of E. Nordell (1951, pp. 14, 18–21, 24–25, 28–29, 32, 67, 70–71, 74–76.)

TABLE 7-2. Taxa Typical of North American Running Waters Receiving Minor Organic and Inorganic Enrichment from Human Perturbations (*faunal elements organized according to compartmentalization discussed in text*). [Reproduced with permission. From *River Ecology and Man*, by Kenneth W. Cummins, © 1972 by Academic Press.]

Vertical compart-mentalization	Faunal elements	Horizontal compartmentalization		
		Erosional	Intermediate	Depositional
Nekton	Invertebrates	Drift: all components of the benthos during catastrophic drift; during normal (diurnal) drift especially Ephemeroptera, e.g., Baetidae; Diptera, e.g., Chironomidae; Simuliidae; and Amphipoda, e.g., *Gammarus;* also "planktonic" forms displaced from depositional areas	Drift: similar to erosional sections but reduced Occasional incursions by certain benthic forms, e.g., Chydoridae	Drift: settling out, especially smaller classes and lighter forms not removed or recolonized in other sections "Planktonic" Protozoa (esp. Ciliata), Rotifera (free-swimming Monogononta), and Microcrustacea (esp. cyclopoid and diaptomid copepods and daphnid and chydroid Cladocera)
	Fish	Salmonidae (esp. *Salmo* and *Salvelinus*); Cyprinidae (*Rhinichthys*) (in large rivers *Esox, Micropterus dolomieui, Catostomus commersoni*)	*Notropus cornutus* (and elements from both erosional and depositional areas	Cyprinidae (e.g., *Notropus*), Centrarchidae, Ameiuridae
Benthos	Invertebrates	Sessile: Porifera (Spongillidae) "Suckers": Hirudinea, Diptera (Blepharoceridae), Gastropoda (Ancylidae), *Hydra*	Elements from both erosional and depositional zones but typical gravel	Burrowers: Annelida (Naididae, Tubificidae), Nematoda, Diptera (Chironomidae), Odonata (Gomphidae), Ephemeroptera (Hexagenia), Pelecypoda (thin-shelled *Anodonta*)
		Marginal contact: Coleoptera (*Psephenus, Ectopria*), Ephemeroptera (*Rithrogena, Epeorus, Ephemerella doddsi*) Fixed retreats: Trichoptera (Hydropsychidae, Philopotamidae, Psychomyiidae, Polycentropidae), Diptera (Orthocladiinae, e.g., *Rheotanytarsus*), Lepidoptera (*Cataclysta*)	fauna of: Ephemeroptera (the burrowers, *Ephemera*); Trichoptera (Limnephilidae Pycnopsyche; Leptoceridae *Mystacides, Oecetis*)	Sprawlers: Ephemeroptera (Caenidae, Ephemerellidae with gill flaps), Amphipoda Climbers: along stream margins in rapid and intermediate flow areas: Plecoptera (*Taeniopteryx*), Odonata (*Calopteryx*), Hemiptera (*Belastoma, Ranatra*)

TABLE 7-2. *(Continued)*

Vertical compartmentalization	Faunal elements	Horizontal compartmentalization		
		Erosional	Intermediate	Depositional
Benthos (*cont'd.*)	Invertebrates (*cont'd.*)	Hooks and claws: Diptera (Simuliidae, Deuterophlebiidae, Anthomyiidae–*Limnophora*), Megaloptera (*Nigronia, Corydalus*), Trichoptera (*Rhyacophilidae*), Coleoptera (*Elmidae*) Dorso-ventral flattening: Turbellaria (*Dugesia, Phagocata*), Ephemeroptera (Heptageneidae, Leptophlebiidae), Plecoptera (Perlidae) "Crevice seekers": Decapoda (e.g., *Orconectes*), Diptera (*Tipulidae*) Cases and heavy shells: Trichoptera (esp. *Helicopsyche, Neophylax, Goera*), Pelecypoda (e.g., Fusconaia) Swimmers: Ephemeroptera (Ephemerellidae, Baetidae, Leptophlebiidae, Siphlonuridae) Leaf packs–entrapment against obstructions: A special fauna with coarse-particle feeders "shredders", e.g., *Tipula, Pycnopsyche, Pteronarcys, Peltoperla, Lepidostoma;* fine-particle feeders, e.g., *Gammarus, Stenomena, Baetis;* and predators, e.g., *Nigronia Acroneuria*	Diptera (*Atherix*, Orthocladiinae, Diamesinae); Megaloptera (*Sialis*); Gastropoda (*Limnaea, Physa*), Pelecypoda (Sphaerium, Pisidium)	Climbers: in vascular plant beds: an essentially lentic fauna, e.g., Odonata (*Ischnura, Enallagma, Anax, Libellula*), Trichoptera (Phryganeidae, Leptoceridae, Limnephilidae) Leaf packs at stream margins: similar fauna to packs in erosional sections

zontal life is either *erosional, intermediate,* or *depositional.* A summary of the faunal elements in these classifications of streams is given in Table 7-2.

7-2. WATER QUALITY STANDARDS

There exist today in the United States two fundamental types of water quality standards in use for the control of water pollution. One type, which considers the quality of the receiving body of water, is commonly referred to as "stream standard." The other type, which deals with the quality of the wastes being discharged from any source, is called "effluent standard." The latter is becoming more prevalent as a guideline for industrial effluents. There are advantages and disadvantages, advocates and opponents of each type of pollution control. Stream standards were approached in two manners. The first involved a standard amount of receiving stream dilution required for a given unit volume of waste discharge. This method is largely outmoded in this country today, because similar volumes of wastes contain vastly different amounts and types of pollutants. The same may be true, but to a lesser extent, for equal volumes of receiving waters. The second and more prevalent stream standard procedure utilizes limiting values for various specific substances when mixed with the stream water. In the past it has been common practice to have a slightly different set of water quality criteria for each beneficial use of the stream, such as drinking, swimming, fishing, irrigation, etc. Since 1968 a new philosophy has begun to influence our criteria: that of establishing one set of water quality standards for all waters, based on the most beneficial use. Each of the 50 states in the United States has, for example, set water quality standards which have been approved by the federal government's agency, the Federal Water Pollution Control Administration. Although there is considerable similarity in these standards from one state to another, there are also some variations. *Chemical Week* (1969) has listed the major standards relating to industrial receiving waters for each of the states (Table 7-3).

Water quality must be related directly to quantity of flow in the stream; the greater the streamflow the more pollutants it can assimilate without contravening water quality standards. Industry, for example, recognizes the value of sites with high minimum streamflows. It was reported (*Chemical Week,* 1968) that there are less than 200 rivers in the United States with minimum flows over 50 cubic feet per second. Although there are some similarities in the water quality criteria established by the 50 states (*Chemical Week,* 1969) there are also considerable variations, especially in dissolved oxygen loads. For example, New York State (*Chemical News,* 1966) established in November of 1968, through its Water Resources Commission, a set of "Classifications and Standards of Quality and Purity for Waters of New York State." Although the system includes tidal water and groundwater criteria, we are primarily interested in surface water

TABLE 7-3. A Summary of Water Quality Standards Set by All 50 States.
[*After Chemical Week, 1969. By permission of publishers.*]

	pH Allowable,		Temperature (°F) allowable		Dissolved oxygen	Other requirements
	Range	Deviation	Maximum	Increase above ambient	(minimum mg/l or %)	
Alabama	6.0–8.5	1.0	90° (24 hr) 93° (any 8 hr) 93° (cooling water)	10% – 10°	2.0 at 5 ft or middepth if less than 10 ft.	Radioactivity, Color, Taste and Odor, Toxic Substances: Only in amounts that would not render waters unsuitable for industrial-cooling process-water supply purposes. Solids: Free from waste materials that cause unsightly or putrescent conditions or interfere directly or indirectly with industrial use.
Alaska	7.0–8.0	0.5	70°	–	5.0	Radioactivity: Not to exceed limits of PHS Drinking Water Standards. Turbidity: No imposed values that would interfere with established levels of treatment. Color: True color less than 50 color units. Taste and Odor: Shall not unreasonably impair esthetic considerations. Solids: No dissolved solids above natural conditions causing corrosion or scaling problems. No visible evidence of other floating solids or sludge deposits. No imposed sediment loads that would interfere with established treatment levels.* Toxic Substances: Chemical constituents should be below concentrations found to be of public health significance.*
Arizona	6.5–8.6	0.5	93°	5°	–	Radioactivity: Not to exceed 1/30 of the MPC_w value given for continous occupational exposure in NBS Handbook 69. Turbidity: 50 JCU (streams); 25 JCU (lakes). Color: Free from waste materials in amounts sufficient to change existing color enough to interfere with industrial use or to create a nuisance. Taste and

						Odor: Free from wastes in amounts sufficient to produce enough taste and odor to create a nuisance or interfere with industrial use. Solids: Free from wastes that would be unsightly, putrescent, odorous, or in amounts that would interfere with industrial use. Toxic Substances: Free from wastes toxic to human, animal, plant or aquatic life or in amounts that would interfere with industrial use.
Arkansas	6.0–9.0	1.0 (24 hr)	95°	5°	4.0 (average) for any cross section)	Radioactivity: Rules and Regulations for the Control of Ionizing Radiation," Arkansas Board of Health, apply. Turbidity: No distinctly visible increases due to wastes. Color: Shall not be increased to the extent that it interferes with industrial use, present or future. Taste and Odor: Must not cause offensive odors or otherwise interfere with industrial use. Solids: No distinctly visible persistent solids, bottom deposits or sludge banks due to wastes. Toxic Substances: Must not be present in amounts toxic to human, animal, plant or aquatic life.
California	6.5–8.6 7.0–8.6 (Coastal waters)	–	71.6°* (fresh) water)	None that would cause ecological change or harm aquatic life (coastal waters)*	6.0 Coastal water: 5.0 (unless naturally lower)	Radioactivity: Shall not exceed 1/10 of the MPC_w values given for continuous occupational exposure in NBS Handbook 69. Turbidity: Free from wastes that could alter water's existing turbidity. Color: Free from substances attributable to wastes that produce detrimental color. Taste and Odor: No substances that impart foreign taste or odor. Solids: Dissolved solids in fresh water must not exceed 300 mg/l at any time; annual mean: 175 mg/l. Settleable solids must not be able to change nature of stream bottom or harm aquatic environment. Toxic Substances: At all times

TABLE 7-3. *(Continued)*

	pH Allowable, Temperature (°F) allowable				Dissolved oxygen (minimum mg/l or %)	Other requirements
	Range	Deviation	Maximum	Increase above ambient		
						free from concentrations harmful to humans, aquatic life or wild or domestic animals.
Colorado	5.0–9.0	–	93°	–	3.0	Radioactivity: Not to exceed 1/30 of the 168-hr week values in NBS Handbook 69. turbidity: Must not interfere with established levels of treatment. Color: Wastes present must not cause appreciable change in color or interfere with industrial use. Taste and Odor: Free from wastes that cause odor or appreciable change in taste. Solids: Free from wastes that are unsightly, putrescent, or odorous or would interfere with use. Toxic Substances: Free from wastes in concentrations or combinations sufficient to harm human or animal life.
Connecticut	6.0–9.0	–	–	None unless it does not exceed recommended limits for industrial use	2.0	Radioactivity: Limits to be approved by appropriate state agency. Turbidity, Color, Taste and Odor: None in such quantities that would impair industrial use. Solids: Limited to small amounts that may result from discharge of appropriately treated wastes. Toxic Substances: Free from chemical constituents in concentrations or combinations harmful to human, animal or aquatic life.
Delaware	6.5–8.5	–	–	5°	50%* or 4.0	Radioactivity: Alpha emitters limited to 3 pc/l; beta emitters, to 1,000 pc/l. Color, Taste and Odor: None

						in concentrations that cause color, taste, or odor. Solids: Free from unsightly and malodorous nuisances due to floating solids or sludge deposits. Toxic Substances: None in concentrations harmful (synergistically or otherwise) to humans, fish, shellfish, wildlife or aquatic life.
Florida	6.0–8.5	1.0	–	–	4.0	Radiation: Gross beta–1,000 pc/l (in absence of Sr-90 and alpha emitters). Turbidity: 50 JCU. Color: Must not render water unfit for industrial-cooling or process-water supply purposes. Taste: Must not render water unfit for industrial use; phenols 0.001 maximum. Solids: Dissolved solids must not exceed 1,000 mg/l; monthly average: 500 mg/l. Must be free from floating wastes that are unsightly or deleterious or other wastes that settle to form putrescent or objectionable sludge deposits. Toxic Substances: Free from wastes harmful to human, animal or aquatic life. Cu, 0.5 mg/l; Zn, 1.0; Cr, 0.05; Pb, 0.05; Fe 0.3; As, 0.05; F, 10.0; Cn, none detectable.
Georgia	6.0–8.5	–	93.2°	10° (above intake)	2.5 3.0 (daily average)	Radioactivity: Must conform to state statutes. Turbidity, Color, Taste and Odor: Free from wastes that cause objectionable conditions or interfere with industrial use. Solids: Free from wastes that are unsightly, putrescent or otherwise objectionable or would interfere with industrial use. Toxic substances: No wastes in concentrations that would prevent fish survival or interfere with industrial use.
Hawaii	6.5–8.5	–	–	–	4.5	Radioactivity: Not to exceed 1/30 of the values given by NBS Handbook 69. Turbidity: Free from soil particles from erosion caused by land development or agricultural use. Taste and Odor: Wastes, after dilution

TABLE 7-3. *(Continued)*

	pH Allowable, Temperature (°F) allowable				Dissolved oxygen	Other requirements
	Range	Deviation	Maximum	Increase above ambient	(minimum mg/l or %)	
						and mixture, must not interfere with industrial use. Toxic Substances: Free from substances in concentrations harmful to human, animal, or marine life or that make waters unsuitable for industrial use.
Idaho	6.5–9.0	0.5	–	2° Only if water 68° or less	75% (at seasonal low)	Radioactivity: Not to exceed limits of '62 PHS Drinking Water Standards. Turbidity: No objectionable turbidity that can be traced to a point source. Solids: No floating or submerged matter; no sludge deposits that could adversely affect industrial use. Toxic Substances: No wastes of other than natural origin in concentrations of public health significance or that could adversely affect industrial use.
Illinois	5.0–9.0	–	95°	–	2.0 3.0 (for 16 hr in any 24-hr period)	Color, Taste and Odor: Free from wastes that produce color, odor, or taste in such a degree as to create a nuisance. Solids: Free from floating wastes that settle and form unsightly, deleterious or putrescent deposits. Toxic Substances: Free from wastes in concentrations or combinations harmful to human, animal, plant or aquatic life.
Indiana	5.0–9.0	–	95°	–	1.0 2.0 (daily average)	Color, Taste and Odor: Free from wastes that produce color, taste or odor in such a degree as to create a nuisance. Solids: Dissolved solids must not exceed 1,000 mg/l; monthly average, 750 mg/l. Must be free from unsightly, putrescent, deleterious or otherwise

						objectionable wastes. Toxic Substances: Free from wastes in concentrations or combinations harmful to human, animal, plant or aquatic life.
Iowa	–	–	–	–	–	Color, Taste and Odor: Free from wastes that produce color, taste or odor in such a degree as to be detrimental to industrial use. Solids: Free from floating wastes in amounts that would be unsightly or deleterious or other wastes that settle to form putrescent or objectionable sludge deposits. Toxic Substances: No wastes in concentrations or combinations detrimental to human, animal or aquatic life or to industrial use.
Kansas	6.5–9.0	–	90°*	–	4.0*	Turbidity: No increase that causes substantial visible contrast with natural appearance or that is detrimental to industrial use. Color: Discharges of color-producing substances limited to concentrations not detrimental to industrial use. Taste and Odor: Concentrations limited to those that would not result in noticeable offensive odors or otherwise interfere with industrial use. Solids: Free from floating debris or material in amounts that would be unsightly or detrimental to industrial use. Toxic Substances: Pollutional substances must be maintained below concentrations detrimental for industrial use.
Kentucky	5.0–9.0	–	95° 73°(Dec.-Feb.)	2°/hr 10°/day	–	Color, Taste and Odor: Wastes must not create a nuisance. Solids: Dissolved solids must not exceed 1,000 mg/l; monthly average: 750 mg/l. No floating wastes in unsightly or deleterious amounts; no other wastes that settle to form putrescent or objectionable

TABLE 7-3. *(Continued)*

	pH Allowable, Temperature (°F) allowable				Dissolved oxygen (minimum mg/l or %)	Other requirements
	Range	Deviation	Maximum	Increase above ambient		
						sludges. Toxic Substances: No wastes in concentrations or combinations harmful to human, animal, plant or aquatic life.
Louisiana	6.0–9.0	–	96.8°	5.4°	50%	Radioactivity: Specific limits set for all radioactive isotopes released as waste. Solids: None that would produce floating masses, sludge banks or beds on bottom, either organic or inorganic. Toxic Substances: No wastes in concentrations or combinations harmful to animal or plant life.
Maine	6.0–9.0*	0.5*	90°*	–	2.0*	Radioactivity: Not to exceed '62 PHS Drinking Water Standards. Turbidity, Color, Taste and Odor: Free from wastes that impart turbidity, color, taste or odor or impair industrial use. Solids: Free from sludge deposits, solid refuse and floating solids. Toxic Substances: No chemical constituents from waste sources harmful to humans or that adversely affect industrial use.
Maryland	5.0–9.0 (unless natural)	–	100°	–	4.0 (unless naturally lower)	Color, Taste and Odor: Free from waste materials that change existing color or produce taste and odor to such a degree as to create a nuisance of interfere with industrial use. Solids: Free from wastes that float, settle to form deposits, create a nuisance or interfere with industrial use and are unsightly, putrescent or odorous. Toxic Substances: Free from toxic wastes

						that interfere with industrial use or that are harmful to human, plant, animal or aquatic life.
Massachusetts	6.0–9.0	–	90°	–	2.0	Radioactivity: None in concentrations harmful to human, animal or aquatic life. Turbidity, Color, Taste and Odor: None in concentrations that would impair industrial use. Solids: None allowed except that which may result from the discharge from waste-treatment facilities providing appropriate treatment. Toxic Substances: None in concentrations or combinations harmful to human, animal or aquatic life.
Michigan	6.5–8.8	0.5	–	10°	Enough to prevent nuisance	Radioactivity: Standards to be established when information is available on deleterious effects. Turbidity, Color: No objectionable unnatural turbidity or color in quantities sufficient to interfere with industrial use. Taste and Odor: Below levels that are or may become injurious to industrial use. Solids: Dissolved solids must not exceed 750 mg/l; monthly average: 500 mg/l. No floating solids or objectionable deposits in quantities that would interfere with industrial use. Toxic Substances: Limited to concentrations less than those that are or may become injurious to this use.
Minnesota	6.0–9.0	–	86°	–	–	Color, Taste and Odor, Solids: Free from wastes that cause nuisance conditions, such as material discoloration, obnoxious odors, significant floating solids, excessive suspended solids or sludge deposits.
Mississippi	6.0–8.5	1.0	93°	10°	3.0	Color, Taste and Odor: Free from wastes that produce color or odor in such a degree as to create a nuisance. Solids: Dissolved solids must not exceed 1,500 mg/l; monthly average 750 mg/l. Must be free from floating

TABLE 7-3. *(Continued)*

	pH Allowable, Temperature (°F) allowable				Dissolved oxygen	Other requirements
	Range	Deviation	Maximum	Increase above ambient	(minimum mg/l or %)	
						wastes that settle to form unsightly, deleterious, objectionable or putrescent deposits. Toxic Substances: No wastes in concentrations or combinations harmful to human, animal or aquatic life.
Missouri	6.5–9.0	–	90°	9° (average of cross section)	4.0*	Radioactivity: Gross beta: 1,000 pc/l (in absence of Sr-90 and alpha emitters). Sr-90: 10 pc/l. Dissolved Ra-226: 3 pc/l. Color: Wastes must not cause substantial visible contrast with natural appearance of stream or interfere with industrial use. Taste and Odor: Limited to concentrations that would not result in noticeable offensive odors or otherwise interfere with industrial use. Solids: No noticeable organic or inorganic deposits or floating materials in unsightly or deleterious amounts. Toxic Substances: Concentrations not detrimental to industrial use or toxic to humans, fish, wildlife. F:1.2.
Montana	6.5–9.5	0.5	–	No adverse change	–	Radioactivity: Not to exceed '62 PHS Drinking Water Standards. Turbidity: Must not interfere with established levels of treatment. Color, Taste and Odor: Water shall be maintained in condition not offensive to sense of sight or smell. Solids: No floating solids and sludge deposits in amounts deleterious to industrial use; no sediments or settleable solids that affect

						treatment levels. Toxic Substances: Amounts present must not adversely affect industrial use.
Nebraska	6.5–9.0	1.0	90°	5° (May-Oct.) 10° (Nov.-Apr.) Rate: 2°/hr	5.0	Radioactivity: Must conform with Radiological Health Regulations (1st. ed.), State of Nebraska, '66. Turbidity: No more than 10% increase above normal level. Color: No evidence of matter that creates nuisance conditions. Taste and Odor: Less than amounts that would degrade water quality for industrial use; phenol: 0.001 mg/l. Solids: Dissolved solids must not exceed 1,500 mg/l. No more than 20% increase (limit 100 mg/l) from any point source. No waste solids that permit deposition or are deleterious to industrial use. Toxic Substances: None in concentrations or combinations that would render water unsuitable for industrial use.
Nevada	6.5–8.5 Annual median: 7.4–8.3	–	77.0° (summer) 57.2° (winter)	–	5.0 6.0 (average, June-Sept.)	Radioactivity: Limited to 1/10 of the 168-hr-week values in NBS Handbook 69. Turbidity, Color, Taste and Odor: Free from wastes in amounts sufficient to change existing turbidity or color enough to create a nuisance or interfere with industrial use, or to produce taste or odor in the water. Solids: Free from floating or other wastes that settle to form sludge banks or deposits in amounts that would be unsightly or odorous or interfere with industrial use. Toxic Substances: Free from wastes in concentrations or combinations toxic to human, animal, plant or aquatic life or that interfere with industrial use.
New Hampshire	6.0–8.5 (unless	–	90°*	No increase that would	5.0	Turbidity, Color, Taste and Odor: None in objectionable amounts. Solids: No floating solids or sludge

TABLE 7-3. *(Continued)*

	pH Allowable, Temperature (°F) allowable				Dissolved oxygen	Other requirements
	Range	Deviation	Maximum	Increase above ambient	(minimum mg/l or %)	
	natural)			interfere with this use*		deposits in objectionable amounts. Toxic Substances: None in toxic concentrations or combinations.
New Jersey	6.5–8.5 (unless natural)	–	87° (unless natural)	5° (up to 87°)	4.0*	Turbidity, Solids: None noticeable in water or deposited along shore. Color, Taste and Odor: None that are offensive to humans or detrimental to aquatic biota. Toxic Substances: None that would affect humans or be detrimental to aquatic biota.
New Mexico	6.6–8.6	–	–	Must not pollute or make water unfit for this use	No oxygen demand that would cause pollution	Radioactivity: Not greater that 1/10 of the 48-hr value in NBS Handbook 69. Turbidity: Shall not cause substantial visible contrast with natural appearance. Color: Should not create an esthetically undesirable condition. Taste and Odor: No odors, other than of natural origin, that are esthetically objectionable or obnoxious. Solids: No objectionable floating solids or debris and sediment that significantly alter properties of bottom. Toxic Substances: No amounts toxic to humans, plants, fish, animals.
New York	6.0–9.5	–	86°	5° (average 7 days) Rate: 2°/hr 9°/24 hr	3.0	Color: No colored wastes that alone or in combinations make water unsuitable for industrial use. Solids: No floating or settleable solids or sludge deposits that are readily visible and attributable to wastes. Toxic Substances: None alone or in combinations that would impair industrial use.

North Carolina	Normal for area, usually 6.0–8.5; as low as 4.3 in swamps	–	95°*	7°	3.0	Color: Must not render water unfit for industrial cooling. Solids: Must not, after dilution and mixture, make water unfit for industrial cooling. Toxic Substances: Must not make water unfit for industrial cooling.
North Dakota	6.0–9.0	0.5	93°	10°	3.0 5.0 (for 16 hr/day)	Radioactivity: No discharge allowed unless materials are readily soluble or dispersible and of quantities acceptable to state health department. Color, Taste and Odor: No wastes that color water or result in objectionable odors to a degree that impairs industrial use. Solids: No unsightly floating wastes that would adversely affect industrial use or wastes that settle to form putrescent or objectionable deposits. Toxic Substances: No concentrations or combinations harmful to human, animal or aquatic life.
Ohio	5.0–9.0	–	95°	–	1.0 2.0 (daily average)	Color, Taste and Odor: Free from wastes that produce color or odor to a degree that creates a nuisance. Solids: Dissolved solids must not exceed 1,000 mg/l; monthly average: 750 mg/l. Must be free from floating or other wastes that settle to form putrescent or objectionable deposits or that are unsightly or deleterious. Toxic Substances: No wastes in concentrations or combinations harmful to human, animal or aquatic life.
Oklahoma	6.5–8.5	–	93°	5°	4.0	Radioactivity: Average concentration at points of controlled release shall not exceed State Board of Health Radiation Protection Regulations. Turbidity: None that causes visible contrast with natural conditions.

TABLE 7-3. *(Continued)*

	pH Allowable, Temperature (°F) allowable				Dissolved oxygen	Other requirements
	Range	Deviation	Maximum	Increase above ambient	(minimum) mg/l or %)	
						Color: None that is persistent in concentrations detrimental to industrial use. Taste and Odor: No concentrations that would cause offensive odors in vicinity of water or otherwise interfere with industrial use. Solids: Free from floating debris, bottom deposits or other materials. Toxic Substances: None in quantities that make water toxic to human, animal, plant or aquatic life.
Oregon	6.5–9.0	–	–	2° (only if water 70° or less)	5.0	Radiation: Shall not pose external hazard. Turbidity: 5 JCU above natural. Color, Taste and Odor: No objectionable discoloration or conditions esthetically offensive to human senses of taste or smell. Solids: No floating solids, organic or inorganic deposits injurious to industry. Toxic Substances: No conditions injurious to public health.
Pennsylvania	6.0–9.0	–	93°	2°/hr	4.0 5.0 (daily average)	Color, Taste and Odor: No wastes that produce colors, tastes or odors in amounts harmful to industrial use. Solids: Dissolved solids must not exceed 750 mg/l; monthly average: 500 mg/l. No floating wastes or substances that settle to form sludge in amounts harmful to industrial use. Toxic Substances: None in amounts harmful to industrial use.
Rhode Island	6.0–8.5	–	–	4°	3.0* 5.0* (16 hr/	Radioactivity, Toxic Substances: No concentrations or combinations harmful to human, animal or aquatic

					day)	life. Turbidity, Taste and Odor: None in concentrations that would impair industrial use. Solids: No solid refuse, floating solids or sludge deposits.
South Carolina	6.0–8.5 5.0–8.5 (swamps)	–	9.32°*	10°*	3.0* 2.5* (swamps)	Turbidity, Color, Taste and Odor: Free from wastes that change the existing turbidity or color or that produce taste or odor to such a degree as to cause a nuisance or interfere with industrial use. Solids: None from waste sources in amounts that are unsightly, putrescent, odorous or that cause a nuisance or interfere with industrial use. Toxic Substances: Free from wastes harmful to human, animal, plant or aquatic life or that interfere directly or indirectly with industrial use.
South Dakota	6.0–9.5	1.0	–	–	–	Radioactivity: None permitted in water unless readily soluble or dispersible and in quantities allowed by federal or state agencies. Color, Taste and Odor: No wastes that produce material discoloration or undesirable odors. Solids: Dissolved solids must not exceed 2,000 mg/l. No wastes producing floating solids, sludge deposits or other offensive effects. Toxic Substances: None in concentrations toxic to human, animal or aquatic life.
Tennessee	6.0–9.0	1.0 (24 hr)	93°	10° Rate: 3°/hr	Enough to prevent offensive conditions	Radioactivity: None that could adversely affect industrial use. Turbidity, Color: None in amounts or concentrations that could not be reduced to acceptable levels by conventional treatment. Taste and Odor: None that would result in taste or odor that would prevent use for industrial processing. Solids: Dissolved solids must not exceed 500 mg/l. No distinctly visible solids, bottom deposit or sludge

TABLE 7-3. *(Continued)*

	pH Allowable, Temperature (°F) allowable				Dissolved oxygen	Other requirements
	Range	Deviation	Maximum	Increase above ambient	(minimum) mg/l or %)	
						banks that could be detrimental to industrial use. Toxic Substances: None that produces toxic conditions that would adversely affect water for industrial use.
Texas	5.0–8.5 5.0–9.0 (cooling water)	–	–	–	4.0	Radioactivity: Regulated by Texas Radiation Control Act and Texas Regulations for Control of Radiation. Turbidity: No substantial increase due to wastes. Color: No substantial visible contrast with natural appearance of receiving waters after wastes receive best practical treatment. Taste and Odor: No concentrations that produce offensive odors. Solids: Dissolved solids must not exceed 1,000 mg/l, unless water used only for cooling. Must be essentially free from floating or settleable suspended solids that would adversely affect industrial use. Toxic Substances: Shall not show acute or chronic toxicity to humans, animals or aquatic life to such an extent as to interfere with industrial use.
Utah	6.5–9.0	–	–	–	–	Radioactivity: Shall not exceed 1/30 of MPC_w values for continuous occupational exposure in NBS Handbook 69. Turbidity, Color: No wastes in amounts that would change existing turbidity or color enough to create public nuisance or interfere with industrial use. Taste and Odor: No wastes in amounts that

						would produce taste or odor. Solids: No floating wastes that are unsightly or that interfere with industrial use; no wastes that settle to form unsightly or odorous sludge or bottom deposits. Toxic Substances: No wastes in concentrations or combinations toxic to human, animal, plant or aquatic life or that would interfere with industrial use.
Vermont	6.0–8.5	–	–	4°	3.0* 5.0* (16 hr/day)	Radioactivity: To be approved by appropriate state agency. Turbidity, Color, Taste and Odor: None in concentrations that would impair industrial use. Solids: No floating solids, sludge deposits or solid refuse. Toxic Substances: No chemical constituents in concentrations or combinations harmful to human, animal or aquatic life.
Virginia	5.0–9.0 (swamps as low as 4.3)	–	95°* (unless naturally higher)	No sudden changes that could harm aquatic life	1.0* 2.0* (daily average)	Color, Taste and Odor: No wastes that change existing color or produce odor to such a degree as to create a nuisance or interfere with industrial use. Solids: No floating wastes that are unsightly or create a nuisance or other wastes that settle to form unsightly, putrescent or odorous deposits. Toxic Substances: No wastes in concentrations or combinations that would interfere directly or indirectly with industrial use.
Washington	6.5–8.5	0.5	70°	t = 110/(T-15)†	6.5 or 70%	Radioactivity, Toxic Substances: Below concentrations that could adversely affect industrial use. Turbidity: Less than 10 JCU over natural conditions. Color, Taste and Odor, Solids: Dissolved, suspended, floating

TABLE 7-3. *(Continued)*

	pH Allowable, Temperature (°F) allowable				Dissolved oxygen (minimum) mg/l or %)	Other requirements
	Range	Deviation	Maximum	Increase above ambient		
						or submerged matter shall not reduce esthetic values so as to affect industrial use.
West Virginia	Process water: 5.5–9.0 Cooling water: 5.0–9.0	–	Cooling water: 93° (May-Nov.) 73° (Dec.-Apr.) Process water must permit fish passage	– 5° Rate: 2°/hr (Dec.-Apr.)	1.0 2.0 (daily average)	Color: None that is objectionable. Taste and Odor: No objectionable odors in vicinity of the water. Solids: No distinctly visible floating, settleable or suspended solids of unreasonable kind or quantity. No objectionable bottom deposits or sludge banks. Toxic Substances: No concentrations of materials poisonous to human, animal or fish life.
Wisconsin	6.0–9.0	0.5	89°	–	1.0 2.0 (daily average)	Color, Taste and Odor: No materials producing color, taste or odor in amounts that would create a nuisance. Solids: Dissolved solids must not exceed 1,000 mg/l; daily average: 750 mg/l. No floating or submerged debris or waste substances that would cause objectionable deposits in amounts to create a nuisance. Toxic Substances: None in concentrations or combinations toxic to humans or of public health significance.
Wyoming	6.5–8.5	–	–	4° 2° (for streams where temp. not over 70°)	6.0	Radioactivity: Not to exceed 1/30 of NBS Handbook 69 values. Turbidity: No more than 15 JCU above natural (when turbidity is 150 JCU or less); otherwise, no more than 10% above natural. Color: Essentially free of wastes that visibly alter natural color of water or impart color to vessels or structures. Taste and

Odor: Essentially free from substances that would produce detectable odor at site of use. Solids: Essentially free from floating or settleable solids that are unsightly or settle to form sludge, bank or bottom deposits. Toxic Substances: Free from toxic substances in concentrations or combinations toxic to human, animal or aquatic life.

---

*Standard reserved from Federal Water Pollution Control Administration approval.

†t = total cumulative heat addition allowed from unnatural waste sources, at any point throughout the given stream reach. T = highest occurring temperature for a given period, in a specific stream reach.

Abbreviations: PHS–Public Health Service; NBS–National Bureau of Standards; JCU–Jackson Candle Units; pc/l–picocuries per liter.

Source: Water Quality Standards of the United States, Territories, and the District of Columbia, American Public Health Assn., Subcommittee on Water Quality Control, and Engineering and Sanitation Section.

NOTE–Specific limits for coliforms, biologic oxygen demand, plant nutrients, oil, grease, scum, bottom deposits, pesticides, specific conductance, carbon chloroform extract, synthetic detergents not included. Some states set standards for each stream reach or river basin; in such cases, table shows least stringent requirement.

TABLE 7-4. Classes and Standards for Fresh Surface Waters Class AA.
[*Chemical News, 1966. By permission of publishers.*]

Best usage of waters. Source of water supply for drinking, culinary or food processing purposes and any other usages.

Conditions related to best usage. The waters, if subjected to approved disinfection treatment, with additional treatment if necessary to remove naturally present impurities, meet or will meet U.S. Public Health Service drinking water standards and are or will be considered safe and satisfactory for drinking water purposes.

Quality standards for class AA waters

Items	*Specifications*
1. Floating solids; settleable solids; oil; sludge deposits; tastes or odor producing substances	None attributable to sewage, industrial wastes or other wastes.
2. Sewage or wastes effluents	None which are not effectively disinfected.
3. pH	Range between 6.5 and 8.5.
4. Dissolved oxygen	For trout waters, not less than 5.0 parts per million; for non-trout waters, not less than 4.0 parts per million.
5. Toxic wastes, deleterious substances, colored or other wastes or heated liquids	None alone or in combination with other substances or wastes in sufficient amounts or at such temperatures as to be injurious to fish life, make the waters unsafe or unsuitable as a source of water supply for drinking, culinary or food processing purposes or impair the waters for any other best usage as determined for the specific waters which are assigned to this class.

NOTE 1–In determining the safety or suitability of waters in this class for use as a sources of water supply for drinking, culinary or food processing purposes after approved treatment, the Water Pollution Control Board will be guided by the standards specified in the latest edition of *Public Health Service Drinking Water Standards* published by the United States Public Health Service.

NOTE 2–With reference to certain toxic substances as affecting fish life, the establishment of any single numerical standard for waters of New York State would be too restrictive. There are many waters, which because of poor buffering capacity and composition will require special study to determine safe concentrations of toxic substances. However, based on non-trout waters of approximately median alkalinity (80 ppm) or above for the State, in which groups most of the waters near industrial areas in this State will fall, and without considering increased or decreased toxicity from possible combinations, the following may be considered as safe stream concentrations for certain substances to comply with the above standard for this type of water. Waters of lower alkalinity must be specially considered since the toxic effect of most pollutants will be greatly increased.

Ammonia or ammonium compounds	Not greater than 2.0 parts per million (NH_3) at pH of 8.0 or above
Cyanide	Not greater than 0.1 part per million (CN)
Ferro- or Ferricyanide	Not greater than 0.4 parts per million ($Fe(CN)_6$)
Copper	Not greater than 0.2 parts per million (Cu)
Zinc	Not greater than 0.3 parts per million (Zn)
Cadmium	Not greater than 0.3 parts per million (Cd)

TABLE 7-5. Classes and Standards for Fresh Surface Waters Class A. [*Chemical News, 1966. By permission of publishers.*]

Best usage of waters. Source of water supply for drinking, culinary or food processing purposes and any other usages.

Conditions related to best usage. The waters, if subjected to approved treatment equal to coagulation, sedimentation, filteration and disinfection, with additional treatment if necessary to reduce naturally present impurities, meet or will meet U.S. Public Health Service drinking water standards and are or will be considered safe and satisfactory for drinking water purposes.

Quality standards for class A waters

Items	*Specifications*
1. Floating solids; settleable solids; sludge deposits	None which are readily visible and attributable to sewage, industrial wastes or other wastes or which deleteriously increase the amounts of these constituents in receiving waters after opportunity for reasonable dilution and mixture with the wastes discharged thereto.
2. Sewage or waste effluents	None which are not effectively disinfected.
3. Odor producing substances contained in sewage, industrial wastes or other wastes	The waters after opportunity for reasonable dilution and mixture with the wastes discharged thereto shall not have an increased threshold odor number greater than 8, due to such added wastes.
4. Phenolic compounds	Not greater than 5 parts per billion (Phenol).
5. pH	Range between 6.5 and 8.5
6. Dissolved oxygen	For trout waters, not less than 5.0 parts per million; for non-trout waters, not less than 4.0 parts per million.
7. Toxic wastes, oil, deleterious substances, colored or other wastes or heated liquids	None alone or in combination with other substances or wastes in sufficient amounts or at such temperatures as to be injurious to fish life, make the waters unsafe or unsuitable as a source of water supply for drinking, culinary or food processing purposes or impair the waters for any other best usage as determined for the specific waters which are assigned to this class.

NOTE–Refer to notes 1 and 2 under class AA, which are also applicable to class A standards.

standards. New York State uses five such standards, on the basis of best uses: (1) drinking water with no treatment; (2) drinking water after complete treatment; (3) bathing; (4) fishing; and (5) agriculture or industrial cooling and processing. The precise standards for each of these five classifications are shown in Tables 7-4 through 7-8. The water quality should be at least comparable to that required for its "best use" classification.

TABLE 7-6. Classes and Standards for Fresh Surface Waters Class B. [*Chemical News, 1966. By permission of publishers.*]

Best usage of waters. Bathing and any other usages except as source of water supply for drinking, culinary or food processing purposes.

Quality standards for class B waters

Items	*Specifications*
1. Floating solids; settleable solids; sludge deposits	None which are readily visible and attributable to sewage, industrial wastes or other wastes or which deleteriously increase the amounts of these constituents in receiving waters after opportunity for reasonable dilution and mixture with the wastes discharged thereto.
2. Sewage or wastes effluents	None which are net effectively disinfected.
3. pH	Range between 6.5 and 8.5
4. Dissolved oxygen	For trout waters, not less than 5.0 parts per million; for non-trout waters, not less than 4.0 parts per million.
5. Toxic wastes, oil, deleterious substances, colored or other wastes, or heated liquids.	None alone or in combination with other substances or wastes in sufficient amounts or at such temperatures as to be injurious to fish life, make the waters unsafe or unsuitable for bathing or impair the waters for any other best usage as determined for the specific waters which are assigned to this class.

NOTE–Refer to note 2 under class AA, which is also applicable to class B standard.

7-3. POLLUTANTS IN WATER

For the purpose of discussion here, the term "pollution" means *too much of any given contaminant such that it renders the receiving water unusable in its existing state for its desired best usage*. It then becomes equally important to identify the major "best usages." The National Technical Advisory Committee on Water Quality Criteria of the Federal Water Pollution Control Administration considered five major water uses important enough to establish criteria (Interim Report, 1967):

1. Aesthetics and recreation
2. Public water supplies
3. Fish, other aquatic life, and wildlife
4. Agriculture
5. Industrial water supplies

TABLE 7-7. Classes and Standards for Fresh Surface Waters Class C. [*Chemical News, 1966. By permission of publishers.*]

Best usage of waters. Fishing and any other usages except for bathing as source of water supply for drinking, culinary or food processing purposes.

Quality standards for class C waters

Items	*Specifications*
1. Floating solids; settleable solids; sludge deposits	None which are readily visible and attributable to sewage, industrial wastes or other wastes or which deleteriously increase the amounts of these constituents in receiving waters after opportunity for reasonable dilution and mixture with the wastes discharged thereto.
2. pH	Range between 6.5 and 8.5
3. Dissolved oxygen	For trout waters, not less than 5.0 parts per million; for non-trout waters, not less than 4.0 parts per million.
4. Toxic wastes, oil, deleterious substances, colored, or other wastes, or heated liquids	None alone or in combination with other substances or wastes in sufficient amounts or at such temperatures as to be injurious to fish life or impair the waters for any other best usage as determined for the specific waters which are assigned to this class.

Aesthetics and Recreation

The National Committee recommended criteria for aesthetic purposes such that "all surface waters should contribute to the support of life forms of aesthetic value." (Interim Report, 1967) This general criterion was implemented by the following specific recommendations:

Surface waters should be free of substances attributable to discharges of wastes.

A. Materials that will settle or form objectionable deposits.

B. Floating debris, oil, scum and other matter.

C. Substances producing objectionable color, odor, taste, and turbidity.

D. Materials including radionuclides in concentrations of combinations which are toxic or which produce undesirable physiological responses in human, fish, and other animal life and plants.

E. Substances and conditions or combinations thereof in concentrations which produce undesirable aquatic life.

The reader may note that no actual numbers are recommended by the Committee. Instead they suggest guidelines from which specific criteria may be

TABLE 7-8. Classes and Standards for Fresh Surface Waters Class D.
[*Chemical News, 1966. By permission of publishers.*]

Best usage of waters. Agricultural or source of industrial cooling or process water supply and any other usage except for fishing, bathing or as source of water supply for drinking, culinary or food processing purposes.

Conditions related to best usage. The waters will be suitable for fish survival; the waters without treatment and except for natural impurities which may be present will be satisfactory for agricultural usages or for industrial process cooling water; and with special treatment as may be needed under each particular circumstance, will be satisfactory for other industrial processes.

Quality standards for class D waters

Items	*Specifications*
1 Floating solids; settleable solids; sludge deposits	None which are readily visible and attributable to sewage, industrial wastes or other wastes or which deleteriously increase the amounts of these constituents in receiving waters after opportunity for reasonable dilution and mixture with the wastes discharged thereto.
2. pH	Range between 6.0 and 9.5
3. Dissolved oxygen	Not less than 3.0 parts per million.
4. Toxic wastes, oil, deleterious substances, colored or other wastes, or heated liquids	None alone or in combination with other substances or wastes in sufficient amounts or at such temperatures as to prevent fish survival or impair the waters for agricultural purposes or any other best usage as determined for the specific waters which are assigned to this class.

NOTE–Refer to note 2 under class AA, which is also applicable to class D standards.

established to suit the local situation. Many of the criteria for aesthetic use depend upon a subjective analysis, and thus will vary geographically and from individual to individual. In addition, as time goes on, and as people become more concerned about the quality of the environment, these criteria may become more stringent and/or more generally enforced. However, enforcement of such guidelines is extremely difficult, since each case must be brought into the judicial branch of government to settle the question of fact–whether the guidelines were violated. The more specific the criteria the more enforceable would be the guidelines, thus enabling administrative procedures to replace judicial ones.

Effect of Stream Pollution upon Recreational Uses of the River Water. Recreational use can be divided into three classifications:

1. Swimming and bathing in natural waters

2. Fishing in natural waters
3. Boating on or camping along natural waters

For (1) the public health aspects are paramount; for (2) the maintenance of a favorable physical, chemical and biological balance of the water, together with the absence of elements which are detrimental to fish propagation, is most important; and for (3) the major consideration is prevention of nuisance conditions (odors, floating matter, deterioration of protective coatings of the hulls of floating craft, or growth of macrophytes which can foul propellers).

Swimming and bathing. The major item of concern in this classification is coliform density, the allowable quantity of which varies from 50 per 100 ml in the Tennessee Valley to 2,400 per 100 ml in New York State. Most states use from 500-1,000 per 100 ml as the dividing line. The American Public Health Association feels that waters showing a concentration of most probable number (mpn) less than 1,000 per 100 ml are fairly acceptable for bathing, and the occasional presence of organisms of the coliform group in waters, even to the extent of 2,400 per 100 ml or slightly higher, does not necessarily preclude their safe use for bathing waters where the sanitary survey suggests no gross pollution from human beings. There are many knowledgeable environmentalists who maintain that limits should be placed upon fecal coliform types only, rather than total coliforms, which include soil types. Emphasis is placed upon a correlative sanitary survey, thus giving the following considerations for acceptability:

1. *Bacteriological*—count of coliform group
2. *Biological*—toxic aquatic plants and algae blooms, protozoa, worms; rotifers, etc.
3. *Chemical*—toxic substances, free acids, core of chemicals which produce objectionable tastes, odors, presence of dissolved oxygen.
4. *Physical*—absence of oils, tars, grease, scum, floating solids, debris (except that which comes from natural sources), absence of preventable sludge deposits, and turbidity

Stevenson (1953) states that the scientific proof of the transmission of disease from bathing water to bather is extremely meager. However, many reports have been made by competent authorities who cite numerous incidents where epidemiological evidence has incriminated particular bathing areas as the source of disease cases and epidemics, Jaeger (1892) has reported several cases of enteric illness caused by bathing in streams. Leptospirosis has also been reported to be caused by bathing in dirty waters. Typhoid fever attributed to swimming in water polluted by sewage has been reported by several persons (Reece, 1908-09; Report of Committee on Bathing Places, 1922; Vaughan, 1922-23; New York State Dept. of Health, 1932; California State Board of Health, 1942). Cholera also has been attributed to bathing in polluted waters (Miller, 1941). The parasitic infection, schistosome dermatitis (swimmer's itch), has been a troublesome problem, especially in cold fresh water (Smith, Woolsey, and Stevenson, 1951).

The following diseases are given as important when dealing with polluted bathing areas:

1. Eye infections–pinkeye and conjunctivitis
2. Ear infections
3. Nasal infections–colds and sinus infections
4. Skin infections–gonococcus infection of the eyes
5. Gastrointestinal–typhoid, etc., paratyphoid

Three studies (U.S. Public Health Service, 1952 and 1961; Stallybrass, 1931) conducted by the Public Health Service, in order to secure epidemiological and correlated bathing water quality data, have been outlined. These studies have furnished information as to swimming and illness experience by age, sex, and other groupings. They have demonstrated that *higher evidence of illness may be expected in the swimming group* than in the nonswimming group, regardless of the water quality. They have shown that in studies of this type, eye, ear, and respiratory ailments will represent more than half of the overall illness incidence, gastrointestinal disturbances up to one-fifth, and skin irritations and others the balance of the illnesses. Two instances of statistically significant correlation between illness incidence and bathing water quality were noted. They conclude also that some of the most rigid of natural bathing water quality requirements could be relaxed without detrimental effect to the health of the bathers. However, they hasten to add that this has always been and will always remain a matter of judgment.

Camp (1963) confirms the widespread suspicion among physicians and health officers that diseases are spread through bathing waters, although convincing epidemiological evidence is lacking in most cases. The types of diseases suspected are (1) ear, eye, nose, and throat infections; (2) skin diseases such as ringworm, eczema, scabies, etc.; (3) venereal infections; and (4) gastrointestinal disorders. Although intestinal disease acquired from swallowing polluted water is of much less significance than that acquired from drinking water, typhoid, dysentery, and even polio and hepatitis are being connected directly to swimming in polluted waters.

The Public Health Service has reported that hepatitis A, a virus, is transmitted through the fecal-oral route, but there is no evidence for a respiratory route. It is possible, therefore, that infectious hepatitis may be contracted at sewage-polluted bathing beaches.

For swimming pools employing recirculation, filtration, and chlorination, the American Public Health Association recommends that the total bacteriological count on standard nutrient agar, incubated 24 hours at 37°C, should not exceed 200/ml or 20,000/100 ml in more than 15 percent of the samples taken over a considerable period of time. No coliforms are allowed (except in 15 percent) and 0.4-1.0 ppm Cl_2 residual maintained.

The National Committee (Interim Report; 1967), probably due to lack of consistent and conclusive data, considers it neither practical nor desirable to recommend an absolute standard of safety for outdoor swimming. Most standards of safety, however, lie between 240/100 ml and 2,400/100 ml. Difficulties arise from:

1. Even naturally contaminated streams having coliform counts of 240/100 ml of streamwater.
2. Effects of sewage treatment and chlorination on coliform bacteria.
3. Effects of salt water on coliform bacteria.

Chlorination of sewage effluents is recommended, since it has been shown to be 1,000-1,700 times as effective against coliform bacteria as conventional treatment without chlorination.

Schistosome dermatitis–swimmer's itch–has increased greatly recently because of flooding of lands by new reservoirs for irrigation pruposes. This disease is commonly referred to as snail fever and often results in fever, rather gradual immobilization, and ultimately, death for the infected person. The larvae of trematode worms (which are parasitic to birds and snails) penetrate a bather's skin. They are not usually found in salt water and are controlled by using $CuSO_4$ and $CuCO_3$.

The National Committee (Interim Report, 1967) more recently recommended that for waters of primary contact sports such as swimming

> Fecal coliform should be used as the indicator organism for evaluating the microbiological suitability of recreation waters. As determined by multiple tube fermentation or membrane filter procedures, and based upon a minimum of not less than five samples taken over not more than a 30 day period, the fecal coliform content shall not exceed a geometric mean of 200/100 ml, nor shall more than 10 percent of total samples during any 30 day period exceed 400/100 ml.

The Committee also recognized the importance of pH in swimming. "When the lacrimal fluid of the human eye, through exhaustion of its buffering capacity, is unable to adjust the immediate contact layer of a fluid to a pH of 7.4, eye irritation occurs. A deviation of no more than 0.1 unit from the normal pH of the eye may cause discomfort."

In view of the above the Committee (Interim Report, 1967) recommended:

> In primary contact recreation waters, the pH should be within the range of 6.5 to 8.3 except when due to natural causes, (some relatively clean forest streams

have naturally occurring low pH values) and in no case shall be less than 5 nor more than 9.0. When the pH is less than 6.5 or more than 8.3, discharge of substances which would increase the buffering capacity of the water should be limited.

The Committee also makes a case for clarity of primary contact waters: "... Clarity should be such that a Secchi disk, a Standard aluminum reflector piece, is visible at a minimum of four feet" and shows a concern over the physiological effects of continuous swimming in overheated waters: "... except where caused by natural conditions, maximum water temperatures should not exceed 85°F (30°C)."

Public Water Supplies

The Committee listed two types of criteria, "permissible" and "desirable," for public water supplies (see Table 7-9–7-11). *Permissible* includes "those characteristics and concentrations of substances in raw surface waters which will allow the production of a safe, clear potable, aesthetically pleasing and acceptable public water supply which meets the Public Health Service Drinking Water Standards 1962 after treatment." The Public Health Service is an agency of the Department of Health, Education, and Welfare, which is entrusted with maintaining drinking water standards. The criterion *desirable* includes "those characteristics and concentrations of substances in the raw surface waters which represent high quality water in all respects for use as public water supplies."

Fish, Other Aquatic Life, and Wildlife

Pertinent factors necessary for the sustenance of native fish life are the presence of adequate concentrations of dissolved oxygen, proper temperature, and adequate sources of fish food, as well as the absence of organic sludge blankets (sediments) which are deteremental to spawning, and the absence of toxic substance.

Minimum dissolved oxygen concentrations of 5 ppm or more have been suggested as standards for fish survival. The exact level depends not only upon the type of fish, but also upon the environmental conditions such as pH, temperature, stress, etc. Much emphasis has been given, of late, to the influence of change of environment on fish deaths rather than the exact degree of the environmental factor. However, Katz and Gaufin (1952) found that mixed fish deaths occur where dissolved oxygen values are often well below 5 ppm for periods of several hours.

Effect of Pollution on Fish. The effect of sewage and wastes on fish life varies greatly with the season. During the winter, when the water is very cold, fish are

TABLE 7-9-7-11. Table of Criteria for Public Water Supplies.
[*Interim Report, 1967. By permission of publishers.*]

Constituent or characteristic	Permissible criteria	Desirable criteria	See paragraph in National Technical Committee Report
Physical			
Color (color units)	75*	<10†	1
Odor	Narrative†	Virtually absent	2
Temperature*	Narrative	Narrative	3
Turbidity	Narrative	Virtually absent	4
Microbiological			
Coliform organisms	10,000/100 ml§	<100/100 ml§	5
Fecal coliforms	2,000/100 ml§	<20/100 ml§	5
Inorganic chemicals	(mg/l)	(mg/l)	
Alkalinity	Narrative	Narrative	6
Ammonia	0.5 (as N)	<0.01	7
Arsenic*	0.05	Absent	8
Barium*	1.0	Absent	8
Boron*	1.0	Absent	9
Cadmium*	0.01	Absent	8
Chloride*	250	<25	8
Chromium*, Hexavalent	0.05	Absent	8
Copper*	1.0	Virtually absent	8
Dissolved oxygen	⩾ 4 (monthly mean) ⩾ 3 (ind. sample)	Near saturation	10
Fluoride*	Narrative	Narrative	11
Hardness*	Narrative	–	12
Iron (filterable)	0.3	Virtually absent	8
Lead*	0.05	Absent	8
Manganese* (filterable)	0.05	Absent	8
Nitrates plus Nitrites*	10 (as N)	Virtually absent	13
pH (range)	6.0–8.5	Narrative	14
Phosphorus*	Narrative	Narrative	15
Selenium*	0.01	Absent	8
Silver*	0.05	Absent	8
Sulfate*	250	<50	8
Total dissolved solids* (filterable residue)	500	<200	16
Uranyl ion*	5	Absent	17
Zinc*	5	Virtually absent	8
Organic chemicals	(mg/l)	(mg/l)	
Carbon chloroform extract* (CCE)	0.15	<0.04	18
Cyanide*	0.20	Absent	8
Methylene blue active substances*	0.5	Virtually absent	19
Oil and grease*	Virtually absent†	Absent	20
Pesticides			
Aldrin*	0.017	Absent	21
Chlordane*	0.003	Absent	21

TABLE 7-9–7-11. *(Continued)*

Constituent or characteristic	Permissible criteria	Desirable criteria	See paragraph in National Technical Committee Report
DDT*	0.042	Absent	21
Dieldrin*	0.017	Absent	21
Endrin*	0.001	Absent	21
Heptachlor*	0.018	Absent	21
Heptachlor epoxide*	0.018	Absent	21
Lindane	0.056	Absent	21
Methoxychlor	0.035	Absent	21
Organic phosphates plus carbamates*	0.1‡	Absent	21
Toxaphene*	0.005	Absent	8
Herbicides			
2, 4-D plus 2, 4, 5-T, plus 2, 4, 5-TP*	0.1	Absent	21
Phenols*	0.001	Absent	8
Radioactivity	(pc/l)	(pc/l)	
Gross beta*	1,000	<100	8
Radium-226*	3	<1	8
Strontium-90*	10	<2	8

*The defined treatment process has little effect on this constituent.
†See paragraph.
‡Microbiological limits are monthly arithmetic averages based upon an adequate number of samples. Total coliform limit may be relaxed if fecal coliform concentration does not exceed the specified limit.
§As parathion in cholinesterase inhibition. It may be necessary to resort to even lower concentrations for some compounds or mixtures. See paragraph 21.
NOTE–Committee could not arrive at a single numerical value for entire country. Narrative in Interim Report (1967) must be used to obtain background of existing data.

extremely resistant to the effects of pollution. This is probably due to their low respiration rate and the higher oxygen content of the water during this season. The rapid rise in water temperature in the spring brings on a critical period when fish are susceptible to unfavorable conditions, and often die. Again, in late summer and early fall, when high water temperatures are attained, the fish show reduced resistance to low oxygen and other effects of pollution. Because of this increased rate of respiration and low dissolved oxygen which occur only under these critical conditions, fish deaths in a stream or lake are usually confined to a very few days of the year when the weather is very warm and the water level low (Hubbs, 1933).

Cairns (1970) discussed various approaches that can be taken to the discharge of heated effluents: uncontrolled discharge without regard to the effects; regulating the discharge to suit the capacity of the receiving water body; finding alternative ways of dissipation or using the waste heat; and modifying the environment to suit the new temperature conditions.

Effect of Time of Day. It is not all infrequent for a stream badly polluted with sewage to show not only saturation with dissolved oxygen, but even 100-200 percent supersaturation on sunny afternoons. This same stream is very apt to show a great reduction–often complete depletion during the night, especially just prior to sunrise.

Most harmful effects of pollution on fish life pass unnoticed or are attributed to other causes than pollution when the fish are slowly killed off. The sewage may change conditions so that only the small ones may be killed, and the dead may not even be seen. Or the spawn may be prevented from hatching, or the development may be made abnormal, so that malformed fish result. Or the natural spawning beds may be covered over a deposit of septic sludge, in which the eggs cannot hatch. The pollution may kill the animal life on which the fish normally live, thus depriving them of nourishment. Fish then tend to become dwarfed in polluted waters. Certain chemicals, even in minute quantities, have been known to affect the taste of fish so as to prevent their sale. Fish unquestionably show a negative reaction toward water badly polluted with organic wastes, for they swim away from such wastes. Extremely small concentrations of certain chemicals have been found to make fish very nervous, rendering them an easy prey for other aquatic life or driving them away from the waters. Sewage conditions tend to favor the coarse fish (and to decrease the better fish) which eat eggs and clean out the shallow weed beds so that the young will be deprived of a shelter, and which compete for food.

The incidence of disease increases among fish in polluted waters. Parasitic worms produce little black cysts on the skin of the fish in polluted waters more commonly than in most unpolluted water. Another bacterial disease causing the fish to be eaten away, called saprollignia or fin and tail rot, has been observed. Gill disease in trout is due to slime bacteria.

The ratio of CO_2 to O_2 has a very definite effect on fish life: the higher the CO_2 the higher the O_2, which is necessary to supply life for fish.

Other Factors Causing Fish Deaths. *pH.* Ellis (1948) states that fish and the common aquatic organisms prefer pH values of 6.5 to 8.4; pH values below 5 or greater than 9 are definitely detrimental or even lethal. Changes in these values within a few hours, coupled with a slight increase in temperature or lowering of the dissolved oxygen, can be fatal to various warm water fishes and trout. Stiemke and Eckenfelder (1947) verified Hubb's observations when they found that fish reacted differently according to either their stage of development or the season. In the spring they were killed at pH values higher than those values producing death in the winter.

Cairns (Cairns, Dickson, and Crossman, 1972) found acute and residual toxicity to aquatic fauna following river pollution by acid and alkaline wastewaters. They describe the effects on populations of fish and benthic fauna of three accidental discharges of H_2SO_4, NaOH, and a mixture of ethyl benzene and creosote to the Clinch and Roanoke Rivers, Virginia.

At a pH of only 6.5, trout of three species (brook, brown, and rainbow) have shown a significant lessened hatching of eggs, and growth. When the pH is lowered to 5.5, bass, walleyed pike, and rainbow trout have been reported to be eliminated (Boyle, 1981) with declines in trout and salmon reported, as well. Below a pH of 5, most fish are unable to survive. This low pH causes female fish to deter laying of their eggs and, if eggs are laid, the fish are very sensitive in the egg, larval, and fish frog stages. Low pH can interfere with the salt balance freshwater species of fish need to maintain their body tissues and blood plasma. Acid ionizes or otherwise activates many metals already present, such as aluminum, which can be toxic to the fish even at pH values normally considered safe.

Acid Rain. Acidic precipitation (acid rain) is generally acknowledged by environmental scientists to be one of the most pressing subjects facing areas such as Eastern North America, Western Europe, and Scandinavia. Sulfur and nitrogen industrial and household emissions are transported by the atmosphere, transformed in chemical composition, and finally, these pollutants are deposited through either precipitation or dry deposition onto our lands and waters.

Unfortunately, once these emissions are released, their ultimate return to the earth's waters is beyond the normal control of mankind. At present, the major concern of scientists, regarding these acidic contaminants, has been their effect on the flora and fauna in the world's lakes. The most direct result is that of preventing the propagation or growth of native fish life.

The biological effects on fish of low pH waters have been described in (Table 7-11A) by the Ontario Ministry of the Environment (1980).

There has been some recent evidence that when the sulfur and nitric oxides in the deposited emissions are combined with dissolved aluminum from soils, the effects are deadly. Aluminum acid actually replaces the sulfuric and nitric acids. If these acids are neutralized by alkaline soils (calcium and magnesium), the lethal effects of acids will be deferred, at least for the present. Where alkaline soils do not exist, lake waters will turn acid sooner, with subsequent prevention of aquatic life of all kinds.

Salt concentration. Ellis (1948) states that changes in salinity as reflected by the specific conductance of a stream water have been found to be critical to some species. Eldridge (1942, p. 335) claims that fish and other aquatic life can stand only a limited concentration of the various minerals contained in brine.

TABLE 7-11A. Biological Effects on Fish of Low pH Waters.

pH	Effect
6.5 or less	Continued exposure results in significant reductions in egg hatchability and growth in brook trout.–*Menendez, 1976.*
6.0	Coupled with high CO_2 concentrations pH's below 6.0 can adversely affect certain trout species.–*Lloyd and Jordan, 1964*
5.5–6.0	Rainbow trout do not occur. Small populations of relatively few fish species found. Fathead minnow spawning reduced. Molluscs rare.–*EPA, 1972*
5.5	Declines in a salmonid fishery can be expected.–*Jensen and Snekvik, 1972*
5.0–5.5	Very restricted fish populations but not lethal unless CO_2 is high. May be lethal to eggs and larvae. Prevents spawning of fathead minnow. Lethal to some mayflies. Bacterial species diversity reduced.–*EPA, 1972, Scheider et al, 1975*
5.0	Tolerable lower limit for most fish.–*Doudoroff and Katz, 1950, McKee and Wolf, 1963*
4.5–5.0	No viable fishery can be maintained. Lethal to eggs and fry of salmonids. Benthic fauna restricted.–*EPA, 1972*
4.5	Flagfish reproduction inhibited and general activity of adults reduced.–*EPA, 1972*
4.0–4.5	Fish population limited–only a few species survive (pike). Flora restricted. –*EPA, 1972.*

–Ontario Ministry of the Environment, Extensive Monitoring of Lakes in the Greater Sudbury Area 1974–76, Ministry of the Environment (Ontario), 1978, p. 20. References shown in italic.

Concentrations up to 5,000 to 10,000 ppm do not permanently affect most types of fish unless these concentrations are maintained for a period of longer than 24 hours. Brines cause wild fowl to seek fresher water. Vegetation is soon destroyed in localities and streams into which there is a continuous discharge of brine. Anderson (1946) explains that all salts are toxic when they are present in concentrations high enough to exert unfavorable osmotic pressure.

Temperature. Dimick and Merryfield (1947) concluded that at the highest temperature recorded in their experiments (89°F) no trout, salmon, or whitefish were found, and that temperatures of this magnitude would probably be lethal to these fish. On the other hand, largemouth bass, pumpkin seed sunfish, blue gill sunfish, white crappie, bullhead, catfish, and carp did not appear to be distressed. The main effects of temperature appear to be the lowering of the dissolved oxygen level and a change in the metabolic rate (a rapid lowering in this level can be lethal).

Mackenthun (1969) presented an excellent chapter entitled "Water-Quality Constituents," in which he discussed the major reasons behind selection of various water quality limits. A section on the effects of temperature on bacteria and fish is especially revealing. Generally, he warns that when stream temperatures approach 90°F, fish food organisms can be altered rather severely. He uses the National Technical Advisory Committee on Water Quality Criteria's recommended provisional maximum temperatures for fish life and associated biota as follows:

93°F—Growth of catfish, gar, white or yellow bass, spotted bass, buffalo, carpsucker, threadfin shad, and gizzard shad

90°F—Growth of largemouth bass, drum, bluegill, and crappie

84°F—Growth of pike, pearch, walleye, smallmouth bass, and sauger

80°F—Spawning and egg development of catfish, buffalo, threadfin, and gizzard shad

75°F—Spawning and egg development of largemouth bass, white, yellow, and spotted bass

68°F—Growth or migration routes of salmonids and egg development of perch and smallmouth bass

55°F—Spawning and egg development of salmon and trout (other than lake trout)

48°F—Spawning and egg development of lake trout, walleye, northern pike, and sauger

Cairns (1968) discussed the extent and effects of cooling water discharges from electricity-generating stations. He recommended that suitable control methods, include monitoring programs.

Toxic chemicals. *Water Quality Criteria* (1952) emphasizes that the time-concentration relationship is very important in all studies of tolerances of aquatic life toward pollutants. Normally a slug of waste would be more deleterious than a steady, uniform discharge with adequate mixing. Some of the variables to be considered in setting up water criteria for fish propagation are:

1. The effects of harmful substances vary with species, size, age, and the physiological condition of the individuals.
2. The effects vary with the physical and chemical composition of the water supply. For example, in soft water, the damaging effects of poisons are generally greater than in hard water. The interrelationships between the dissolved constituents of the water supply are also extremely important. By synergistic action, the combined influence of several substances simultaneously may result in greater damage to fish life than the sum of the individual effects taken independently. On the other hand, certain combinations of compounds act to repress the injurious effects.
3. Hydrographic features of water courses and fluctuating water levels, particularly in impoundments, may also modify the effects of pollutants on fish

in their natural habitats. While probable safe limits of concentrations of various materials can serve as a helpful guide for industry, it is generally a good policy to conduct supplemental biological tests of bioassays upon the organisms involved, or upon suitable indicator organisms.

Extensive fish mortality in streams in the cotton-growing section of the country, which apparently was associated with crop dusting operations, led to a study of this matter by the Public Health Service (Dusdoroff, Katch, and Tarzwell, 1951). They drew up the following table from their results. The results of their tests with goldfish indicate that toxaphene is the most toxic of the insecticides tested. The tests of soils from tested fields indicate that stream waters can be rendered toxic to fish by drainage from such fields. Adsorption of

*Commercial mix	Fish were observed to die in 10 days	Fish were observed to turn over in 24 hours
Toxaphene dust	.025 ppm	.125 ppm
Aldrin	.8	2.00
BHC-DDT dust	3.0	25.00

*These are the mixtures commonly sold for field use as follows: Toxaphane dust (20 percent toxaphene), Aldricide (2.5 percent Alrin equivalent), BHC-DDT dust (13 percent BHC-DDT).

the insecticide on the soil apparently cannot always either prevent its being washed down into streams by rainwater, render it innocuous, or ensure its effective removal from the water through sedimentation of the coarser, settleable solids.

Cairns and Sparks (1971) showed that the presence of zinc in concentrations ranging from 2.55 to 8.7 ppm in dechlorinated tap water could be detected by an increase in breathing rate or a change in breathing rate variance of bluegills. Trace amounts of zinc did not affect reproduction (.025-.075 ppm), but 0.25 ppm inhibited spawning and killed newly hatched fry. They recommend a system for monitoring fish response to water quality.

Cairns, Cromer, Bahns, and Waller (1971) showed that statistical analysis of results obtained for fish exposed to but not killed by zinc, fish killed by zinc, and fish not exposed to zinc, confirmed the usefulness of the Mounts Autopay technique.

Cairns, Hokanson, and Smith (1971) found that dissolved oxygen, water hardness, and acclimation to linear alkylate sulphurate (LAS) were the most important factors affecting toxicity of LAS to early stages of bluegill. A concentration of 1 ppm of LAS is thought to be a fairly safe level for bluegills in good quality Mississippi River water.

Cairns and Dickson (1970) found that with exposures of up to 24 ppm copper, the number of species in Douglas Lake, Michigan, was reduced from 46 to 7, after 24 hours, and subsequently increased to 14 species after 144 hours. Exposure to 24 ppm zinc reduced the number of species after 144 hours. Exposure to 24 ppm zinc reduced the number of species from 35 to 11 after 24 hours, followed by a rapid recovery to 20 species after 48 hours and to 34 after 120 hours. They concluded that copper has a more lasting effect than zinc pollution on freshwater protozoa.

Ciguatera poisoning occurs in humans following the eating of fish with sufficient concentration of toxic chemical, especially in the fish liver. The poisoning causes severe abdominal cramps followed by nausea, vomiting, diarrhea, headache, and difficulty in swallowing. In servere cases, there are sharp and shooting leg pains and muscle spasms. It is believed that the poisoning starts with a fish eating a toxic food, probably algae. Larger fish accumulate the toxin as they feed on smaller fish that have eaten the algae.

Shellfish culture. The requirements for good yields of marine shellfish include favorable bottom, food, and overlying water conditions. The bottom must be firm enough to give support to the young and keep the adults from sinking, and must be about six feet from the surface. Water containing an abundance of microscopic vegetation and a minimum of suspended solids is best. Characteristics of the water that significantly affect shellfish production include: (1) salinity, (2) pH, (3) oxygen concentration, (4) copper concentration, and the presence of substances deleterious to shellfish development or hazardous to shellfish consumers. Limits of some of these areas are:

1. Salinity: 7,000-35,000 ppm is necessary.
2. pH: neutral and slightly alkaline is desirable.
3. DO: variable amounts required by different species but important to be present at all times.
4. Copper: optimum .05 ppm; > 1 ppm is toxic BOD quality–a limiting mpn of 70 coliform per 100 ml is placed upon water for oyster production that will receive no Cl_2 and will be eaten raw.

Shellfish growing areas. Coliform limit for these areas is about 70 mpn/100 ml. The state supervises all operations, which include the growing, harvesting, and transporting of oysters, clams, or mussels that are shipped interstate. The Public Health Service oversees the state's program. The major problems with disease of shellfish are:

1. Infectious hepatitis
2. Paralytic poisoning
3. Insecticide poisoning (more recently)

Coliforms are a poor measure of these three hazards, but no others have been developed. Viruses such as hepatitis are probably more resistant to death than coliform, especially in salt water. Paralytic poisons are accumulated by some shellfish from certain algae used as food by the fish. Poison concentrations of 80 micrograms per 100 grams of edible portions of raw shellfish appear to be the toxic limit.

A study of sources of pollution, adequately of waste treatment, and tidal and other currents will usually indicate, when coupled with a bacterial analysis, whether the shellfish are safe for human consumption. Growing areas are usually classed as either

1. Approved: < 70 coliforms/100 ml, median
2. Conditionally approved: < 70 coliforms/100 ml, but depends upon season or operation of existing plants
3. Restricted: > 70 and < 700 coliforms/100 ml } median
4. Prohibited: > 700 coliforms/100 ml } median

The Committee (Interim Report, 1967), under the Subcommittee chairmanship of Clarence Tarzwell, recommended water quality criteria for (a) freshwater organisms, (b) marine and estuarine organisms, and (c) wildlife. The Subcommittee findings are presented in tabular form for easier reference in Table 7-12.

Among other interesting and rather detailed literature reviews of contaminants that have been found to affect water quality when related to biological stream life, Mackenthun (1969) includes one of siltation. He reports some findings of Bennett, Thompson, and Parr (1940) in which turbidity was found to reduce the the enjoyment of fishing and limit fishing success; fewer fish were caught per man hour when the water transparency was decreased. He also records an anonymous report of the European Inland Fisheries Advisory Commission (FAO) in 1965, which stated generally that no evidence exists that suspended solids concentrations of less than 25 ppm have any harmful effects on fisheries, whereas waters containing over 80 ppm are unlikely to support good freshwater fisheries. Although sometimes the solids have a direct effect on the fish themselves, the major effect is an indirect one of either reducing fish food or harming spawning grounds.

Agriculture

Since most water used by agriculture is used for crop irrigation, in this section we will concern ourselves primarily with irrigation water quality requirements. Irrigation water is usually judged by three criteria:

1. Boron atom concentration: Compounds of boron are toxic to plants in extremely minute amounts.

TABLE 7-12. Water Quality Recommended by National Technical Advisory Committee for Fish, Other Aquatic Life, and Wildlife. [*Interim Report, 1967. By permission of publishers.*]

	Dissolved materials	pH, alkalinity, acidity	Temperature	Oxygen	Oil	Turbidity	Settleable solids
Freshwater organisms	Increase above natural concentration limited to 33%; max. concentration is 50 million mhos	6–9 Alkalinity > 20 mg/l; bioassays used to determine max. acids and alkalhes	*Warm waters* Max. Rise of 5°F of monthly max. avg. temperature peak; 83°F–86°F in smaller streams to 90°F–96°F in large streams *Cold waters* No heated wastes in spawning areas	*Warm water biota* > 5 ppm 4–5 ppm under extreme cases *Cold water biota* At or near saturation > 7 ppm in spawning areas; 4–6 ppm in extreme cases	No oily film 25 mg/l No oily odor No coated banks or bottoms	< 50 Jackson units in warm waters and < 10 units in cold water < 25 units in warm water lakes and < 10 units in cold water or oli gatrophic lakes	None so as to affect natural biota
Wildlife		7.0–9.2 Alkalinity 35–200 mg/l		Aerobic bottom areas to prevent botulinus organisms	Oil and petrochemical must be excluded	10% of incident light of surface should reach 6-foot depth	None
Marine and estuarine organisms		6.5–8.5 No additions such as to cause a greater pH change than 0.1 unit	Max 4°F rise Dec.–Feb., Mar.–May, and Sept.–Nov. and max. 1.5°F during summer June–Aug. < 1°F per hour and max. 24 hr. Change of 7–9°F	Surface coastal water > 5 mg/l Estuaries & tidal trib. > 4 mg/l	No detectable film or sheen no tainting of fish No oil sludge deposits	No change in turbidities or color unless it can be shown to be non-detrimental to aquatic life	None which may either precipitate or settle out so as to affect these biota adversely

Freshwater organisms	None so as to impart ordor or taste to fish	None so as to restrict use of harvested organisms as per Rad. Prot. Grades of Fed. Rad. Council or in drinking water as per USPHS Pub. No. 956 revised 1962	Controlled so as to prevent undesirable aquatic vegetation and nuisance organisms. None so as to decrease O_2 in hypolimnion or to increase it in zone of anaerobic decomposition	None such as shown by bioassay tests to be harmful	No chlorinated hydrocarbons None so as to contravene known toxicity as shown in reference cited.	*ABS* < 1/7 48 hr. TL_m < 1 mg/l *LAS* < .2 mg/l or < 1/7 48 TL_m
Wildlife				Excluded from wildlife habitat to the degree that they affect the health and well-being of wildlife		Efforts to be made to find a substitute for lead shot
Marine and estuarine organisms	None so as to produce off-flavors in fish	Same as for fresh-water organisms	Naturally occurring ratio of NO_3-N to PO_4-P should be maintained. No artificial enrichment so as to cause any major quantitative change or qualitative alteration of the flora	All foreign matter considered harmful unless bioassays show otherwise	A maximum of 50 parts per trillion for many organochlorides and phosphates and 10 parts per billion of a group of 8 other pesticides	Based on 96 hr. TL_m values of local typical organisms. Application factors of 1/100 for metals, 1/20 for NH_3, 1/10 for Cn, and 1/20 for sulfide

2. Ratio of Na^+ to $Na^+ + Mg^{++} + Ca^{++} + K^+$: Too much Na^+ makes the soil sticky when wet and forms clods when dry. These effects are counteracted by Ca^{++}, Mg^{++}, and K^+.

3. Total dissolved chemical content: A large fraction of the water poured on soil evaporates and deposits its salts into the ground. In regions of low rainfall these salts build up to a very high level. Owing to osmotic pressure, the plants are not able to absorb water through their roots, and in fact are often dehydrated or undergo plasmolysis by this same effect. (Water passes from plants to equalize salt concentration.) Plants, therefore, wilt from lack of water even though the ground is wet. The U.S. Department of Agriculture Regional Salinity Laboratory (1946) estimated that half of the irrigated land in the United States has had its productivity impaired or threatened by the accumulation of salts.

Boron. Boron occurs naturally in most rivers. It can also be found in fertilizers and certain wastewaters, such as those from citrus washing. Although traces of boron are essential for all plant growth, it is doubtful whether more than 0.5 ppm can be applied continuously to soils without ultimately producing some plant injury.

Na^+. In most normal soils or in semiarid regions, Ca^{++} and Mg^{++} are the principal cations held by the soil in replaceable form, with Na consisting of a small percentage, that is, about 3-7 percent. Such soils, when not misused, represent a favorable physical condition for root and water penetation. An increase of the exchangeable sodium to as much as 12 or 15 percent causes the granular soil to begin to break down when the soil is moistened. It has been widely recommended that the percentage of Na

$$\frac{\text{Na}}{\text{Na} + \text{Ca} + \text{Mg} + \text{K}} \times 100$$

in irrigation water should not exceed 50-60. Where the soil has a high cation

Permissible Limits for Concentration of Boron in Several Classes of Water for Irrigation

Class of water	Concentration of Boron in ppm for crops that are: *Sensitive*	*Semi-tolerant*	*Tolerant*
Excellent	<.33	<.67	<1.0
Good	.33-0.67	.67-1.33	1.0-2
Permissible	.67-1.0	1.33-2.0	2.0-3
Doubtful	1.00-1.25	2.00-2.5	3.0-3.75
Unsuitable	>1.25	>2.5	>3.75

exchange capacity, and where the irrigation water is very dilute, values above 50 may be within safe limits.

Salts. According to authorities in western Australia, water containing up to 1,000 ppm of salinity (dissolved solids) is suitable for growing all types of plants, including the salt-susceptible ones, provided the drainage is good. Water containing up to 2,150 ppm is suitable for most plants except sensitive ones, and water containing up to 3,150 ppm has been used for growing tomatoes, cabbages, and other salt-resistant plants. Generally, 3,150 ppm is near the maximum for the safe watering of any plant; in such instances drainage must be excellent, and each watering should permit leeching from the root zone.

Limits of toxicity. Absolute limits to the permissible concentration of salts in irrigation waters cannot be fixed for several reasons: (a) It is almost universally true that the soil solution is at least 3–8 times as concentrated as the water that replenishes it because of the evaportion of water from the soil surface, transpiration of plants, and the selective absorption of salts by the plants. (b) There is apparently no definite relationship between the concentration and composition of the irrigation water and those of the soil solution, which in some cases may be as much as 100 times more concentrated than the water. (c) Plants vary widely in their tolerance of salinity as well as of specific salt constituents. (d) Soil types, climatic conditions (such as temperature, rainfall, and humidity), and irrigation practices may all influence the reactions of the crop to the salt constituents. (e) Interrelationships between and among constituents may be highly significant; the effect of one ion may be modified by the presence of another. (Such antagonistic influences operate between calcium and sodium, boron and nitrate, selenium and sulfate) (Ehlers, 1935). *The following represent some of the effects of salts on plants and the soil:*

1. Direct effects of salts in preventing water uptake by plants (osmotic effects).
2. Direct metabolic effects upon metabolic reactions of plants.
3. Indirect effects through changes in soil structures, permeability, and aeration.

The utilization of sewage effluent for crop irrigation and fertilization was studied over a three-year period by means of lysimeter (soil prototypes) and field irrigation (Henry, Moldenhauer, Englebert, and Truog, 1954). The purpose of the study was to determine how much effluent could be disposed of through irrigation; the benefits to the crop of both the additional water and fertilizer supplied by the wastewaters; and the effect of irrigation with effluent on the chemical content and microbiological population of the percolating waters. The authors found that: (1) crop and soil are effective in removing virtually all of the nitrogen, phosphorus, and potassium from the percolating waters. When there is a high concentration of sodium in the effluent the losses of calcium and magnesium to the percolate are increased. Sodium causes soil particles to become dispersed and allows the calcium and magnesium to percolate through. If too much calcium and magnesium are lost and if the sodium continues to accumulate in the soil, the physical condition of the soil will be harmed; (2) the crop yields are substantially increased by plant nutrients in the effluent. The increases amounted to one ton per acre on Miami salt loam (a mineral soil) and to about two tons per acre on peat (a soil high in organic matter); (3) drainage waters from the soil irrigated with sewage effluent did not increase the coliform index of the nearby creek. However, the chloride, content, and to a lesser extent the sodium content of the creek waters, was increased. The effectiveness of the soil in reducing the coliform numbers of percolating waters was substantiated by analysis of the groundwater. The coliform index, except in one instance, never exceeded 100, in contrast to an index of about 10,000,000 in the effluent applied to the soil. Toxic metals contained in sewage effluents possessing industrial wastes as well will accumulate in the soil and plants, and affect men and animals that eat these plants. It is apparent that soil of certain types will effectively retain and remove coliform-type bacteria from unchlorinated sewage effluents applied to them.

Effect of salt on oxygen solubility. Based on the assumption that the solubility of oxygen changes linearly with salinity, the solubility in saline waters can be expressed by the equation

$$C_s = 14.161 - .3943T + .007714T^2 - .0000646T^3 - S(.0841 - .00256T + .0000374T^2)$$

where C_s is solubility (ppm) of oxygen in water where salinity is 10 × % salt, or $S\%$ and $T = {}^\circ\text{C}$. Less accurate but with 0.1 ppm is the following formula:

$$C_s = \frac{47.5 - 2.65S}{33.5 + T} \quad \text{(after Truesdale and Knowles, 1956)}$$

The 1967 Department of Interior Interim Report separates agriculture uses into (a) farmstead water supplies, (b) livestock, and (c) irrigation. For farmstead uses any water that meets the U.S. Public Health Service drinking water standards (1962) will be "generally safe and acceptable to the user." Because of many important variables, including nature and intake of dietary dry matter, species, age, and productivity, the establishment of a single set of water-purity values for livestock is unfeasible. Table 7-13 does give some major guidelines to acceptable livestock water quality. Variations and interactions of soils, plants, water, and climate also preclude the establishment of a single set of criteria to

TABLE 7-13. Key Water Criteria for Livestock Use

Characteristic	Recommendations
Total dissolved solids (TDS)	< 10,000 mg/l
Hazardous trace elements	
Arsenic	< 0.05 mg/l
Cadmium	< 0.01 mg/l
Chromium	< 0.05 mg/l
Fluorine	< 2.40 mg/l
Lead	< 0.05 mg/l
Selenium	< 0.01 mg/l
Organic substances	
Algae ("water bloom")	Avoid abnormally heavy growth of blue-green algae
Parasites and pathogens	Conform to epidemiological evidence
Dissolved organic compounds	Biological accumulation from environmental sources, including water, shall not exceed established, legal limits in livestock products.
Radionuclides	Conform to recommendations for farmstead water supplies.

evaluate all water quality characteristics for irrigation purposes. The Interim Report, however, summarized suggested guidelines for major criteria. The writer has taken the liberty of extracting these and consolidating them in a convenient table (Table 7-13). Since irrigation water is recognized as the largest single-purpose beneficial consumptive use, water quality criteria for its use become more and more important (see Table 7-14). For example, in 1966 a total of 458 million acres of cropland existed in the United States (Interim Report, 1967), of which about 44 million acres were irrigated (about 10 percent of the total cropland). Two other major sources of information about irrigation waters are given by Ellis (1954) and by the state of California (1952).

Industrial Water Quality

In 1964 the total water intake of both industrial manufacturing plants and investor-owned thermal electric utilities was about 48,900 billion gallons, with over 90 percent being used for either cooling or condensing purposes. A rather complete breakdown of the water used by industry for the major needs is shown in Table 7-15. For our purpose in this section, the significance of this table is the

TABLE 7-14. Guidelines for Water Quality Characteristics for Irrigation Purposes.
[Derived from interim Report, 1967. By permission of publishers.]

Characteristics	Salinity: Total dissolved solids (mg/l)	Salinity: Conductivity (mmhos/cm)	Sodium (sodium absorption ratio)*	Chlorides (me/l)	Trace elements (in ppm): Coarse soils	Trace elements (in ppm): Firm soils	Radionuclides (pc/l)	Microorganisms no's/100 ml	pH	Temperature (°F)	Pesticides† mg/l
No detrimental effects usually noted	< 500	> 0.75	< 4	Maximum level of 1–20 for chloride sensitive crops	0.75 Boron	2.0	3 Sr-90	< 5,000 on a monthly average Total coliforms	4.5–9.0	55–85	2,4D
					1.0 Arsenic	10.0					
					5.0 Chromium	20.0					
Can have detrimental effects on sensitive crops	500–1,000	0.75–1.50			0.2 Copper	5.0	10 Ra-226	< 1,000 on a monthly average fecal coliforms			3.5–10 field beans
					1.0 Aluminum	20.0					0.7–1.5 grapes
					Iron No problem	No problem	1,000 Gross beta activity	< 20,000 on any one sample total coliforms			3.5 sugar beets
											Dimethylamines
May have adverse effects on many crops	1,000–2,000	1.50–3.00			5.0 Zinc	10.0		< 4,000 on any one sample fecal coliforms			< 25 corn
											< 25 soybean
											25 sugarbeets
Used only on salt-tolerant plants on permeable soils with careful management practices	2,000–5,000	3.00–7.50		Salinity itself usually controls for chloride tolerant crops							

*Sodium absorption ratio = $SAR = \frac{Na^{+}}{\sqrt{\frac{Ca^{++} + Mg^{++}}{2}}}$ expressed at Me/l.

†See Table 3 in Interim Report (1967) for more trade pesticides and concentrations.

TABLE 7-15. Industrial and Investor-Owned Thermal Electric Plant Water Intake, Reuse, and Consumption, 1964.

SIC	Industrial group	Water intake (billion gallons per year [bgy])				Water recycled (bgy)	Gross water use, incl. recycling (bgy)	Water consumed (bgy)	Water discharged (bgy)
		Purpose							
		Cooling & condensing	Boiler Feed san. serv., etc.	Process	Total				
20	Food and kindred products	392	104	264	760	520	1,280	72	688
22	Textile mill products	24	17	106	147	163	310	13	134
24	Lumber and wood products	71	24	56	151	66	217	28	123
26	Paper and allied products	607	120	1,344	2,071	3,945	6,016	129	1,942
28	Chemicals and allied products	3,120	202	564	3,886	3,688	7,574	227	3,659
29	Petroleum and coal products	1,212	99	88	1,399	4,763	6,162	81	1,318
31	Leather and leather products	1	1	14	16	2	18	1	15
33	Primary metal industry	3.387	195	996	4,578	2,200	6,778	266	4,312
	Subtotal	8,814	762	3,432	13,008	15,347	28,355	817	12,191
	Other industries	571	197	271	1,039	1.207	2,246	71	968
	Total industry	9,385	959	3,703	14,047	16,554	30,601	888	13,159
	Thermal electric plants	34,849	*	–	34,849	5,815	40,664	68	34,781
	Total	44,234	959†	3,703	48,896	22,369	71,265	956	44,940

*Boiler-feed water use by thermal electric plants estimated to be equivalent to industrial sanitary service, etc., water use.

†Total boiler-feed water.

Source: 1963, Census of Manufacturers, Water Use in Manufacturing (U.S. Dept. Commerce, 1966) and Water Resources Activities in the United States–Electric Power in Relation to the Nation's Water Resources (U.S. Senate, 1960)

great variety of uses for water, and the diverse assortment of industries that use it. It is also an established fact that raw water of almost any existing quality can be utilized with treatment at certain costs by any industry. This fact has made pollution abatement difficult. The National Technical Advisory Committee for Water Quality Requirements for Industrial Water Supplies (Interim Report, 1967) reached the following conclusions regarding industrial water quality and requirements.

1. The quality characteristics of the water supply for an established industry at a given site, if allowed to deteriorate from the range usually experienced for those characteristics of significance to that industry, can cause an undesirable increase in the cost of treatment. To the contrary, an improvement in the quality of the same supply will not significantly decrease the cost of treatment at an existing installation.

2. Marked variations in the quality of an industrial water supply that continually require adjustment of treatment are undesirable, as they increase the cost of treatment and the cost of the facilities required, and may result in deterioration of product quality for some industries.

3. The water quality requirements at the point of use in each process in each industry, as distinguished from the quality characteristics at the point of supply, are generally well established for each existing industrial process use. These water quality requirements, however, vary considerably even for the same process, depending upon the technological age of the design, and other factors.

4. The quantity of water employed for process use by different plants in the same industry may vary considerably between plants, depending on the cost of treatment, the age of the plant design, operating practices, and the quality and quantity of the available supply.

5. Studies should be undertaken to assemble information on the costs of altering water quality characteristics so that the future needs of industry can be assessed accurately both as to the quality and quantity requirements.

All water supplies for industrial use should, in general, meet the following basic three minimum standards for all uses:

1. Free from substances attributable to municipal, industrial, or other discharges, or agricultural practices that will settle to form putrescent or otherwise objectionable sludge deposits;

2. Free from floating debris, oil, scum, and other floating materials attributable to municipal, industrial, or other discharges, or agricultural practices in amounts sufficient to be unsightly or deleterious;

3. Free from materials attributable to municipal, industrial, or other discharges, or agricultural practices producing color, odor, or other conditions in such a degree as to cause a nuisance.

In addition, specific industries require specific qualities for certain plant production uses. These—as pointed out previously—are extremely variable. However, Table 7-16 summarizes acceptable qualities for many industries. It can be used as a guide to reasonable industrial water quality.

7-4. PRIORITY POLLUTANTS

Water Quality Criteria

In addition to the promulgation of effluent limitations, considering each of the 65 pollutants whenever they occur in the effluents from the 21 designated industrial categories, stream water quality criteria were required to be developed for each of the pollutants. Such criteria are guidelines that specify the maximum permissible pollutant concentrations consistent with the protection of aquatic organisms, human health, and recreational activities. As of October 1, 1979, a criterion document for each of the 65 pollutants (see Table 7-16A) was made available by the EPA for public comment. When those documents were published complete in 1980, they contained, wherever data were available, the Federally-recommended pollutant concentrations in ambient water that should not be exceeded to maintain water quality associated with fishing and other recreation. (See Table 7-16B.)

The water quality criteria themselves may not affect all managers of environmental control programs, but the expected use to which the criteria will be put has an impact. On July 10, 1978, EPA suggested in the Federal Register that it expected to require use of some of the criteria (guidelines only) by states in setting water quality standards (having the force and effect of law). When established in water quality standards, such criteria for toxic pollutants could be used through the process of wasteload allocation to require more stringent treatment than that required through the effluent guidelines described earlier.

It is recognized that there are now many unresolved issues associated with the development of both effluent limitations and water quality criteria. The prudent managers of environmental control programs should, however, plan now for what seems to be a certainty of the future—that effluent limitations and ambient water quality standards will exist for many toxic pollutants.

On November 28, 1980, the EPA published in the *Federal Regiser Part V* water quality criteria for 64 pollutants listed in Table 7-16A. Recommended limits are given for three uses: (a) freshwater aquatic life, (2) saltwater aquatic life, and (3) human health. Your author has extracted data from the *Register* material to form the basis of Table 7-16B.

TABLE 7-16. Summary of Specific Quality Characteristics of Raw Waters That Have Been Used as Sources for Industrial Water Supplies (Unless otherwise, indicated, units are mg/l and values are maximums. No one water will have all the maximum values shown.). [*Ater Interim Report, 1967. By permission of publishers.*]

Characteristic	Boiler make-up water		Cooling water			
			Fresh		Brackish*	
	Industrial 0–1,500 psig	Utility 700–5,000 psig	Once-through	Make-up recycle	Once-through	Make-up recycle
Silica (SiO_2)	150	150	50	150	25	25
Aluminum (Al)	3	3	3	3		
Iron (Fe)	80	80	14	80	1.0	1.0
Manganese (Mn)	10	10	2.5	10	0.02	0.02
Copper (Cu)						
Calcium (Ca)			500	500	1,200	1,200
Magnesium (Mg)						
Sodium & potassium (Na + K)						
Ammonia (NH_3)						
Bicarbonate (HCO_3)	600	600	600	600	180	180
Sulfate (SO_4)	1,400	1,400	680	680	2,700	2,700
Chloride (Cl)	19,000	19,000	600	500	22,000	22,000
Fluoride (F)						
Nitrate (NO_3)			30	30		
Phosphate (PO_4)		50	4	4	5	5
Dissolved solids	35,000	35,000	1,000	1,000	35,000	35,000
Suspended solids	15,000	15,000	5,000	15,000	250	250
Hardness ($CaCO_3$)	5,000	5,000	850	850	7,000	7,000
Alkalinity ($CaCO_3$)	500	500	500	500	150	150
Acidity ($CaCO_3$)	1,000	1,000	0	200	0	0
pH, units			5.0–8.9	3.5–9.1	5.0–8.4	5.0–8.4
Color, units	1,200	1,200		1,200		
Organics						
Methylene blue active substances	2§	10	1.3	1.3		1.3
Carbon tetrachloride extract	100	100	No floating oil	100	No floating oil	100
Chemical oxygen demand	100	500		100		200
Hydrogen sulfide (H_2S)					4	4
Temperature, °F	120	120	100	120	100	120

NOTE–Application of the above values should be based on analytical methods in ASTM Manual on Industrial Water and Industrial Waste Water (2) or APHA Standard Methods for the Examination of Water and Waste Water (1).

*Water containing in excess of 1,000 mg/l dissolved solids.

†May be ≤ 1,000 for mechanical pulping operations.

‡No large particles ≤ 3 mm diameter.

§One mg/l for pressures up to 700 psig.

Process water							
Textile industry SIC-22	Lumber industry SIC-24	Pulp & paper industry SIC-26	Chemical industry SIC-28	Petroleum industry SIC-29	Prim. metals industry SIC-33	Food & kindred products SIC-20	Leather industry SIC-31
		50		50		In general, the quality of raw surface supply should be that prescribed by the NTA Committee on water quality requirements for public water supplies.	
0.3		2.6	5	15			
1.0			2				
0.5							
			200	220			
			100	85			
				230			
			600	480			
			850	570			
		200†	500	1,600	500		
				1.2			
				8			
150		1,080	2,500	3,500	1,500		
1,000	‡		10,000	5,000	3,000		
120		475	1,000	900	1,000		
			500		200		
					75		
6.0–8.0	5–9	4.6–9.4	5.5–9.0	6.0–9.0	3–9		
		360	500	25			
					30		
		95			100		

TABLE 7-16A. The Toxic Pollutant List

1. Acenaphthene
2. Acrolein
3. Acrylonitrile
4. Aldrin/Dieldrin
5. Antimony and compounds
6. Arsenic and compounds
7. Asbestos
8. Benzene
9. Benzidine
10. Beryllium and compounds
11. Cadmium and compounds
12. Carbon tetrachloride
13. Chlordane (technical mixture and metabolites)
14. Chlorinated benzenes (other than dichlorobenzenes)
15. Chlorinated ethanes (including 1,2-dichloroethane, 1,1,1-trichloroethane, and hexachloroethane)
16. Chloroalkyl ethers (chloromethyl, chloroethyl, and mixed ethers)
17. Chlorinated naphthalene
18. Chlorinated phenols (other than those listed elsewhere, includes trichlorophenols and chlorinated cresols)
19. Chloroform
20. 2 Chlorophenol
21. Chromium and compounds
22. Copper and compounds
23. Cyanides
24. DDT and metabolites
25. Dichlorobenzenes (1,2-, 1,3-, and 1,4-dichlorobenzenes)
26. Dichlorobenzidine
27. Dichloroethylenes (1,1- and 1,2-dichloroethylene)
28. 2,4 Dichlorophenol
29. Dichloropropane and dichloropropene
30. 2,4 Dimethylphenol
31. Dinitrotoluene
32. Diphenylhydrazine
33. Endosulfan and metabolites
34. Endrin and metabolites
35. Ethylbenzene
36. Fluoranthene
37. Haloethers (other than those listed elsewhere, includes chlorophenyl-phenyl esters, bromophenyl, phenyl ether, bis(dichloroisopropyl) ether, bis(chloroethoxy)methane, and polychlorinated diphenyl ethers)
38. Halomethanes (other than those listed elsewhere, includes methylene chloride, methyl-chloride, methyl-bromide, bromoform, dichloro-bromomethane, trichlorofluoro-methane, dichlorodifluoromethane)
39. Heptachlor and metabolites
40. Hexachlorobutadiene
41. Hexachlorocychlorhexane (all isomers)
42. Hexachlorocyclopentadiene
43. Isophorone
44. Lead and compounds
45. Mercury and compounds
46. Naphthalene
47. Nickel and compounds
48. Nitrobenzene
49. Nitrophenols (including 2,4-dinitro-phenol, dinitrocresol)
50. Nitrosamines
51. Pentachlorophenol
52. Phenol
53. Phthalate esters
54. Polychlorinated biphenyls (PCB)
55. Polynulcear aromatic hydrocarbons (including benzanthracenes, benzo-pyrenes, benzofluoranthene, chrysenes, dibenzanthracenes, and indenopyrenes)
56. Selenium and compounds
57. Silver and compounds
58. 2,3,7,8-Tetrachlorodibenzo-p-dioxin (TCDD)
59. Tetrachloroethylene
60. Thallium and compounds
61. Toluene
62. Toxaphene
63. Trichloroethylene
64. Vinyl chloride
65. Zinc and compounds

TABLE 7-16B. Water Quality Limits for Toxic Pollutants for Three Uses

Toxic chemical	Concentration of toxic material considered limit (ug/ℓ)				
	Freshwater aquatic life		*Saltwater aquatic life*		
	Acute	*Chronic*	*Acute*	*Chronic*	*Human health*
1. Acenaphthene	1,700	–	970	710	20 (est.)
2. Acrolein	68	21	55	–	320
3. Acrylonitrile	7,550	–	not available	–	058–.006 lifetime
4. Aldrin-Dieldrin	.0019 (24 hr avg.)	–	0.0019 (24 hr avg.)		.0071 ng/ℓ–.71 ng/ℓ
	2.5 maximum conc.	–	0.71 (max. conc.)	–	
4A. Aldrin	3.0	–	1.3	–	.0074 ng/ℓ–.74 ng/ℓ
5. Antimony	9,000	1,600	not available	–	146
6. Arsenic	440	40	508	–	.22 ng/ℓ–22 ng/ℓ
7. Asbestos	not available	–	not available	–	3,000–300,000 fibers/ℓ
8. Benzene	5,300	–	5,100	700	.066–6.6
9. Benzidine	2,500	–	not available	–	.01 ng/ℓ–1.2 ng/ℓ
10. Beryllium	130	5.3	not available	–	.37 ng/ℓ–37 ng/ℓ
11. Cadmium	3.0 (100 ppm hardness) max.		59 (maximum)	–	10
	.025 (100 ppm hardness) avg		4.5 (avg.)	–	
12. Carbon tetrachloride	35,200	–	500,000	–	.04–4.0
13. Chlorodane	2.4 max	–	.09 max	–	.046–4.6
	0.0043 (24 hr. avg.)	–	.0040 (24 hr. avg.)		
14. Chlorinated benzenes	250	50 (fish 7.5 days)	160	129	hexachlorobenzene .072 ng/ℓ–7.2 ng/ℓ tetrachlorobenzene 38 ug/ℓ–48 pentachlorobenzene 74–85 monochlorobenzene 488

TABLE 7-16B. *(Continued)*

Toxic chemical	*Concentration of toxic material considered limit (ug/ℓ)*				
	Freshwater aquatic life		*Saltwater aquatic life*		
	Acute	*Chronic*	*Acute*	*Chronic*	*Human health*
	Acute		*Acute*		
15. Chlorinated Ethanes	118,000 (1,2 dichloroethane)		113,000		.094–9.4 (1,2 dichloro-ethane)
	18,000 (two trichloroethane)		31,200		18.4 mg/ℓ–1.03 g/ℓ
	9,320 (two tetrachloroethanes)		9,020		.017–1.7 ug/ℓ
	7,240 (pentachloroethane)		390		–
	980 (hexachloroethane)		940		.19–19 ug/ℓ
	Chronic		*Chronic*		
	20,000		–		
	9,400		–		
	2,400 Same as above		–		
	1,100		261		
	540		–		
	Acute	*Chronic*	*Acute*	*Chronic*	
16. Chlorinated Napthalenes	1,600	–	7.5	–	not available
17. Chlorinated Phenols	30 to 500,000	970	440 to 29,000	–	0.1 (3 monochlorophenol) 0.1 (4 monochlorophenol) 0.04 (2,3 dichlorophenol) 0.5 (2,5 dichlorophenol) 0.2 (2,6 dichlorophenol) 0.3 (3,4 dichlorophenol)

					1.0 (2,3,4,6 tetrachloro-phenol) 2.6 mg/ℓ (2,4,5 trichloro-phenol) .12–12 ug/ℓ (2,4,6 trichlorophenol) 1,800 ug/ℓ (2 methyl 4 chlorophenol) 3,000 (3 methyl, 4 chlorophenol) 20 (3 methyl, 6 chloro-phenol)
18. Chloroalkyl Ethers	238,000	–	not available	–	.00038 ng/ℓ to .038 ng/ℓ (for bischloro-methyl ether) .003 ug/ℓ–.3 ug/ℓ (for bis 3 chloroethyl ether) 34.7 ug/ℓ (for bis-2-chloro-isopropyl ether)
19. Chloroform	28,900	1,240	not available	–	.019–1.9 ug/ℓ
20. 2 Chlorophenol	4,380	2,000 (one fish species)	not available	–	.01 ug/ℓ
21. Chromium	21 (max) .29 (avg. 24 hr.)	hexavalent	1260 (max) Cr^{vi}	–	170 mg/ℓ Cr^{iii}
	4700 ug/ℓ (100 ppm trivalent hardness) 44 ug/ℓ chronic toxicity		18 (24 hr. avg.) 10,300 (Cr^{+3}) chronic toxicity)	–	50 ug/ℓ (Cr^{vi})
22. Copper	5.6 (24 hr. avg.) 22 ug/ℓ (100 ppm hardness) max	–	4.0 (24 hr. avg.) 23 (max)	– –	1 mg/ℓ (for taste and odor) none other available
23. Cyanide	3.5 (24 hr. avg.) 52 (max.)	–	2–30 ug/ℓ	–	200 ug/ℓ
24. DDT and Metabolites	1.1 max .001 (24 hr. avg.) DDT 0.6 acute toxicity TDE 1050 acute toxicity DDE	–	0.13 (max DDT) .001 (24 hr. avg.) 3.6 acute toxicity TDE 14 acute toxicity DDE	– –	.0024 ng/ℓ to .24 ng/ℓ for DDT

TABLE 7-16B. *(Continued)*

Toxic chemical	*Concentration of toxic material considered limit (ug/ℓ)*				
	Freshwater aquatic life		*Saltwater aquatic life*		
	Acute	*Chronic*	*Acute*	*Chronic*	*Human health*
25. Dichlorobenzenes	1,120	763	1970	–	400
26. Dichlorobenzidines	not available	–	not available	–	.00103–.103
27. Dichloroethylenes	11,600	–	224,000	–	.0033–.33
28. 2,4 Dichlorophenol	2,020	365	not available	–	.3 ug/ℓ (for taste and odor)
					3.09 mg/ℓ (toxicity)
29. Dichloropropanes	23,000	5,700	10,300	3,040	87 ug/ℓ
Dichloropropenes					
30. 2,4 Dimethylphenol	2,120	–	not available	–	400 ug/ℓ (taste and odor)
31. 2,4 Dinitrotoluene	330	220	590	370	.001–1.1 ug/ℓ
32. 1,2-Diphenylhydrazine	270	–	not available	–	4–422 ng/ℓ
33. Endosulfan	.056 (24 hr. avg.)	–	.0087 (24 hr. avg.)	–	74 ug/ℓ
	.22 (maximum)	–	.034 (maximum)	–	
34. Endrin	.0023 (24 hr. avg.)	–	.0023 (24 hr. avg.)	–	1 ug/ℓ
	.18 (maximum)	–	.037 (maximum)	–	
35. Ethylbenzene	32,000	–	430	–	1.4 mg/ℓ
36. Fluoranthene	3,980	–	40	18	42 ug/ℓ
37. Haloethers	360	122	not available	–	not available
38. Halomethanes	11,000	–	12,000	6,400	.019–1.9 ug/ℓ
39. Heptachlor	.0038 (24 hr. avg.)	–	.0036 (24 hr. avg.)		.028–2.78 ng/ℓ
	.52 (maximum)	–	.053 (maximum)	–	
40. Hexachlorobutadiene	90	9.3	32	–	.045–4.47 ug/ℓ
41. Hexachlorocyclo-hexane (Lindane)	.080 (24 hr. avg.)	–	0.16	–	.92–9.2 ng/ℓ
	2.0 (maximum)	–			
BHC	100	–	0.34	–	1.63–163 ng/ℓ
42. Hexachlorocyclo-pentadiene	7.0	5.2	7.0	–	206 ug/ℓ
					1.0 ug/ℓ (taste and odor)

43. Isophorone	117,000	–	12,900	–	5.2 mg/ℓ
44. Lead	3.8 ug/ℓ (100 ppm hardness and 24 hr. avg.)				
45. Mercury	.00057 ug/ℓ (24 hr. avg.)		.025 (24 hr. avg.)	–	144 ng/ℓ
	.0017 ug/ℓ (maximum)		3.7 (maximum)	–	
46. Naphthalene	2,300	620	2,350	–	not available
47. Nickel	96 (100 ppm hardness and 24 hr. avg)		7.1 (24 hr. avg.)	–	13.4
	1,800 (maximum)	–	140 (maximum)	–	
48. Nitrobenzene	27,000	–	6,680	–	19.8 mg/ℓ 30 mg/ℓ (taste and odor)
49. Nitrophenols	230	150	4,850	–	13.4 (for 2,4 dinitrocresol) 70 (for dinitrophenol)
50. Nitrosamines	5,850	–	3,300,000	–	.14–14 ng/ℓ (for n-nitrosodimethylamine) .08–8.0 ng/ℓ (for n-nitrosodiethylamine) .064–64 ng/ℓ (for n-nitrosodi-n butylamine) 490–49,000 ng/ℓ (for n-nitrosodiphenylamine) 1.60–160 ng/ℓ (for n-nitrosopyrolidine)
51. Pentachlorophenol	55	3.2	53	34	1.01 mg/ℓ 30 ug/ℓ (for taste and odor)
52. Phenol	10,200	2,560	5,800	–	3.5 mg/ℓ 0.3 mg/ℓ (for taste and odor)

TABLE 7-16B. *(Continued)*

Toxic chemical	Concentration of toxic material considered limit (ug/ℓ)				
	Freshwater aquatic life		Saltwater aquatic life		
	Acute	Chronic	Acute	Chronic	Human health
53. Phthalate Esters	940	3	2,944	3.4	313 mg/ℓ (dimethylphthalate) 350 mg/ℓ (diethylphthalate) 34 mg/ℓ (dibutylphthalate) 15 mg/ℓ (di-2-ethyl-hexyl-phthalate)
54. Polychlorinated Biphenyls	0.014 (24 hr. avg.) 2.0 (maximum)	– –	0.30 (24 hr. avg.) 10 (maximum)	– –	.0079–.79 ng/ℓ
55. Polynuclear Aromatic Hydrocarbons (PAH's)	not available	–	300	–	.28–28 ng/ℓ
56. Selenium	35 (24 hr. avg.) 260 (maximum) 760 (inorganic selenate)	–	54 (24 hr. avg.) 410 (maximum)	–	10 ug/ℓ
57. Silver	4.1 (maximum 100 ppm hardness)	0.12 (average chronic)	2.3	–	50 ug/ℓ
58. Tetrachloroethylene	5,280	840	10,200	450	.08–8 ug/ℓ
59. Thallium	1,400	40	2,130	–	13 ug/ℓ
60. Toluene	17,500	–	6,300	5,000	14.3 mg/ℓ
61. Toxaphene	.013 (24 hr. avg.) 1.6 (maximum)	–	.070 (maximum)	–	.07–7.1 ng/ℓ
62. Trichloroethylene	45,000	21,900	2,000	–	.27–27 ug/ℓ
63. Vinyl chloride	not available	–	not available	–	.20–20 ug/ℓ
64. Zinc	47 ug/ℓ (24 hr. avg.) 320 (max. at 100 ppm hardness)	–	58 (24 hr. avg.) 170 (maximum)	– –	5 mg/ℓ

Source: Adapted From *Federal Register* 1980.

7-5. OVERALL POLLUTION INDEX–A MEASURE OF WATER QUALITY

The desire to solve water pollution control problems on a nationwide scale resulted in passage of the Water Quality Act of 1965. The establishment of water quality standards for interstate water is the most important part of the act influencing the activities of pollution control. Measurement and expression of a polluted water quality are important factors in controlling pollution. At present, water quality is usually expressed by several parameters such as BOD (Biochemical Oxygen Demand), DO (Dissolved Oxygen), coliform bacteria count, and so on. However, "pollution" and a "polluted water" are relative terms, and expressions to describe them have been arbitrarily chosen in each instance. These expressions, chosen to differentiate a polluted water from a clean water, have been mainly based on the characteristics of the water uses. As an example, the item of fecal coliform count is selected for a regional recreational use to express one form of a pollution. On the other hand, the six iterms of pH, hardness, suspended solids, copper, iron, and manganese concentrations have been selected for a textile industry. Even though agreement may be reached on the proper expression of pollution, persons may not agree on the critical values of those expressions.

More recently, and especially in regional systems, benefits of pollution control and damages of pollution are being considered. A regional, rather than local or separate control system, has great merit. A major characteristic of the regional system is that a region is often composed of many different kinds of waste discharges and water uses. Therefore, the most basic and decisive problem is how to quantify pollution by a common measurement for the society affected by it. This is made difficult because of the many different methods of quantifying pollution, as previously mentioned. Up to now, in regional cases, one representative item, such as BOD or DO, has often been selected as an overall common expression of pollution, neglecting the effects of the other items of water qualities, or assuming that those effects may be substituted by the representative item. It is also common to use multiple items of water qualities to determine a minimum quality level of a pollution. These single-item (first group) and multiple-item (second group) common expressions of pollution are not always satisfactory. A single item cannot always serve as a substitute for other important items, especially in a large regional problem in which many kinds of water uses coexist; and also, in the multiple-item method, the quantified values are often expressed as discrete values rather than useful moving plots of pollution characteristics. The mutual relations of those multiple items of water quality such as pH, BOD, and DO are also unclear at this stage. Therefore, it is most urgent that a simple, reasonable, and useful method be developed for expressing quality of polluted water. The major objective in this section is the development of one such common quality expression for relative states of pollution of a watercourse.

Water Quality Expressions in Water Pollution Control Problems

Pollution control problems may be classified quite generally into two groups, based on the type of the problem. In one group, pollution control is concerned with a single item of water quality, and in the other group, multiple items of water qualities are used. For instance, a specific control problem of either iron, color, or temperature in water supply is an example of the first group. On the other hand, a problem of determining whether a water is suitable for swimming requires the use of multiple items such as coliform bacteria, pH, color, and turbidity. Most of the problems, precisely speaking, are classified into the second group, in which it is difficult to express an absolute measure of pollution. But some problems in this group can often be transferred into the first group, in practice. One representative item is often selected from the multiple items. Stream quality description by DO sag curve and the decision of BOD removal efficiency at sewage treatment plants are typical examples of the transfer. In the above case, the concentration of total biodegradable organic matter is used as an index of pollution, neglecting the other items of qualities, and the purpose of sewage treatment is practically defined as a removal of the organic matter expressed as BOD.

However, when considering a specific water use, such simplification as in the above example is not realistic, since water quality must satisfy other contaminant criteria simultaneously. When water is used for swimming and cooling, for example, multiple items such as coliform bacteria, temperature, pH, and so on should be used rather than a general expression of pollution such as BOD. Single-item quality control (such as BOD) can only be used when the receiving water is not being used for a specific purpose. In certain cases when data are incomplete or water use is not highly specialized, single-item, quality control may still be useful. For these ever disappearing cases, the values expressed for water quality can be employed as a measure of pollution without any special problems. In the second group, a direct expression of all the multiple items concerned could be employed as a measure of pollution–if discrete values can measure pollution for the existing expression of a minimum quality level, and if each item is employed quite independently among multiple items themselves. Quality standards for classified waters in New York State, such as a range of 6.5 to 8.5 in pH and 4.0 to 5.0 in DO, are examples of discrete and independent expressions of a pollution using multiple items. It should be emphasized here that pollution effect on a stream can be quantified quite reasonably by the existing quality expression methods in many cases, as shown above. In these instances, additional discussions of a general expression method of pollution are not required.

Maximum treatment efficiency and/or minimum treatment cost have been mainly used to control pollution. Little consideration has been given to the specific human use of the receiving water. In other words, pollution control problems have been considered as problems in the first group (single items), most of which are transferred from the second group (multiple items). Therefore, one of the major items such as BOD, COD, or DO has been reasonably employed for those cases as a general expression of pollution. More recently in regional problems, uses of the water body into which the treated wastes are discharged are usually taken into account to arrive at an optimum method of controlling the regional pollution. This means that pollution control criteria have been changing from the minimization of the treatment costs to some sort of maximization of the total benefits of water uses in an entire region, including the treatment costs as negative benefits. In the discussion of the many kinds of regional uses of a receiving water, the use of multiple-item expression of pollution is inevitable.

A brief procedure to select the appropriate water quality expression method in water pollution control problems is suggested as follows:

1. Clarify the major purpose of the pollution control and the definition of the pollution, and classify the problem into one of two groups, based on single or multiple water quality items concerned. (Even after the classification, the possible transfer from the multiple-item group to the single-item group must be considered.)
2. Check the scale of the control problem to determine whether it is local or regional.
3. Determine whether the pollution control objective involves the uses of the receiving water or is only concerned with the waste treatments.

The proper quality expression method can now be determined for each control problem by using Table 7-17. From Table 7-17 the water pollution control problems are more generally defined as follows:

1. For local and individual problems, the existing methods of single items or multiple discrete items are applicable.
2. For regional problems, the quality expressions suggested are as follows:
   (*a*) The existing expressions using a representative item, such as BOD or DO, are advantageous, if only treatments for pollution control are discussed, without considering the individual uses of the water body (e.g., cost analysis of regional waste treatment).
   (*b*) A new expression using multiple items should be developed if treatments are discussed, considering the effects of the treatments for individual uses of the receiving water, for example, benefit analysis of a regional waste treatment (Nemerow, 1967).

TABLE 7-17. Quality Expression Methods of a Pollution for Pollution Control Problems

Number of items concerned with the problem	Single		Multiple	
Size of the problem / Purpose of the problem	Local	Regional	Local	Regional
Discussions of discharges (Removal efficiency and cost of waste treatment) e.g., Cost and efficiency analysis analysis	*Trad.* ex: (BOD removal, etc.)	*Trad.* ex: (DO sag curve, etc.)	*Trad.* (Ind.)	*Trad.* (Ind.) or *Index*
Discussions of discharges and uses of the receiving water. (Effects and damages of waste treatment) e.g., Benefit analysis	*Trad.* ex: (Toxic ion removal, Temperature increase, etc.)	*Trad.* ex: (Spread of a disease germ, toxic ion removal, etc.)	*Trad.* (Ind.)	*Trad.* (Disc.) and *Index*

NOTE–*Trad.*, Traditional expression by the existing water quality expression method; *Trad.* (Ind.), Independent expression of each item using the existing water quality expression method; *Trad.* (Disc.), Discrete value expression by traditional expression such as stream standards; *Index*, A new continuous quantification method of a pollution considering multiple items of water qualities.

(*c*) The existing expression by multiple items is useful for a discrete quality expression of a pollution (e.g., stream quality standards expressed using multiple items).

A Proposal of Pollution Index

A new quality expression method of pollution is required, especially for regional benefit analysis of pollution control, as mentioned in the previous section. On a regional basis, all effects of a pollutant on the water uses should be taken into account at the same time to evaluate total damages of the pollution. In the rest of this chapter, a common and overall measuring unit of pollution will be discussed, considering overall effects of the multiple items of pollutants for specific uses.

An expression of pollution quality will be developed by a single value for each major classified water use, which is arbitrarily referred to as "Pollution Index." An overall expression of pollution quality for all uses will be derived from these index values. A certain precision is sacrificed in developing an overall index value. But such an overall quality expression of pollution is expected to be useful for many practical purposes. For example, it may offer means for measuring pollution to be used for administrative purposes and as communication with the public (Horton, 1965).

Basic Consideration for the Index. The primary problem in developing the index is how to integrate the independent multiple items of pollutants in a water into a common expression. There is often little correlation between the significance and the dimension of each pollutant: for instance, 1,000 (MPN/100 ml) in coliform count, 6 mg/l in DO, and 80°F in temperature. Each value may be compared only with the same item of quality. When those values are related to some kind of standard values of the respective items such as coliform number of 500/100 ml, DO of 5 mg/l, and temperature of 50°F, the relative value can be expressed as 2.0 (1,000/500 = 2.0), for coliforms, 1.2 ($\frac{6}{5} = 1.2$) for DO, and 1.6 ($\frac{80}{50} = 1.6$) for temperature, respectively, as nondimensional relative values.

It is possible to derive a meaningful, comparable, and relative value, if reasonable standard values are chosen. If these standard values are the permissible quality levels, the Pollution Index can be defined as a truly relative term. The permissible pollutant level at a location of a water use is recommended here as the standard value for the index development. An index for a major specific water use will be developed first, and then the overall index, which is a common measure of a pollution for all uses, will be discussed (Sumitomo and Nemerow, 1967).

When the multiple items of water qualities are expressed as C_i's and the permissible levels of the respective items for a use are expressed as L_{ij}'s, the

Pollution Index for the use j, PI_j, may be expressed as a function of the relative values (C_i/L_{ij})'s. Here, i is the number of the i-th item of water quality, and j is the number of the j-th water use

$$PI_j = \text{A function of } \left[\left(\frac{C_i}{L_{ij}}\right)\text{'s}\right] \quad \text{or} \tag{1}$$

$$PI_j = f\left(\frac{C_1}{L_{1j}}, \frac{C_2}{L_{2j}}, \frac{C_3}{L_{3j}}, \ldots \frac{C_i}{L_{ij}}\right)$$

$$(i\ 1, 2, 3, \ldots i, \ldots 1 \quad j\ 1, 2, 3, \ldots j, \ldots J)$$

This approach is similar to that used in describing toxicity of multiple items of toxic materials. When several toxic materials $(T_1, T_2, T_3, \ldots)$ coexist in a water, it is suggested (Eq. [1]) that the total toxicity may be evaluated by the next relation, applying the respective permissible levels $[(TL_1), (TL_2), (TL_3) \ldots]$

$$\text{Total toxicity} = T_1/(TL_1) + T_2/(TL_2) + T_3/(TL_3) + \cdots \tag{2}$$

A Development of the Pollution Index. The index may be expressed by the relative value (C_i/L_{ij}) as shown in the relation above (1). Each value of (C_i/L_{ij}) shows the relative pollution contributed by the single item. A value of 1.0 is the critical value for each (C_i/L_{ij}). Values greater than 1.0 indicate that the water requires some treatment prior to use for that specific purpose. Likewise, when combining the individual values of (C_i/L_{ij})'s into a common index, values over 1.0 signify a critical condition under which a proper treatment is necessary for the water use. The major problem here is visualizing how to get an index value from the multiple relative values. Theoretically speaking, there is no absolute solution among the numerous possible methods. However, we propose a reasonable method for an overall expression of pollution. The average value of all the calculated (C_i/L_{ij}) values may be recommended as one of the most important parameters for the PI_j.

For example, when the quality of a water is expressed as BOD $(C_1) = 10$ mg/l, coliform bacteria $(C_2) = 1{,}300/100$ ml, and hardness $(C_3) = 80$ mg/l, and their permissible levels for a use j are given as BOD $(L_{1j}) = 20$ mg/l, coliform $(L_{2j}) = 1{,}000/100$ ml, and hardness $(L_{3j}) = 100$ mg/l, then the (C_i/L_{ij}) values are expressed as follows: $(C_1/L_{1j}) = 0.5$, $(C_2/L_{2j}) = 1.3$, and $(C_3/L_{3j}) = 0.8$. The average value of the above three values is about 0.9, which may generally indicate that the water is just under the critical condition for the use j with no treatment of the water. However, the average value may not satisfactorily measure pollution, because the necessity of water treatment for a use is often determined by the maximum value of (C_i/L_{ij}) rather than the average value. Even if the mean value is very small, say, under 1.0, the water cannot be used for use j without treatment, if one of the (C_i/L_{ij}) values is over 1.0. As shown in the above example in which the mean value (0.9) is under 1.0, the water should be treated for the use, because one item, coliform bacteria, is over the permissible level, $(C_2/L_{2j} = 1.3)$. Therefore, the maximum of the (C_i/L_{ij}) values is proposed as another decisive factor to be included in computation of PI_j, in addition to the mean value. The index may be expressed using the maximum and mean values of (C_i/L_{ij}), as shown in the next relation.

$$Pi_j = f\,[\text{max. of } (C_i/L_{ij})\text{'s and mean of } (C_i/L_{ij})\text{'s}] \tag{3}$$

The relation (3) shows that the index PI_j is expressed by the maximum and mean (C_i/L_{ij}) values, although the relation among those three contaminants is still unknown. One expression of the relations (3), when the abscissa is the maximum (C_i/L_{ij}) and the ordinate is the mean (C_i/L_{ij}), shows the PI_j value as a point somewhwere in the space between the two axes, as shown in Fig. 7-1. It may be generally agreed that the larger the values of maximum (C_i/L_{ij}) and/or mean (C_i/L_{ij}), the more the water is polluted. Therefore, the length of the line from the origin to the point PI_j in Fig. 7-1 may be proposed as a significant factor to express pollution. Another factor in Fig. 7-1 is determined by the ratio of both values of the maximum and mean (C_i/L_{ij})'s (θ). But it is difficult from a practical standpoint to discuss the significance of the ratio of the maximum and mean values of (C_i/L_{ij}) or to determine which is relatively more important in regard to pollution. Many existing data plotted according to the above graphic considerations are distributed within a limited angle, as seen in Fig. 7-2. It is proposed that the general quality expression of pollution for use j is related to

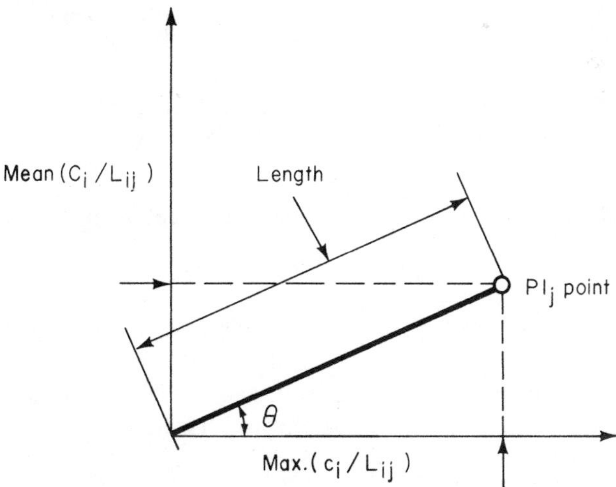

FIG. 7-1. Relationship between mean and maximum ratios of contaminant concentration and allowable concentration.

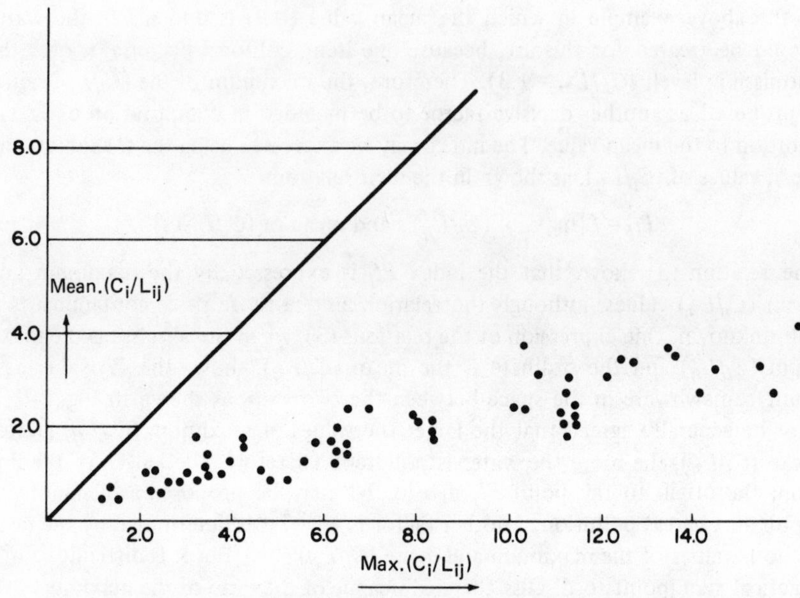

FIG. 7-2. An example of data distribution for human contact use.

the length of a line between the origin and each point. The length is determined by the values of the maximum and mean of (C_i/L_{ij}) in Fig. 7-1. In our computation of a Pollution Index, we propose to neglect the effect of the angle (θ). The Pollution Index, for use j, PI_j, is measured by the lengths of the radii of the concentric circles, as shown in Fig. 7-3. Therefore, the relation (3) is expressed as shown in Fig. 7-3 as follows:

$$PI_j = m\sqrt{\text{max.}\,(C_1/L_{ij})^2 + \text{mean}\,(C_i/L_{ij})^2} \tag{4}$$

Here, m = the proportionality constant.

A critical condition to determine the coefficient m is recommended as follows:

$$PI_j = 1.0, \text{ when max. } \frac{C_i}{L_{ij}} = 1.0 \quad \text{and mean} \quad \frac{C_i}{L_{ij}} = 1.0 \tag{5}$$

This means that the index for use j is expressed as 1.0 when all items of water qualities are just equal to their respective permissible levels for the use. The relation (4) is as follows under relation (5):

$$1.0 = m\sqrt{1^2 + 1^2}, \therefore m = \frac{1}{\sqrt{2}}$$

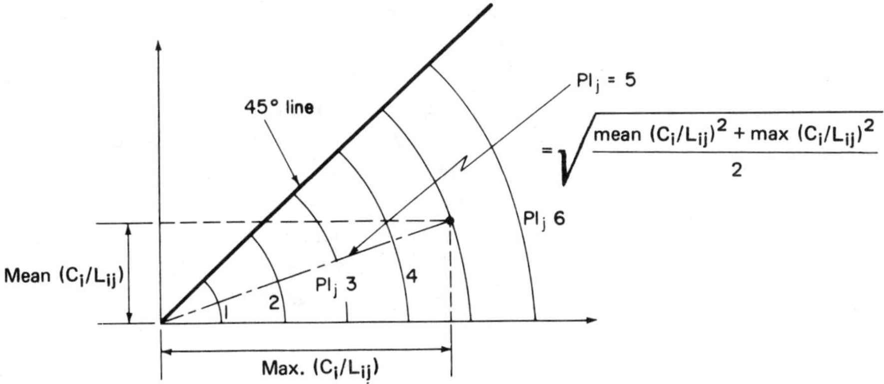

FIG. 7-3. Relationship between mean and maximum ratios of contaminant concentration and allowable concentration.

Therefore, PI_j is proposed as follows, and the relation among those three is expressed graphically, as shown in Fig. 7-3.

$$PI_j = \sqrt{1/2} \cdot \sqrt{\text{max.}\,(C_i/L_{ij})^2 + \text{mean}\,(C_i/L_{ij})^2}$$

$$= \sqrt{\frac{\text{max.}\,(C_i/L_{ij})^2 + \text{mean}\,(C_i/L_{ij})^2}{2}} \qquad (6)$$

Overall pollution index. The index in Equation (6) is proposed only for a particular use j. However, with some modification, it may be employed for some sort of grouped uses, depending upon what kind of permissible levels are employed as L_{ij}'s. If the L_{ij}'s are determined from the general permissible levels for swimming, boating, sport fishing, and hiking, as an example, the index PI_j, may be defined as the Pollution Index for outdoor recreation use. All existing water uses in the region should be taken into consideration in regional benefit analyses of pollution control problems—and an overall Pollution Index PI can be used for such problems. If an overall permissible level for all uses in a region can be determined, the overall Pollution Index could be easily determined by Equation (6), employing the overall permissible level for L_{ij}'s, (j = all uses.) However, such permissible levels have not yet been developed. The overall index PI is recommended for general use as a kind of mean value of the calculated respective PI_j's for all uses in the region.

When pollution is quantified for each use by the calculated PI_j's such as, in one case, for example, 5.0 for drinking, 1.0 for swimming, and 0.9 for an industrial use, these values will indicate a more general evaluation of the pollution for all uses, if the relative importances of those water uses are quantified in the region. If the swimming and the industrial uses are negligible or not *that* important in the region, the overall evaluation of pollution may depend mainly upon drinking use. In this case, the overall evaluation of this polluted water for the entire region may be quantified primarily according to the index value for drinking use. However, the relative effects or importances of many regional uses mentioned above are not easily quantified. One possible procedure for establishing the overall index PI is proposed here. This procedure assumes that the relative effects are determinable as simple constant numbers in an overall estimation.

Judging from the fact that each PI_j value is a relative value, nondimensional, and that the relative importance of each use may be generally determined as a constant value, the overall index PI is proposed here as a weighted average value of all the PI_j's as follows

$$PI = \sum_{j=1}^{j=n} (w_j, PI_j) \tag{7}$$

Here w_j = Weight coefficient (constant value), which is determined by the relative importances of the water use j in the region or society

n = Number of water uses

and

$$\sum_{j=1}^{j=n} (w_j) = 1.0 \tag{8}$$

Additional consideration for pollution index. Equation (6) is proposed as a general form of Pollution Index for water use j. However, some modifications of the general equation are necessary in order to complete the actual index.

First of all, calculation of the (C_i/L_{ij}) value will be discussed in more detail. Every item of water quality (C_i) does not always increase as pollution increases. Most organic and inorganic ions increase in concentration as pollution increases, such as the increase from 20 mg/l to 50 mg/l. But, on the other hand, pH values vary up or down with pollution, within a range of from 0 to 14. Dissolved Oxygen (DO), for example, also usually decreases within a limited range as pollution mounts. In addition, it is common practice to specify the permissible level by a range of values, such as pH from 6.4 to 7.5. In these special cases, the (C_i/L_{ij}) value cannot be calculated in the same ways as previously suggested. Some general practical methods are recommended for some of these special cases.

1. For the contaminant that decreases in value as pollution increases, such as transparency, DO, etc.: the theoretical or practical maximum value (C_{im}) of the (C_i) value should be determined, such as DO concentration at saturation. The usual (C_i/L_{ij}) value may then be replaced by a new (C_i/L_{ij}), where

$$\frac{\text{new } C_i}{L_{ij}} = \frac{C_{im} - C_i}{C_{im} - L_{ij}} \tag{9}$$

For example, when maximum DO = 8.0, the minimum permissible DO level is 4.0, and the existing DO level is 5.0,

$$\frac{C_1}{L_{1j}} = \frac{8.0 - 5.0}{8.0 - 4.0} = \frac{3}{4} = 0.75$$

2. For the contaminant qualities that have permissible levels ranging from L_{ij} min. to L_{ij} max.: the mean value of the ranged levels $\bar{L}_{ij}$ should be calculated

$$\bar{L}_{ij} = \frac{L_{ij.\text{min.}} + L_{ij.\text{max.}}}{2} \tag{10}$$

and the (C_i/L_{ij}) value may be replaced by the following:

$$\frac{C_i}{L_{ij}} = \frac{C_i - \bar{L}_{ij}}{L_{ij.\text{min.}}{}^{*} \text{ or } L_{ij.\text{max.}}{}^{*} - \bar{L}_{ij}} \tag{11}$$

For example, if the maximum allowable pH level is 8.5, and the range is 6.5 to 8.5 when the existing pH is 10.0, then

$$\frac{C_i}{L_{ij}} = \frac{10.0 - 7.5}{8.5 - 7.5} = \frac{2.5}{1.0} = 2.5$$

The (C_i/L_{ij}) value indicates relative pollution as compared to the respective permissible level. At the same time, this value also may express how damaging the water may be for use j, especially when the value is over 1.0. To illustrate the possibility, let us compare the (C_i/L_{ij}) values of 0.9 and 1.1. Both waters may be considered as almost equally polluted in a numerical meaning. However, this is not always valid in practice, because a treatment facility is necessary only for the latter water. Damages of the pollutants for uses, which may be expressed in the necessary expenses for the facilities, are generally quite significantly different from the numerically expressed values of 0.9 and 1.1. Likewise, when (C_i/L_{ij}) values of 5.0 and 10.0 are compared, the difference of the damages caused by both contaminants may not necessarily be so large as the values show. The necessary expenses for treatment are not always proportional to the quality of the raw water, but generally diminish in an increasing rate at *PI* values over 1.0. When the proposed index is expected to reflect the relative damage of pollution rather than a simple numerical expression of the pollutants, some kind of calculation method of the (C_i/L_{ij}) value should be considered to express the relative damage, as shown in Fig. 7-4. Assuming that the shape of the curve in Fig. 7-4

---

*Use whichever value is nearer to existing C_i value.

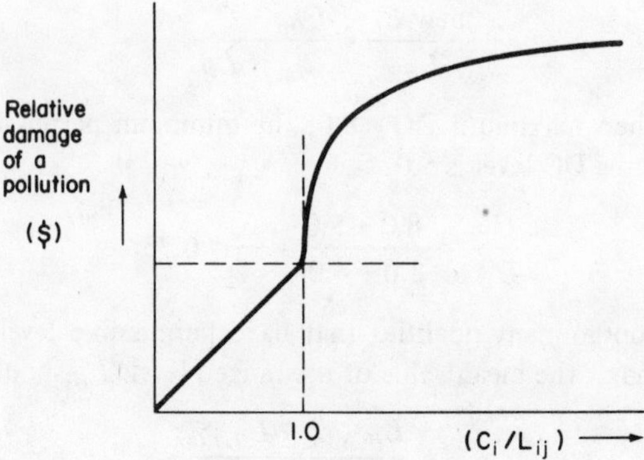

FIG. 7-4. Effect of increasing concentration of pollutant on the damage costs.

is valid, the following calculation method is proposed here for the substitution of (C_i/L_{ij}) values. The existing data of the treatment efficiencies and the treatment costs are used. The following calculation of the (C_i/L_{ij}) value for each (C_i/L_{ij}) may be substituted for the (C_i/L_{ij}) in Equation (6) for PI_j.

When $(C_i/L_{ij}) \leqslant 1.0$,

$$\frac{C_i}{L_{ij}} \leqslant 1.0, \qquad \frac{C_i}{L_{ij}} = \text{the actual value of } \frac{C_i}{L_{ij}}$$

When $(C_i/L_{ij}) > 1.0$,

$$\frac{C_i}{L_{ij}} > 1.0, \qquad \frac{C_i}{L_{ij}} = 1.0 + p \cdot \log_{10} \frac{C_i}{L_{ij}} \tag{12}$$

Here, p = a constant value. (As a standard value for a relative comparison, 5.0 is arbitrarily employed for the p value in the later applications of the proposed indices for the existing pollutions.)

Consideration is now given to the practical grouping of water uses (j's) and the reasonable items of water qualities (i's) for the practical application of the proposed indices. In the previous discussions, PI_j is proposed for each use j, $(j = 1, 2, 3, \ldots j \ldots n)$, but the number of water uses (j's) in a society is so numerous that a collective general grouping of water uses may be used for practical purposes. If a reasonable common permissible level can be determined for similar water uses, grouping of water uses may be possible, facilitating calculation of the Pollution Indices. The following three groups are recommended for separate index expressions.

1. *Human Contact Use* ($j = 1$), which includes drinking, swimming, beverage manufacturing, etc. (see Table 7-18).

TABLE 7-18 Permissible Quality Levels for Human Contact Water Use ($j = 1$).
[*Interim Report, 1967. By permission of publishers.*]

Use	1 Te °F	2 Col unit	3 Tur	4 pH	5 Coli/100 ml	6 TS ppm	7 SS ppm	8 $T(NO_3)$ ppm	9 Alk ppm	10 Har ppm	11 Cl ppm	12 Fe ppm	13 SO_4 ppm	14 DO ppm
Drinking	–	15	5	–	5	500	–	45	–	–	250	0.35	250	
Swimming	85	–	–	6.5	–200	–	–	–	+	+	+	–+	–	–
Beverage manufacturing	–	10	–	–	–	–	–	–	85	–	250	0.35	–	–/
Average; L_{ij}	85	13	5	6.5 8.3	–103	500	–	45	/	/	/	/	250	4.0*

NOTE–1. (Te); Temperature; 2. (Col), Color; 3. (Tur), Turbidity; 5. (Coli), Coliform bacteria count; 6. (TS), Total Solids; 7. (SS), Total Suspended Solids; 8. (TN), Total Nitrate Nitrogen; 9. (Alk), Alkalinity; 10. (Har), Hardness; 11. (Cl), Chloride; 12. (Fe), Iron and Manganese; 13. (SO_4), Sulfate; (–) Now under discussion; (+) No special limit; (/) L_{ij} is not determined because of the presence of (+) mark; (*) Assumed.

2. *Indirect Contact Use* ($j = 2$), which includes fishing, industrial food preparation, agricultural use, etc. (see Table 7-19).

3. *Remote Contact Use* ($j = 3$), which includes industrial cooling, aesthetic uses (picnicking, hiking, and plain visitation), navigation, etc. (see Table 7-20).

The overall permissible levels (L_{ij}'s) should be determined for the three grouped uses by applying the present permissible contamination levels for each particular use. An example of the determination of the overall permissible levels for the grouped uses will be shown later in this chapter.

The following contaminant items (i's) are recommended for the index discussions and computations, although as many items of water qualities as possible should be utilized.

1. Temperature
2. Color
3. Turbidity
4. pH
5. Fecal coliform bacteria
6. Total dissolved solids
7. Suspended solids
8. Total nitrogen
9. Alkalinity
10. Hardness
11. Chloride
12. Iron and manganese
13. Sulfate
14. Dissolved oxygen

Suitable contaminant items should be determined after considering the objective of the grouped use. In addition, one must make or have available water quality data including the data for the permissible levels.

Applications of the proposed indices. As an application of the proposed indices, pollution of the surface water resources in New York State will be quantified by the following procedure. These resources will be evaluated for human contact use (by PI_1), and indirect contact use (by PI_2), remote contact use (by PI_3), and also for general overall use (by PI). The permissible quality levels for the above three grouped uses, (L_{i1}, L_{i2}, and L_{i3}) are determined as average permissible levels shown in Tables 7-18, 7-19, and 7-20. Permissible levels have been selected after careful study of the Federation of Water Pollution Control Association Interim Report of the National Technical Advisory Committee on Water Quality Criteria, June 30, 1967.

An example calculation of the Pollution Index (for human contact use, PI_1) of Oneida Lake in New York State is described in detail. First, the water qualities (New York State Department of Health, 1960–1964) of the Oneida Lake, which are shown in Table 7-21, are divided by the respective permissible levels in Table 7-18, and those calculations are summarized in Table 7-22. From the Table, Max. (C_1/L_{ij}) and Mean (C_i/L_{ij}) are given as 2.9 and 0.87, respectively.

Therefore

$$PI_1 = \sqrt{\frac{2.9^2 + 0.87^2}{2}} = 2.1$$

TABLE 7-19. Permissible Quality Levels for Indirect Contact Use ($j = 2$).
[*Interim Report, 1967. By permission of publishers.*] **(See Table 7-18 for numerical column titles.)**

Use	1	2	3	4	5	6	7	8	9	10	11	12	13	14
Fishing	55	-	30	6.0 9.0	-2,000	-	-	-	-	+	-	-	-	-
Agricultural	-	+	-	6.0- 8.5	-	500	-	45	-	+	+	1.0	-	-
Fruit and vegetable Industrial	+	5	5	6.5- 6.8	-	500	10	10	250	250	250	0.4	250	-
Average; L_{ij}	/	/	18	6.2 8.1	-2,000	500	10	28	250	/	/	0.7	250	3.0*

NOTE–(-) Now under discussion; (+) No special limit; (/) L_{ij} is not determined because of the presence of (+) mark; (*) Assumed. Agricultural use is Farmstead and Irrigation.

TABLE 7-20. Permissible Quality Levels for Remote Contact Use (j = 3).
[*Interim Report, 1967. By permission of publishers.*] **(See Table 7-18 for numerical column titles.)**

Use	1	2	3	4	5	6	7	8	9	10	11	12	13	14
Iron & steel (cooling)	100	+	+	5–9	+	-	10	+	-	-	-	-	-	-
Cement	-	-	+	6.5 9.0	+	600	500	+	400		250	25.5	250	-
Petroleum				6.0- 9.0	+	1,000	10	+	-	350	300	1.0	+	-
Pulp[a]	95	10	-	6.10	+	-	10	+	-	100	200	1.1	+	-
Textile[b]	-	5	-	6.4- 10.3	+	100	5	+	-	25	-	0.2	-	-
Chemical[c]	-	5	-	6.5 8.1	+	338	5	-	145	210	28	0.2	85	-
Navigation	-	-	-	-	-	-	-	-	-	-	-	-	-	-
Aesthetic	-	-	-	-	-	-	-	-	-	-	-	-	-	-
Average; L_{ij}	98	/	/	6.1- 9.1	/	510	90	/	274	17.1	195	5.6	/	2.0*

NOTE–(-) Now under discussion; (+) No special limit; (/) L_{ij} is not determined because of the presence of (+) mark; (*) Assumed. For Navigation and Aesthetic uses, data for permissible levels are not available yet.

[a]Bleached.

[b]Average of sizing, scouring, bleaching, and dyeing.

[c]Average of organic and inorganic matters.

In the same way as with the above, the PI_2 and PI_3 are also calculated, using L_{i2} and L_{i3} in Table 7-18 and 7-16 instead of L_{i1} in Table 7-17.

Then

$$PI_2 = 0.8$$

$$PI_3 = 0.6$$

The overall Pollution Index PI is calculated by equation (7), applying the calculated PI_1, PI_2, and PI_3 values. But the relative weight values w_j's in the equation are quite difficult to determine precisely from the existing available data. Therefore, the relative weights may be determined quite tentatively as follows, assuming that all water uses are equally important in our society:

$$w_1 = w_2 = w_3 = \frac{1}{3} \tag{13}$$

Current water uses, however, in Oneida Lake may be estimated quite roughly as follows: 40 percent for swimming use, 40 percent for fishing use, and 20 percent for navigation, aesthetic uses, and others. From these rough estimations, relative weight values may possibly be determined as follows: $w_1 = 0.4$, $w_2 = 0.4$, and $w_3 = 0.2$. Then, the overall Pollution Index for Oneida Lake is determined from data shown in Tables 7-21 and 7-22. as follows.

$$\begin{aligned} PI &= w_1 \cdot PI_1 + w_2 \cdot PI_2 + w_3 \cdot PI_3 \\ &= 0.4 \times 2.1 + 0.4 \times 0.8 + 0.2 \times 0.6 = 1.28 \cong 1.3 \end{aligned}$$

The Pollution Indices for other surface water resources in New York State are quantified in a manner similar to that used for Oneida Lake. In the calculations, relation (12) is applied for relative weights (w_j's). The water quality data for those resources are obtained from "Periodic Report of Water Quality Surveillance Network, 1960 through 1964" by the New York Department of Health (1960–1964). The results shown in Table 7-23 generally indicate the relative pollution of each water resource for our water uses. However, it should be emphasized that these are preliminary results, and are quantified here quite tentatively and mechanically as a trial application of the proposed index method.

The Pollution Index method does require that individual contaminant determinants must be made correctly. It requires the establishment of a new concept or yet another set of values. However, your author believes that once the general public understands and accepts the idea that many factors are involved in classifying a stream in relative terms of pollution, and further, that all of these factors can be integrated into one general expression, evaluation in overall terms will become relatively simple and helpful.

TABLE 7-21. Water Quality in Oneida Lake (Average data, 1960–1964). [*New York State Dept. Health, 1960–1964. By permission of publishers.*]

Item		Quality	Item		Quality
Temperature	C_1	63.3°F	Total nitrogen	C_8	-
Color	C_2	12.0 unit	Alkalinity	C_9	86 ppm
Turbidity	C_3	12.0	Hardness	C_{10}	128 ppm
pH	C_4	8.2	Chloride	C_{11}	26.0 ppm
Coliform	C_5	72 MPN	Fe, Mn	C_{12}	neg.*
Total solids	C_6	209 ppm	Sulfate	C_{13}	84.4 ppm
Suspended solids	C_7		Dissolved oxygen	C_{14}	8.0 ppm

*neg., Negligible.

TABLE 7-22. Calculations of Index PI_1 for Oneida Lake

i	(C_i/L_{ij})	$\log_{10}(C_i/L_{ij})$	$(C_i/L_{ij}) = (C_i/L_{ij})$ or $= 1.0 + 5 \times \log_{10}(C_i/L_{ij})$
1	63.3/85.0 = 0.75	- -	0.75
2	12.0/12.5 = 0.96	- -	0.96
3	12.0/5.0 = 2.40	0.38	2.9 (max.)
4	* = 0.89	- -	0.89
5	72/103 = 0.7	- -	0.7
6	209/500 = 0.42	- -	0.42
7	-/-	-	-
8	-/4.5	-	-
9	86/-	-	-
10	128/-	-	-
11	26.0/-	-	-
12	0/-	-	-
13	84.4/250 = 0.34	- -	0.34
14	† = 0.01	- -	0.01
			Total 6.97
			Average (C_i/L_{ij}) = 6.97/8 = 0.87

NOTE–(-) Datum is not available; (- -) Calculation is not necessary.
*(8.2 – (6.5 + 8.3)/2)/(8.3 – (6.5 + 8.3)/2) = 0.8/0.9 = 0.89.
†(8.4‡ – 8.0)/(8.4‡ – 4.0) = 0.01.
‡Saturated concentration.

TABLE 7.23 Water Pollution of Surface Water Resources in New York State Expressed by Pollution Indices.
[From New York State Dept. of Health, 1960–1984. By permission of publishers.]

Surface water	Location	PI_1	PI_2	PI_3	PI
Niagara Ri.	Youngstown	6.0	1.4	0.2	2.5
Niagara Ri.	Buffalo	7.9	2.5	0.6	3.6
Erie L.	Buffalo	4.5	2.4	0.6	2.5
Cattaraugua Cr.	Gowanda	2.4	0.7	0.8	1.3
Buffalo Ri.	Buffalo	10.6	6.1	2.2	6.3
Cazenovia Cr.	Buffalo	9.4	4.8	1.0	5.1
Buffalo Ri.	W. Seneca	9.3	6.0	2.3	5.9
Buffalo Ri.	W. Seneca	11.8	7.8	0.9	6.8
Tonawanda Cr.	W. Seneca	3.2	1.7	2.1	2.3
Allegheny Ri.	Indian Res.	0.7	0.6	0.6	0.6
Conewango	Carroll	3.3	2.2	0.8	2.1
Cassadaga	Palconer	4.3	3.3	0.4	2.7
Ontario L.	Rochester	8.1	3.5	0.6	4.1
Allens Cr.	Brington	3.0	2.6	0.8	2.1
Ontario L.	Oswego	4.5	0.7	0.6	1.9
Genesee Ri.	Rochester	2.6	1.2	0.6	1.5
Genesee Ri.	Rochester	2.3	0.8	0.8	1.3
Genesee Ri.	Chili	2.6	1.3	1.5	1.8
Genesee Ri.	Wellsville	8.1	2.5	0.5	3.7
Chemung Ri.	Elmira	8.2	4.0	0.6	4.3
Cohocton Ri.	Campbell	3.4	0.8	0.7	1.6
Tioga Ri.	Lindley	2.7	0.6	0.7	1.3
Susquehanna Ri.	Binghamton	9.2	4.6	0.5	4.8
Chenango Ri.	Chenango	2.3	0.6	0.6	1.2
Toughnioga Ri.	Barber	2.6	0.7	0.6	1.3
Chenango	Chenango	0.8	0.6	0.6	0.7
Susquehanna	Unadilla	0.4	0.5	0.4	0.4
Seneca Ri.	Waterloo	4.7	0.7	0.6	2.0
Seneca Ri.	Geneva	1.1	0.8	0.7	0.9
Cayuga L.	Cayuga	8.1	1.5	0.6	3.4
Cayuga L.	Fayette	1.8	0.6	0.7	1.0
Owasco L.	Fleming	2.0	0.8	0.5	1.1
Skaneateles L.	Skaneateles	1.7	0.6	0.5	0.9
Canandaigua L.	Canandaigua	0.9	1.2	0.6	0.9
Seneca Ri.	Montezuma	4.7	0.8	0.6	2.0
Oswego Ri.	Oswego	10.3	5.5	4.0	6.6
Oneida L.	Cicero	2.1	0.8	0.6	1.1
Seneca Ri.	Clay	4.9	2.6	2.9	3.5
Black Ri.	Watertown	8.2	3.6	0.4	4.1
Black Ri.	Lyons Fall	8.3	3.7	0.4	4.1
Lawrence Ri.	Massena	4.7	2.6	0.6	2.6
Lawrence Ri.	Cape Vincent	0.8	1.1	0.4	0.8

TABLE 7-23. *(Continued)*

Surface water	Location	PI_1	PI_2	PI_3	PI
Raquette Ri.	Massena	5.2	0.6	0.6	2.1
Gross Ri.	Massena	5.9	1.2	0.6	2.6
Oswegatchie Ri.	Ogdensburgh	6.5	2.0	0.4	3.0
Saranac Ri.	Shuyler Falls	8.3	3.7	0.5	4.2
Ausable Ri.	Ausable	6.1	1.4	0.5	2.7
Hudson Ri.	Waterford	8.1	3.5	0.6	4.1
Hosic Ri.	Schaghticoke	2.3	0.8	0.6	1.2
Hudson Ri.	Fort Edward	7.9	3.3	0.7	4.0
Battenkooll	Greenwich	5.9	1.4	0.6	2.6
Hudson Ri.	Orinth	6.1	1.5	0.6	2.7
Mohawk Ri.	Cohoes	4.3	1.3	0.5	2.0
Mohawk Ri.	Schenectady	0.6	0.7	0.4	0.6
Sheoharie Cr.	Frerida	7.4	2.8	0.5	3.6
Mohawk Ri.	Frerida	7.5	2.8	0.5	3.6
E. Canada Cr.	Manheim	8.3	3.7	0.3	4.1
Mohawk Ri.	St. Johnsville	8.1	3.6	0.5	4.1
W. Canada Cr.	Herkimer	8.3	3.7	0.5	4.2
Mohawk Ri.	Shuyler	1.8	0.8	0.7	1.1
Hudson	Poughkeepsie	4.6	2.7	0.7	2.7
Frishkill Cr.	Beacon	11.4	6.8	0.7	6.3
Wappinger Cr.	Lagrange	3.8	0.5	0.6	1.6
Wallkill	Rosendale	8.9	4.2	0.6	4.6
Hudson Ri.	Poughkeepsie	9.9	5.4	0.4	5.2
Hudson Ri.	Bethlehem	11.4	6.6	0.3	6.1
Delaware Ri.	Port Jervis	0.4	0.1	0.3	0.3
Neversink Ri.	Deer Park	0.4	0.2	0.2	0.3
Delaware Ri.	Deposit	2.8	0.8	0.5	1.4
Hackensack Ri.	Orange Town	2.7	0.5	0.5	1.2
Pascack Cr.	Ramapo	10.2	5.3	0.5	5.3
Pascack Cr.	Clarkstown	11.9	7.3	0.5	6.6

NOTE–Cr., creek; Ri., river; L., lake.

7-6. RECLAIMED WASTEWATER QUALITY

The *Water Reuse* publication (No. 29, Feb. 1980, p. 10) reports that there were 536 wastewater reuse projects in the United States, totaling 760 acre-feet per year of wastewater.

Municipal effluent characteristics are given as typical and are compared to stream waters and to the maximum contaminant levels allowable in drinking waters, in Table 7-24. In reusing the effluents one must be concerned primarily with the organic content which exceeds that allowed in drinking waters.

TABLE 7-24. Comparison of Secondary Effluent Quality to Stream Water Quality and Drinking Water Standards

	(mg/l except as noted)				Municipal secondary effluent
	Drinking Water	Stream			
Parameters	MCL	Low	High	Typical	Typical
Primary Regulations					
Arsenic	0.05	<0.005	0.340	0.005	0.005
Barium	1	0.03	0.340	0.15	0.1
Cadmium	0.010	<0.002	0.120	0.002	0.001
Chromium	0.05	<0.005	0.120	0.005	0.05
Fluoride	1.4 to 2.4[1]	0.1	1.6	0.2	0.7
Lead	0.05	<0.005	0.140	0.015	0.02
Mercury	0.002	<0.0005	0.005	0.0005	0.0002
Nitrate (as N)	10	<1	3.4	1	1
Selenium	0.01	<0.001	0.060	0.001	0.001
Silver	0.05	<0.001	0.038	0.003	0.005
Endrin	0.0002	ND	0.00006	<0.00001	ND
Lindane	0.004	ND	0.0002	<0.00001	ND
Methoxychlor	0.1	ND	0.0001	<0.00001	ND
Toxaphene	0.005	ND	ND	ND	ND
2, 4-D	0.1	ND	0.0031	<0.00005	ND
2, 4, 5-TP Silvex	0.01	ND	0.0017	<0.00003	ND
Turbidity, TU	1[2]	<15	1×10^7	50	25
Coliform Bacteria (colonies/100 ml)	1[3]	<100	1.5×10^5	1,000	1×10^7
Secondary Regulations					
Copper	1	<0.0002	3.8	<0.0003	0.1
Iron	0.3	<0.05	7.8	0.12	0.2
Manganese	0.05	<0.1	36.0	0.25	0.05
Zinc	5	<0.02	6.5	0.07	0.5
Color, units	15				60
Foaming Agents (as MBAS)	0.5				1.5
Odor, TON	3				40
Other					
Trihalomethane	0.10[4]	ND	0.0011	ND	0.005
COD	None	2.5	125	30	100
TOC	None	1	50	10	40

[1]Varies with average annual maximum daily air temperature.
[2]Monthly average.
[3]Monthly average, membrane filter technique.
[4]Proposed MCL.

Source: "Guidelines for Planning the Location of Water Supply Intakes Downstream from Municipal Wastewater Treatment Facilities, EPA Office of Water Programs, Washington, D.C., April 1978.

ND = None Detected

Irrigation

Irrigation requirements were divided into two categories: agriculture and landscape. There are several important water quality parameters for irrigation, including salinity, sodium relationships, boron, and toxic metals. Additional parameters important in considering the use of wastewater are organics, suspended solids, nutrients, and pathogens. The suitability of water for irrigation depends upon the effects of the constituents in the water on both the crop and the soil.

Recommended criteria developed by the National Academy of Sciences are shown in Table 7-25. The recommended values do not include criteria for biochemical oxygen demand (BOD) or for suspended solids (SS) concentrations. BOD is not a pollutant in itself and exercises no direct harm, but can create an indirect effect by depressing the dissolved oxygen concentrations and causing odors. Based on operating systems and pilot operations, a BOD value in the range of 30 to 40 mg/l should not be harmful for general irrigation purposes. Suspended solids have not been a problem in the range of 30 to 50 mg/l. However, SS concentrations greater than 20 to 40 mg/l can result in inefficient disin-

TABLE 7-25. Water Quality Criteria for Agricultural Irrigation. (Unrestricted irrigation for waters used continuously on all soil)

Parameter	Concentration mg/l, except bacteria
Aluminum	5.0
Arsenic	0.10
Bacteria, Fecal Coliform, MPN/100 ml	1,000
Beryllium	0.10
Cadmium	0.010
Chromium	0.10
Cobalt	0.050
Copper	0.20
Fluoride	1.0
Iron	5.0
Lead	5.0
Lithium	2.5*
Manganese	0.20
Molybdenum	0.010
Nickel	0.20
Selenium	0.02
Vanadium	0.10
Zinc	2.0

*Value is 0.075 mg/l for citrus crops.

TABLE 7-26. Summary of Water Quality Criteria for Landscape Irrigation

Parameter	Concentration (mg/l, except as noted)
Aluminum	5.0
Arsenic	0.1
Bacteria, Fecal Coliform, MPN/100 ml	2.2
Beryllium	0.1
Biochemical Oxygen Demand	20
Cadmium	0.01
Chloride	100–200
Chromium	0.1
Cobalt	0.05
Copper	0.2
Fluoride	2.0
Iron	5.0
Lead	5.0
Lithium	2.5*
Manganese	0.2
Molybdenum	0.01
Nickel	0.2
Nitrogen	
Oil and Grease	Nil
pH, Units	6.0–9.0
Phenols	50
Selenium	0.02
Sodium Adsorption Ratio	8–18
Suspended Solids	15
Sulfate	200–400
Vanadium	0.1
Zinc	2.0

*0.075 mg/l for citrus crops.

fection, and the recommended value of 1,000/100 ml would be difficult to attain. Therefore, for irrigation of forage corps and pasture land, SS concentrations of less than 40 mg/l should be attained.

The use of treated municipal effluent for irrigation represents the single largest reuse practice in the United States. Reclaimed wastewaters have been successfully used for irrigation for many years. Health considerations are minimal for irrigation of nonfood crops, but the hazards are greater for food crops and pasturelands unless the water is adequately treated. The degree of treatment provided prior to irrigation with effluent varies from primary treatment only, to additional treatment after secondary treatment (tertiary treatment). Although no regulations currently exist, in some states the general trend has been toward

TABLE 7-27. Municipal Wastewater Effluent Quality Used for Power Plant Cooling (Concentrations in mg/l, except as noted)

Parameter	Nevada Power Co. Sunrise Station Las Vegas, Nevada	Clark County Sanitation District Las Vegas, Nevada	City of Denton, Denton, Texas	Southwestern Public Service Co. Lubbock, Texas	City of Burbank, California	City of Colorado Springs, Colorado
BOD	21	30	10	15	2	8
Suspended Solids	24	30	38	10	2	2
TDS	940	1,250–1,500	127	1,250	500	650
Sodium	–	–	–	–	88	50
Chloride	–	315	70	345	82	20
pH, units	7.7	7.5	7.2	7.3	7.0–7.2	6.9
Coliforms, MPN/100 ml	10	–	16,000	–	2–62	225
Total Hardness	–	–	–	250	160	240
Phosphate	19	–	–	21	20	1
Organic Nitrogen	1.0	–	–	–	39	1–5
Heavy Metals	–	–	trace	trace	trace	trace
Color, units	–	–	–	–	1	5
MBAS	–	–	–	–	0.5	0.15
Ammonia	–	–	–	–	6	27
Nitrate	1.0–3.4	–	–	–	8	0.5
Total Average Reuse* mgd	5.2	4.3	varies	20	6	2.4

*Not all use is for cooling–some reused for irrigation and other purposes.

TABLE 7-28. Municipal Effluent Used for Cooling (Concentrations in mg/l except as noted)

Parameter	Effluent Quality Before Treatment by User			
	El Paso Products Company Odessa, Texas	Champlin Refinery Enid, Oklahoma	DOW Chemical Company Midland, Michigan	Texaco, Inc. Amarillo, Texas
BOD	8	28	20–30	10
Suspended Solids	14	28	20–30	15
TDS	–	600	400–500	1,400
Sodium	–	–	–	300
Chloride	570	160	200–300	300
pH, units	7.6	7.2	7.6	7.7
Coliforms, MPN per 100 ml	600,000	–	1,000	2
Total Hardness	240	–	–	300
Phosphate	26	–	–	15
Organic Nitrogen	–	–	–	–
Heavy Metals	–	–	none	–
Color, units	–	15	–	–
MBAS	–	–	–	–
Ammonia	–	4	–	–
Nitrate	0.6	–	–	–
Total Average Reuse,* mgd	4.4	3.0	7.0	12.0

*Not all use is for cooling—some reused for irrigation and other purposes.

a requirement for secondary treatment as a minimum for all land discharges. This secondary treatment standard would include a maximum BOD of 30 mg/l, and maximum fecal coliform MPN of 200/100 ml.

Although turf and landscape vegetation are generally more resistant to harmful contaminants than agricultural crops, there are several water quality parameters that are important in considering the use of reclaimed wastewater for landscape irrigation. The most important consideration is the potential aerosol conveyance of bacteria and virus; others include suspended solids, oil, and grease. The suspended solids, concentration through clogging, can also affect the efficient operation of sprinkler nozzles. Therefore, to ensure coliform MPN of less than 2.2/100 ml, and nonclogging of sprinkler nozzles, the suspended solids concentration should be less than 15 mg/l at all times. Approximate guidelines for maximum contaminant concentrations for most types of landscape vegetation are summarized in Table 7-26.

Some states, such as California, have additional requirements for coliforms, turbidity, and treatment methods. Although no disease transmission has been reported for golf course irrigation with municipal effluent, California does require effluent to be chlorinated in order to keep coliform counts within the standards. No adverse effects have been noted unless excesively high chlorine dosages are used. Overchlorination, however, can cause the turf to become bleached and discolored with yellow streaks.

Water Reuse (No. 29, Feb. 1980, p. 15) also gives acceptable water quality for reuse of effluents for power plant cooling based upon six plants that are using the wastewater (Table 7-27). Similar data is given for industrial cooling waters at four locations (Table 7-28.)

QUESTIONS

1. Give the general characteristics of BOD, DO, pH, color, turbidity, and chlorides of an "contaminated" river water.
2. What is meant by establishing stream standards? What is meant by effluent standards? Give two ways they could be used.
3. What are the two current methods of establishing stream standards?
4. What is the philosophy of the National Technical Advisory Committee on Water Quality Criteria?
5. What four major characteristics of a stream are surveyed for swimming use?
6. What major characteristics should be considered when the receiving water is used for a municipal water supply?
7. List some of the major factors affecting fish life and hence fishing-water quality.
8. What are the three river water qualities which most directly affect use for irrigation?
9. Why is it desirable to compute a Pollution Index for a receiving water?
10. What two types of pollution indices have been attempted?
11. Explain in summary form what the author has proposed as Pollution Index.
12. What is the difference between the "Pollution Index" and the "overall Pollution Index"?

REFERENCES

Anderson, B. G.: The Toxicity Thresholds of Various Sodium Salts Determined by the Use of Daphnia Magna, *Sewage Works J.*, **18**(1):82, 1946.

Bennett, Thompson, and Parr: (1940) ref.

Boyle, R. H.: An American Tragedy, *Sports Illustrated* **55**(13):75, September 21, 1981.

Cairns, J.: We're in Hot Water, *Scientist and Citizen*, **10**:187–198, 1968.

——: Ecological Management Problems Caused by Heated Waste Water Discharge Into the Aquatic Environment, *Water Resources Bull.*, **6**:868–878, 1970.

——, and K. L. Dickson: Reduction and Restoration of the Number of Freshwater Protozoan Species Following Acute Exposure to Copper and Zinc, *Trans. Kans. Acad. Sci.*, **73**:1–10, 1970.

——, and R. E. Sparks: The Use of Bluegill Breathing to Detect Zinc, U.S. Environmental

Protection Agency, Water Pollution Control Res. Series 18050EDQ 12/71, U.S. Government Printing office, Washington, D.C., 1971.

——, W. H. Cromer, T. K. Bahns, and W. T. Waller: A Confirmation of Mount's Autopsy Technique for Zinc-Caused Fish Mortality, *Water Resources Bull.*, 7:956–968, 1971.

——, K. L. Dickson, and J. S. Crossman: The Response of Aquatic Communities to Spills of Hazardous Materials, *Proc. Nat. Conf. Hazardous Materials Spills*, pp. 179–197, 1972.

——, K. E. F. Hokanson, and L. L. Smith: Some Factors Influencing Toxicity of Linear Alkylate Sulphonate (LAS) to the Bluegill, *Trans. Am. Fish Soc.*, **100**:1–12, 1971.

California State Department of Health: *Report on a Pollution Survey of Santa Monica Bay Beachers*, 1942.

Camp, Thomas: *Water and Its Impurities*, Reinhold, New York, 1963.

Chemical News, September 15, 1966, pp. 504–508.

Chemical Week, October 5, 1968, pp. 94–95.

Chemical Week, October 29, 1969, pp. 80–86.

Cummins, Kenneth W.: in *River Ecology and Man*, edited by Oglesby, Carlson, and McCann, Academic Press, New York, 1972.

Dimick and Merryfield: Abstract in *Sewage Works J.*, **19**(5):958, Sept. 1947.

Duodoroff, P., M. Katch, and C. H. Tarzwell: The Toxicity of Some New Organic Insecticides to Fish, Activity Report No. 7, 13 pp., Federal Security Agency, Public Health Service, Environmental Health Center, Cincinnati, Ohio, January–March, 1951.

Ehlers, V. M.: Experiences with Sewage Farming in Southwest U.S., *Am. J. Public Health*, **25**:119, 1935.

Eldridge, E. F.: *Industrial Waste Treatment Practice;* McGraw-Hill Book Company, New York, 1942.

Ellis, M. M.: Industrial Wastes and Fish Life, *Sewage Works J.*, **18**(4):764, 1948.

Ellis: *Fresh Water From the Ocean*, The Conservation Foundation, Inc., Ronald Press, New York, 1954.

H. Jaeger Ztschr. Die Aetiologie des infectisen fieberhaften, Octerus, *Fortschr. Hyg. u. Infectionskrankh*, **12**:525, 1892.

Henry, C. D., R. E. Moldenhauer, L. E. Engelbert, and E. Truog: Sewage Effluent Disposal Through Crop Irrigation, *Sewage Ind. Wastes*, **26**(2):123 1954.

Horton, Robert K.: An Index-Number System for Rating Water Quality, *J. Water Poll. Control Fed.*, **37**(3): 1965.

Hubbs, Carl L.: Sewage Treatment and Fish Life, *Sewage Works J.*, **5**(6):1033–1040, 1933.

Interim Report of the National Technical Advisory Committee on Water Quality Criteria to the Secretary of Interior, June 30, 1967, Washington, D.C.

Mackenthun, K. M.: *The Practice of Water Pollution Biology*, U.S. Department Interior, Federal Water Pollution Control Administration, Div. of Technical Support, Washington, D.C., 1969.

Miller, J. E.: Schistosome Dermatitis as a Bathing Place Problem, *Am. J. Public Health*, **31**:305, 1941.

Nemerow, Nelson L.: "Economics of Waste Treatment," *Proceedings of the 1st Mid-Atlantic Industrial Waste Conference*, University of Delaware, Dover, Del., Nov. 13, 1967.

New York State Department of Health: *Periodic Report of Water Quality Surveillance Network*, 1960–1964.

New York State Department of Health: *Quarterly Report*, 1932.

Nordell, E.: *Water Treatment for Industrial and Other Uses*, Reinhold Publishing Company, New York, 1951.

Rand, M. C.: Unpublished data, Syracuse University, New York, 1969.

Reece, R. J.: *Report of Medical Officer to Local Government Board* (London), **38**:90, 1908–09.

Report of Committee on Bathing Places, *Am. J. Public Health*, **12**:121, 1922.

Smith, R. S., T. D. Woolsey, and A. H. Stevenson: *Bathing Water Quality and Health. I Great Lakes*, U.S. Public Health Service, Cincinnati, 1951.

Stallybrass, C. O.: *The Principles of Epidemiology*, MacMillan Company, New York, 1931.

Steimke, R., and W. Eckenfelder: *N.C. State College Engineering Research Bull. No. 33*, Raleigh, N.C., 1947.

Sumitomo, Hisashi, and N. L. Nemerow: Unpublished research, Syracuse University, New York, 1967.

Swimming Pools and Bathing Beaches. *Proceedings of Seventh Florida Public Health Engineering Conference, Bull. No. 62*, March 24, 25, 1953.

Truesdale, G. A. and —— Knowles: *Water Poll. Res. Lab. Reprint No. 295*, 1956.

U.S. Dept. of Commerce, Bureau of Census: *1963 Census of Manufacturers, Water Use in Manufacturing MC 63*(*1*)-10, Washington, D.C., 1966.

U.S. Public Health Service, *Drinking Water Standards*, P.H.S. Pub. No. 956, 1962.

U.S. Senate, 86th Congress, 2nd Session, Select Committee on National Water Resources 1960: *Water Resources Activities in the U.S., Electric Power in Relation to the Nation's Water Resources Committee*, Print No. 10, Washington, D.C., 1960.

Vaughn, V. C.: *Epidemiology and Public Health*, Kimpton, London, 1922–23.

Water Quality Criteria, State Water Pollution Control Board, Pub. No. 3, Sacramento, California, 1952.

U.S. Public Health Service, Bathing Water Quality and Health. II Inland River, Cincinnati Ohio, 1952.

U.S. Public Health Service, Bathing Water Quality and Health. III Tidal Water, Cincinnati, Ohio, 1952.

Additional Reading

1. *Scientific Monthly* 74, 1, January 1952.
2. Stevenson, Albert H.: Water Quality Requirements for Recreational Uses. Presented at Annual Conference of the Pennsylvania Sewage and Industrial Waste Association, August 26, 1948.
3. Stevenson, Albert H.: A Statistical Study of Illness in Relation to Natural Bathing Water Quality, presented at *American Public Health Assoc.* 76th Annual Meeting, NYC, October 27, 1949.
4. Stevenson, Albert H.: Studies of Bathing Water Quality and Health, *American Journal of Public Health* 43, 5, 529, May 1953.
5. Stiemke, R. E. and W. W. Eckenfelder: A Practical Method for Predicting the Effects of Common Acids and Alkalies on the Survival of Fish" *Eng. Research Bull. No. 33*, N.C. State College 46, 5, January 1947.
6. The Case Against the Rain," Ministry of the Environment, Ontario, Canada, 1980.
7. The Effects of Sewage Pollution on the Fish Population of a Midwestern Stream Katz, Max, and Gaufin, Arden R. *Trans. Amer. Fisheries Society*, 82, 156 (1952).

8
ESTUARINE ANALYSIS

8-1. INTRODUCTION

An estuary can be defined as the confluence of a freshwater river with salt water, where mixing occurs under the tidal action previously described in Chapter 2, Section 6. According to Singer (1969) there are between 850 and 900 estuarine systems around the periphery of our country. The Clean Water Restoration Act of 1966 defines "estuarine zones" as "environmental systems consisting of an estuary and those transitional areas which are consistently influenced or affected by water from an estuary such as but not limited to salt marshes, coastal and intertidal areas, bays, harbors, lagoons, in-shore waters and channels" and defines "estuaries" as "all or part of the mouth of a navigable or inter-state river or stream or other body of water having an unimpaired natural connection with the open sea and within which the sea water is measurably diluted with fresh water derived from land drainage."

Estuaries are very sensitive and their ecosystems can be very easily upset. For example, when the floodgates were opened on Lake Okeechobee, Florida on February 18, 1983 allowing tons of water to gush into the sensitive St. Lucie estuary, conservationists agreed that the grass beds which provided food and shelter for fish, and subsequently, the fish themselves, would die (*The Post*, West Palm Beach, Fla., A-1 Feb. 19, 1983).

We cannot use the exact mathematical procedures given in Chapters 4 to 6 to solve estuary problems because of the different physical and chemical conditions in tidal and nontidal watercourses. The protection of our water resources is largely for economic purposes, as shown by an enumeration of some of the major activities occurring in estuaries: (1) mining and petroleum, (2) marine engineering, (3) recreation, (4) health and welfare, (5) transportation, (6) food and agriculture, (7) defense and space, and (8) research and development. The U.S. Department of Commerce in 1964 made an intensive search of the value of estuaries, and estimated the level of economic activity to be \$21.4 billion. As recently as June 1971 (Janson, 1971) the state of Delaware enacted a new law to restrict new wet industries from locating anywhere along their coastline from Pennsylvania to the Maryland border. The decision was made in order to preserve at least the existing water quality of these resources.

The Governor of Delaware, Russell Peterson, feared that permitting any heavy industry on the marshy bay coast would only lead to more development and to pollution of the popular beaches of Delaware's lower bay and Atlantic Coast. In fact, he was quoted as saying that "jobs are very important to our people, but so is the over-all quality of our environment" (Janson, 1971, p. 1). He said, "southern Delaware chose to retain its natural beauty, to serve millions of vacationers from the eastern megalopolis and to thrive on tourism, a major industry, rather than yield to the threat of oil spills and air pollution" (ibid.). He does not believe that heavy industry is compatible with recreation. This represents a bold step taken by state government. Delaware hopes that the law will serve as a landmark for the nation.

The legal question of ownership, and especially the use of estuarine waters, has been plaguing society for a long time. Since the use of these waters is so closely tied to the use of adjoining land, some controversy must occur. Questions arise such as, how far can government go in telling people how they can use their land? At what point do restrictions amount to confiscation? When does government have to pay property owners if it is to keep them from doing what it doesn't want done?

A recent First District Court of Appeals decision in Tallahassee, Florida (*Miami Herald*, July 27, 1980, p. 14C) declared that the state of Florida, and the county "have taken that property (Estero Bay, Florida) for a public purpose without full compensation in violation of the United States and Florida constitutions." It is said that private property was being unduly burdened to provide public benefit. This and other similar issues are far from settled, but clearly, contamination of estuaries cannot any longer be prevented simply by disallowing private development on the land.

It is not the objective of this chapter to discuss the controversial basic theories that may apply to estuaries. In the author's opinion, the field is still too poorly defined to form conclusions on whether the tidal prism theory or diffusion theory is more suitable, or whether either yields the precise and existing conditions sought. The *National Council for Stream Improvement Technical Bulletin* prepared by Dr. Howard Edde (1970) and Chapter 6 of Dr. Donald O'Connor's lecture notes (1968) provide a detailed and up-to-date review of the subject. This chapter only attempts to demonstrate the use of certain techniques commonly used in solving oxygen sag predictions in estuaries.

Statistical Technique

Your author has used a statistical-averaging method in estuarine sampling for many years. It describes what is taking place at the time of estuary sampling and may be used (with care) to predict what may occur at some future time. However, the method is only valid for future predictions if estuary conditions remain relatively constant, or relatively the same as those under which previous analyses were made. The procedure involves repetitive sampling three times on each tide cycle: (1) high tide; (2) mean tide (half the time interval between high and low tide); and (3) low tide. This is repeated daily for the entire 28-day tide period. The proper depth to select a sample is beneath the surface, about $\frac{1}{4}$ to $\frac{1}{3}$ of the distance to the bottom. It is presumed that adequate mixing has occurred in the estuaries. Concentration of contaminants in an estuary varies considerably during the range between high and low diurnal tides. As pointed out in Chapter 2, most estuaries in the northern hemisphere in waters surrounding the United States exhibit two high and two low tides daily. These are illustrated graphically in Fig. 8-1. As shown, eight samples would be collected in a 24-hour period; two represent *low tide* conditions, (1) and (5); two represent *mean high tide*, (2) and (6); two represent high tide, (3) and (7); and two represent *mean low tide*, (4) and (8).

Table 8-1 illustrates typical results that may be expected when sampling for chloride ion content and dissolved oxygen in a typical estuary eight times daily.

The reader can observe that low tide values for the chloride analyses of the sampling point (Table 8-1) varied from 1,000 to 1,200 just within one daily tide cycle. In a 28-day period the range of variation will be even more extreme because of the varying and relative positions of the earth and moon. Dissolved oxygen values (3.0 and 2.8 ppm) will also vary considerably, not only at low tide but also during mean and high tide periods. Low tide represents the critical time, since, in general, the greatest ratio of river contaminants to seawater dilution exists then. For example, the chloride ion concentration varied in our illustration (Table 8-1) from 1,000 ppm at low tide to 25,000 at high tide. The writer suggests that all data for comparable tide periods (i.e., all low tides, high tides,

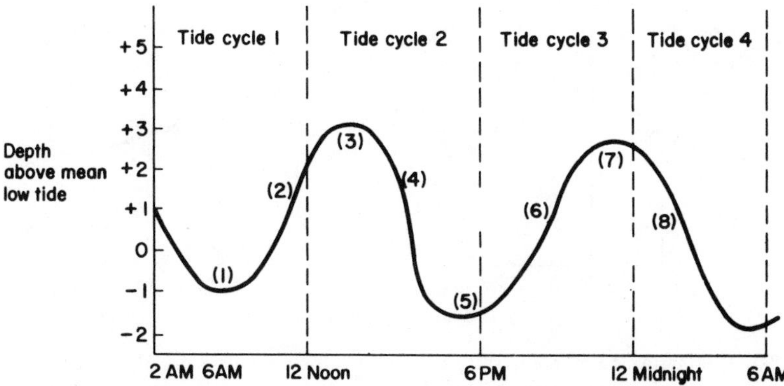

FIG. 8-1. Hypothetical tide cycles and sampling times. (1) through (8) = samples.

TABLE 8-1. Typical Sampling and Chloride Ion Analysis in a Typical Estuary over 24-Hour Period

Sample no.	Time	Description of tide situation	Chloride ion (ppm)	Dissolved oxygen (ppm)
1	6:00 AM	Low	1,000	3.0
2	9:00 AM	Mean high tide	7,500	5.0
3	12:15 PM	High	25,000	8.0
4	3:30 PM	Mean low tide	9,000	5.2
5	6:45 PM	Low	1,220	2.8
6	10:00 PM	Mean high tide	8,000	4.5
7	12:45 AM	High	20,000	7.8
8	3:55 AM	Mean low tide	8,500	5.6

etc.) be statistically analyzed separately. For example, all low tide (critical) analytical results can be arranged in order of magnitude, assigned an intensity number, and computed and placed in plotting position on normal probability paper, such as that shown in Fig. 8-2. Then, the desired risk frequency can easily be calculated graphically.

Suppose we decide that we wish to predict the estuary situation that will exist 95 percent of the time, and have the following data obtained at low tide at one sampling point (Table 8-2). The computed data shown in Table 8-2 are then

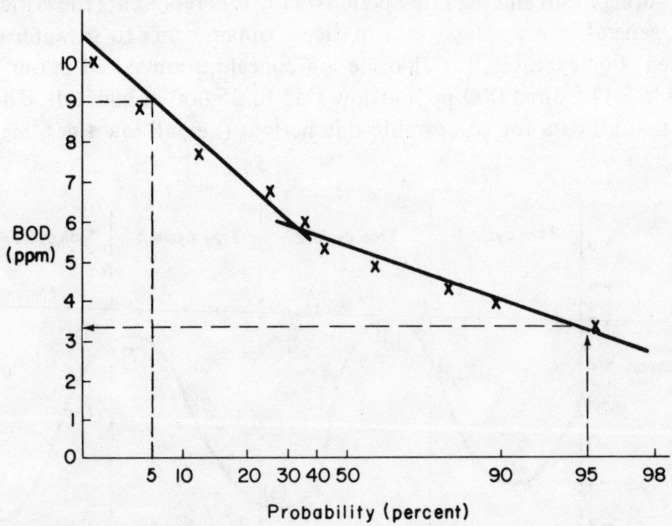

FIG. 8-2. Contaminant probability graph. (A plot of extreme data usually yields a straight line on normal probability paper.)

TABLE 8-2. Illustrative Low Tide Estuary BOD (Data collected over one complete 28-day moon cycle)

Date	Low tide	BOD (ppm)	Magnitude (intensity number)	BOD in order of magnitude	M	Plotting position ($M/N + 1$)
June 1	1	6.5	14	10.1	1	.0175
	2	4.5	45	9.6	2	.035
June 2	3	7.0	12	9.0	3	.0525
	4	5.0	32	9.0	4	.0700
June 3	5	5.8	20	8.2	5	.0875
	6	4.2	48	8.2	6	.1050
June 4	7	6.0	16	8.0	7	.1225
	8	3.8	52	7.6	8	.1400
June 5	9	7.6	8	7.6	9	.1575
	10	5.5	24	7.4	10	.1750
June 6	11	10.1	1	7.0	11	.1925
	12	8.0	7	7.0	12	.2100
June 7	13	7.6	9	6.6	13	.2275
	14	3.0	56	6.5	14	.2450
June 8	15	9.0	4	6.2	15	.2625
	16	5.7	21	6.0	16	.2800
June 9	17	4.6	43	6.0	17	.2975
	18	3.9	51	6.0	18	.3150
June 10	19	5.2	29	6.0	19	.3325
	20	4.5	46	5.8	20	.3500
June 11	21	5.5	25	5.7	21	.3675
	22	4.9	38	5.6	22	.3850
June 12	23	5.0	33	5.5	23	.4025
	24	4.6	42	5.5	24	.4200
June 13	25	5.2	28	5.5	25	.4375
	26	4.8	40	5.4	26	.4550
June 14	27	5.0	34	5.2	27	.4725
	28	5.0	35	5.2	28	.4900
June 15	29	5.2	27	5.2	29	.5075
	30	5.4	25	5.1	30	.5250
June 16	31	5.1	31	5.1	31	.5425
	32	6.0	17	5.0	32	.5600
June 17	33	6.0	18	5.0	33	.5775
	34	7.0	11	5.0	34	.5950
June 18	35	4.2	47	5.0	35	.6125
	36	5.6	22	5.0	36	.6300
June 19	37	4.8	39	4.9	37	.6475
	38	6.2	15	4.9	38	.6650

TABLE 8-2. *(Continued)*

Date	Low tide	BOD (ppm)	Magnitude (intensity number)	BOD in order of magnitude	M	Plotting position ($M/N + 1$)
June 20	39	5.1	30	4.8	39	.6825
	40	7.4	10	4.8	40	.7000
June 21	41	4.0	49	4.7	41	.7175
	42	8.2	5	4.6	42	.7350
June 22	43	4.7	41	4.6	43	.7525
	44	9.0	3	4.5	44	.7700
June 23	45	3.5	55	4.5	45	.7875
	46	5.5	23	4.5	46	.8050
June 24	47	3.8	53	4.2	47	.8225
	48	6.0	19	4.2	48	.8400
June 25	49	3.7	54	4.0	49	.8575
	50	6.6	13	4.0	50	.8750
June 26	51	4.5	44	3.9	51	.8925
	52	4.9	37	3.8	52	.9100
June 27	53	5.0	36	3.8	53	.9275
	54	8.2	6	3.7	54	.9450
June 28	55	4.0	50	3.5	55	.9625
	56	9.6	2	3.0	56	.9800

NOTE–N = number of observations (56).

plotted in Fig. 8-2. From the figure the reader can observe that a BOD value (at low tide at that sampling point) of 3.5 or greater will occur at least 95 percent of the time, based upon past results. If this degree of reliability is sufficient for the design objectives, then the value of 3.5 can be utilized in computations. The same procedure can be utilized with dissolved oxygen values to obtain the oxygen sag curve. The shape of the sag curve thus obtained will be different (dip deeper) at low tide than at high tide (show less sag). The stream analyst can then determine the sag curve under all tide conditions and decide which to use for design purposes. Generally, as mentioned before, the low tide conditions are used for design because of the higher contaminant levels. Stormy weather conditions are usually avoided, since sampling is extremely hazardous, and critical, low tide excess pollution rarely exists during these periods.

8-2. MODIFIED STREETER-PHELPS TECHNIQUE IN ESTUARY ANALYSIS

In general, the estuary study consists of *three basic in situ undertakings*, *four plottings of data*, and *six computations.*

Three Basic In-Stream Studies

1. Measure the following parameters in chronological order of estuary conditions:

   a. Rate of flow of outflow from land drainage Q in mgd
   b. Temperature of outflow from land T in °C
   c. Average cross-sectional velocity U in miles per day
   d. Average depth of water H in feet
   e. Cross-sectional area A in square feet

2. Release a flow tracer, usually at low tide water or high tide water, and usually sodium chloride, in the reach where the minimum dissolved oxygen (bottom of sag) exists.

3. After a given number (three to five) of tidal cycles, measure the five-day, 20°C, BOD, DO, and tracer concentration at various stations both upstream and downstream of the point where the tracer was added at low tide.

Plot the Following Data and Relationships

1. Log of BOD versus stations on the estuary. A graph such as the one shown in Fig. 8-3 will be obtained.

2. Plot the dissolved oxygen versus stations on the estuary. A typical curve as shown in Fig. 8-4 is usually obtained.

3. Plot log C/C_0 versus distance from the peak concentration squared, where C is the tracer concentration and C_0 is the initial tracer concentration. A straight-line relationship such as shown in Fig. 8-5 is usually obtained.

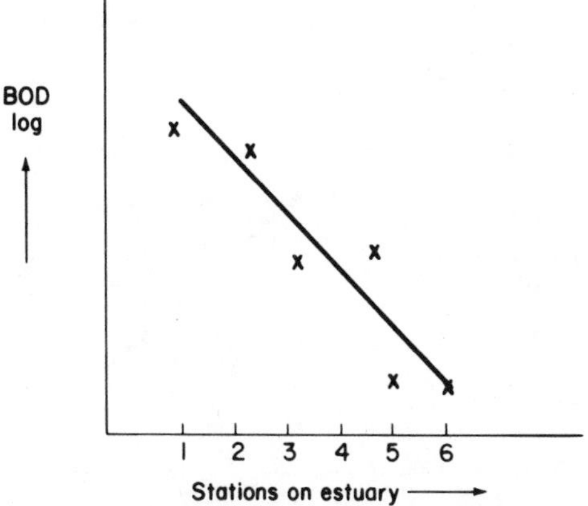

FIG. 8-3. BOD versus stations on estuary.

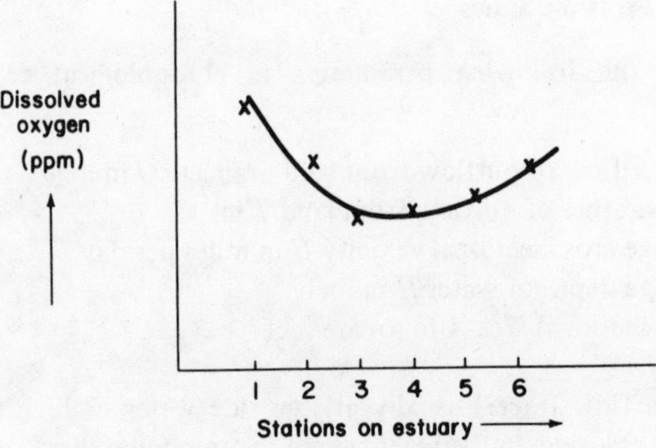

FIG. 8-4. Dissolved oxygen versus stations on estuary.

4. Plot tracer concentration versus stations on estuary. A peak concentration of tracer is usually found at one station as shown in Fig. 8-6.

Six Computations Used in the Modified Streeter-Phelps Technique

1. Compute the Dispersion Coefficient E from tracer data and the formula

$$\frac{1}{(\text{distance from peak})^2} \ln \frac{C}{C_0} = -\frac{1}{4Et}$$

where t = time in days since release of tracer

2. Compute the *estuary decomposition rate* J_1, which can really be considered the BOD reaction coefficient, from the plot, Fig. 8-3, or from

$$L = L_0 e^{-J_1 x}$$

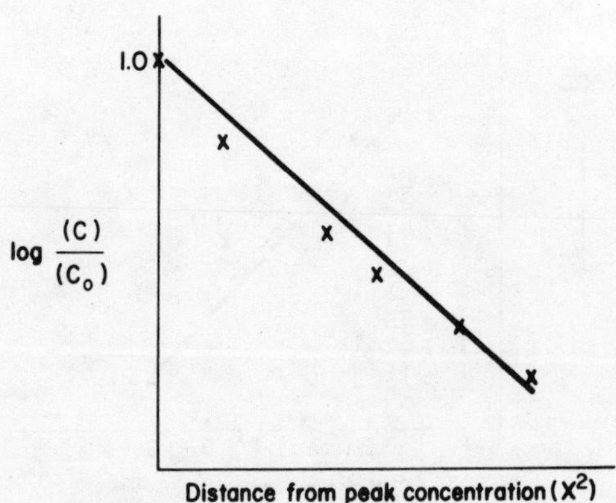

FIG. 8-5. Log C/C_0 versus distance from peak concentration squared.

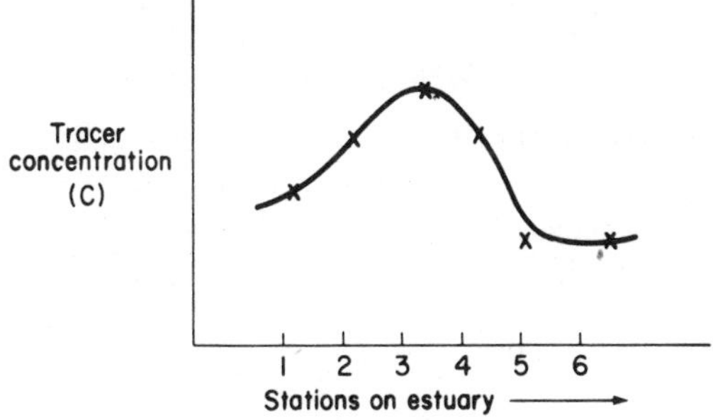

FIG. 8-6. Tracer concentrations versus stations on estuary.

3. Compute the *stream deoxygenation* coefficient constant k_d from the following equation

$$J_1 = \frac{U}{2E}\left[1 - \sqrt{\frac{1 + 4k_d E}{U^2}}\right]$$

4. Compute the *stream reaeration rate* k_2 from the following equation

$$k_2 \text{ base } e = \frac{(D_L U)^{1/2}}{H^{3/2}}$$

5. Compute J_2, *the estuary reaeration constant* (dissolved oxygen reaction coefficient), from the following equation

$$J_2 = \frac{U}{2E}\left[1 - \sqrt{1 + \frac{4k_2 E}{U^2}}\right]$$

6. Now it remains only for the analyst to use all the information obtained from the above computations to compute the sag curve profile or the dissolved oxygen profile from the modified Streeter-Phelps equation

$$D = \frac{k_d}{k_2 - k_d} L_0(e^{-J_1 x} - e^{-J_2 x}) + D_0 e^{-J_2 x}$$

An illustrative solution to an estuary pollution problem is given in the following actual case history. The data and solution have been provided by the engineering firm of O'Brien and Gere of Syracuse, New York (Report of Kent County, 1967). They were asked to design a treatment plant for the city of Dover, Delaware, to determine the assimilation capacity and the dissolved oxygen profile of the St. Jones River from the Dover sewage treatment plant to a point six miles below the discharge.

Kent County Comprehensive Sewerage Study (Report of Kent County, 1967)

Assimilation Capacity of the St. Jones River

It was assumed that the flow from the Dover Treatment Plant to a point six miles downstream is constant. The area contributing runoff to this section of the St. Jones River was calculated to be 38 square miles. Using a runoff figure of 0.3 cfs/sq mi the flow in this section was calculated as follows:

$Q = (38)(.3) = 12.8$ cfs = 8.3 mgd

General at a condition of Low Water Slack in the river

Min. allowable DO	5.0 mg/l
Critical stream temperature	25°C
DO at saturation	8.4 mg/l
Allowable oxygen deficit	3.4 mg/l
Tributary watershed	38 sq mi
Streamflow @ 0.3 cfs/sq mi	8.3 mgd
Cross-sectional area	790 ft^2

$$U = \frac{12.8 \text{ cfs}}{790 \text{ ft}^2} = 58.5 \frac{\text{ft}}{\text{hr}} = 0.26 \frac{\text{mi}}{\text{day}}$$

Evaluation of the Dispersion Coefficient (E) from the Cl^- profiles illustrated in Figs. 8-7 and 8-10:

$$\ln \frac{\frac{C_2}{C_1}}{X_2 - X_1} = \frac{U}{E}$$

Notations:

Q	–	Flow
BOD	–	Biochemical oxygen demand
DO	–	Dissolved oxygen
U	–	Fresh water velocity
E	–	Dispersion coefficient
J_1	–	BOD reaction coefficient
K_d	–	Deoxygenation constant
D_1	–	Oxygen diffusivity constant
K_2	–	Reoxygenation constant
J_2	–	DO reaction coefficient
L	–	Pollutant substance
D	–	Oxygen deficit

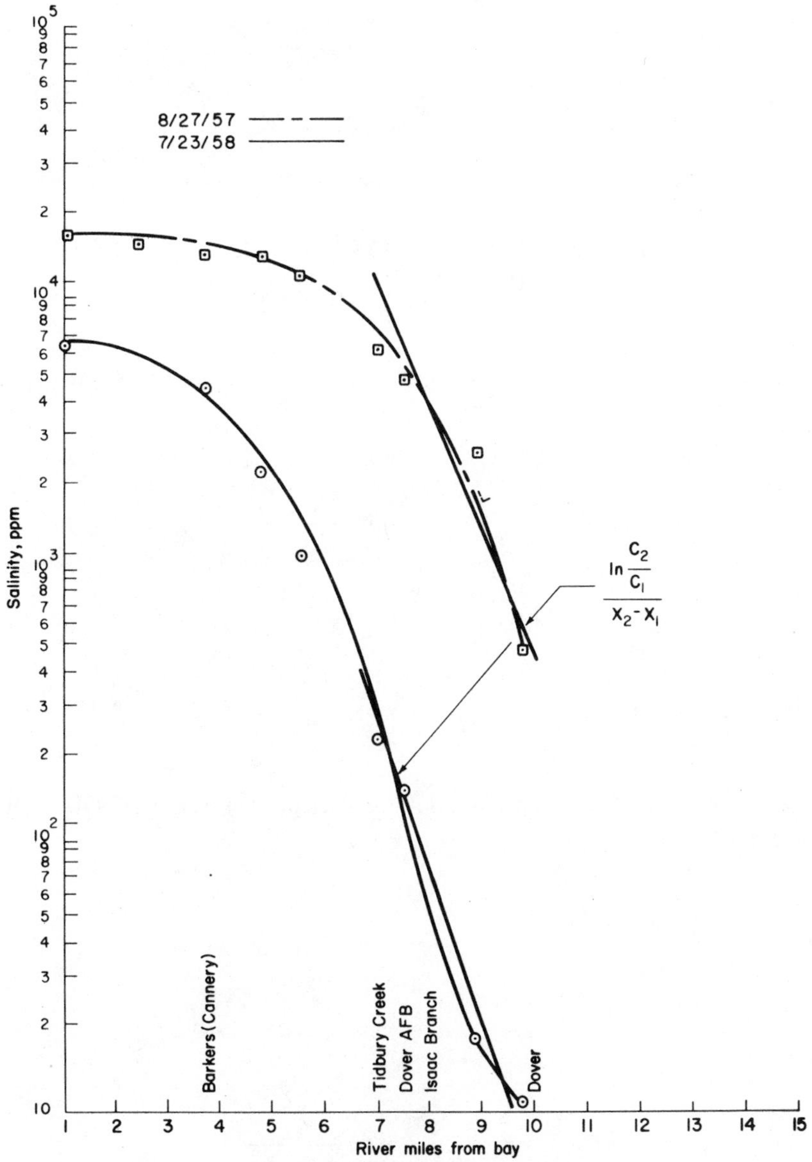

FIG. 8-7. St. Jones River HWS salinity.

D_c – Critical oxygen deficit
X – Distance downstream from Dover Treatment Plant
X_c – Critical distance

Values

@ *LWS*

Year	C_2	C_1	$X_2 - X_1$	$\text{Ln}\dfrac{\frac{C_2}{C_1}}{X_2 - X_1}$
1957	1×10^3	1×10^2	1.40	1.64
1958	1×10^3	1×10^2	1.45	1.58
1959	1×10^2	1×10^1	2.70	0.85
@ *HWS*				
1957	1×10^4	1×10^3	1.95	1.18
1958	1×10^3	1×10^2	1.80	1.28

$$\text{Avg. } Ln \frac{\frac{C_2}{C_1}}{X_2 - X_1} = 1.30$$

$$1.30 = \frac{26}{E}$$

$$E = +\dot{0}.2 \frac{\text{sq mi}}{\text{day}}$$

Evaluation of the BOD River Reaction Coefficient (J_1) from the BOD profiles as illustrated in Fig. 8-8:

$$J_1 = Ln \frac{\frac{C_2}{C_1}}{X_2 - X_1}$$

@ *LWS*

Date	C_2	C_1	$X_2 - X_1$	$Ln \dfrac{\frac{C_2}{C_1}}{X_2 - X_1}$
6/9/66	40	9	2	0.750
7/27/66	40	10	1.4	0.994

$$J_1 \text{ Avg.} = +0.872$$

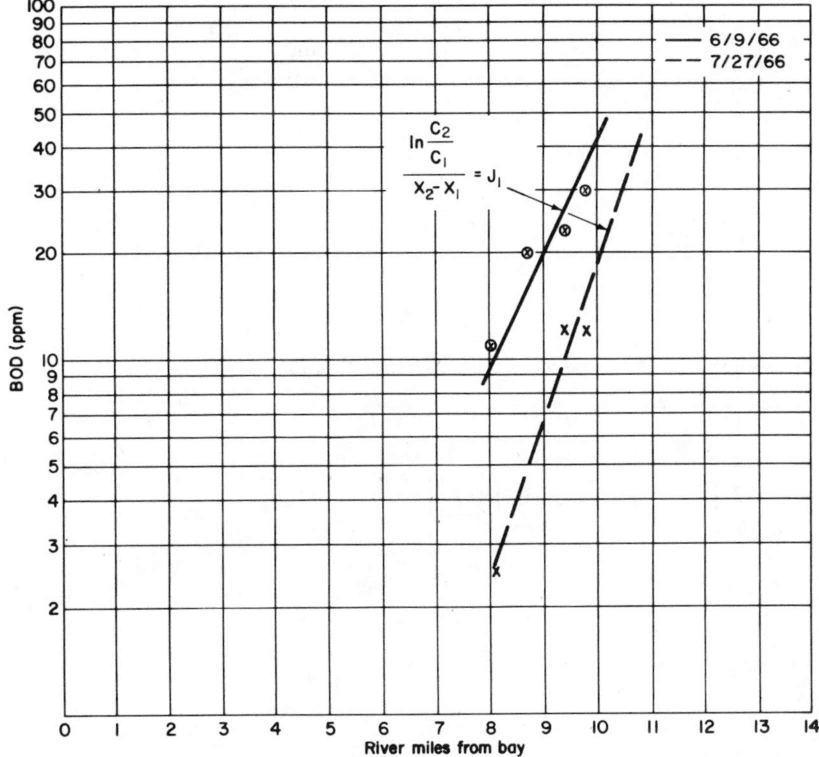

FIG. 8-8. St. Jones River LWS BOD ppm.

Evaluation of the River Deoxygenation Constant (K_d):

$$J_1 = \frac{U}{2E}\left[1 - \sqrt{1 + \frac{4K_d E}{U_2}}\right]$$

$$-0.872 = \frac{0.262}{2(.2)}\left[1 - \sqrt{1 + \frac{4K_d(0.2)}{(0.26)^2}}\right]$$

$$K_d = \frac{0.402}{\text{day}}$$

Evaluation of the River Reoxygenation Constant (K_2):

$$K_2 = \frac{(D_L U)^{1/2}}{H^{3/2}}$$

D_L = Oxygen Diffusivity Constant = $0.81 \times 10^{-4} \dfrac{\text{ft}^2}{\text{mi}}$ at 20°C

$$K_2 = \frac{(0.81 \times 10^{-4})(58.4 \text{ ft/hr})^{1/2}}{5^{3/2}}$$

$$K_2 = \frac{0.147}{\text{day}} \text{ @ } 20°\text{C}$$

$$K_2 = 0.147(1.02)^5$$

$$K_2 = \frac{0.168}{\text{day}} \text{ @ } 25°\text{C}$$

Evaluation of the DO River Reaction Coefficient (J_2):

$$J_2 = \frac{U}{2E}\left[1 - \sqrt{\frac{1 + 4K_2E}{U^2}}\right]$$

$$J_2 = \frac{0.262}{2(.2)}\left[1 - \sqrt{1 + \frac{4(0.168)(0.2)}{(0.26)^2}}\right]$$

$$J_2 = -0.465$$

Summary of Constants:

$$K_d = \frac{0.402}{\text{day}} \text{ @ } 25°\text{C}$$

$$K_2 = \frac{0.168}{\text{day}} \text{ @ } 25°\text{C}$$

$$J_1 = \frac{-0.872}{\text{mi}}$$

$$J_2 = \frac{-0.465}{\text{mi}}$$

Calculation of Dissolved Oxygen Distribution:

$$D = \frac{K_d}{K_2 - K_d} L_0(e^{-J_1X} - e^{-J_2X}) + D_0e^{-J_2X}$$

L (ult) at Dover Treatment Plant effluent = 100 ppm
L_0 (ult) in River at Treatment Plant = 24 ppm
DO in River at Treatment Plant = 2.5 ppm
D_0 = 8.4 - 2.5 = 5.9

$$D = \frac{(0.402)(24)}{0.168 - 0.402}(e^{+0.872x} - e^{+0.465x}) + 5.9e^{+0.465x}$$ and plotted in Fig. 8-9

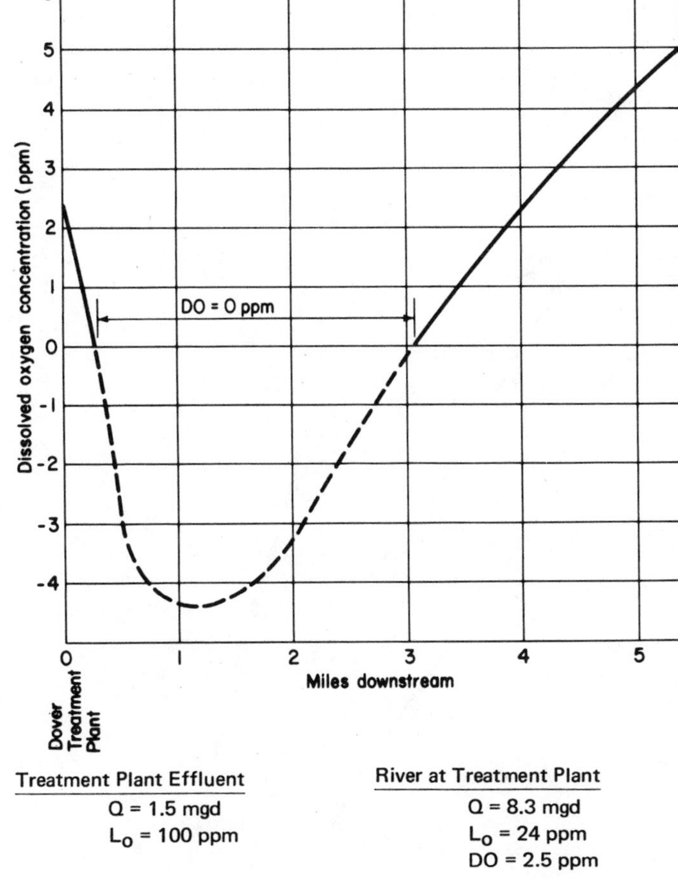

FIG. 8-9. St. Jones River dissolved oxygen versus river miles.

Distance from T.P. X (mi)	Dissolved oxygen deficit D (ppm)	Dissolved oxygen DO (ppm)
0.5	11.0	-2.6
1.0	12.8	-4.4
1.5	12.6	-4.2
2.0	11.7	-3.3
3.0	8.8	-0.2
4.0	6.1	+2.3
5.0	4.1	+4.3

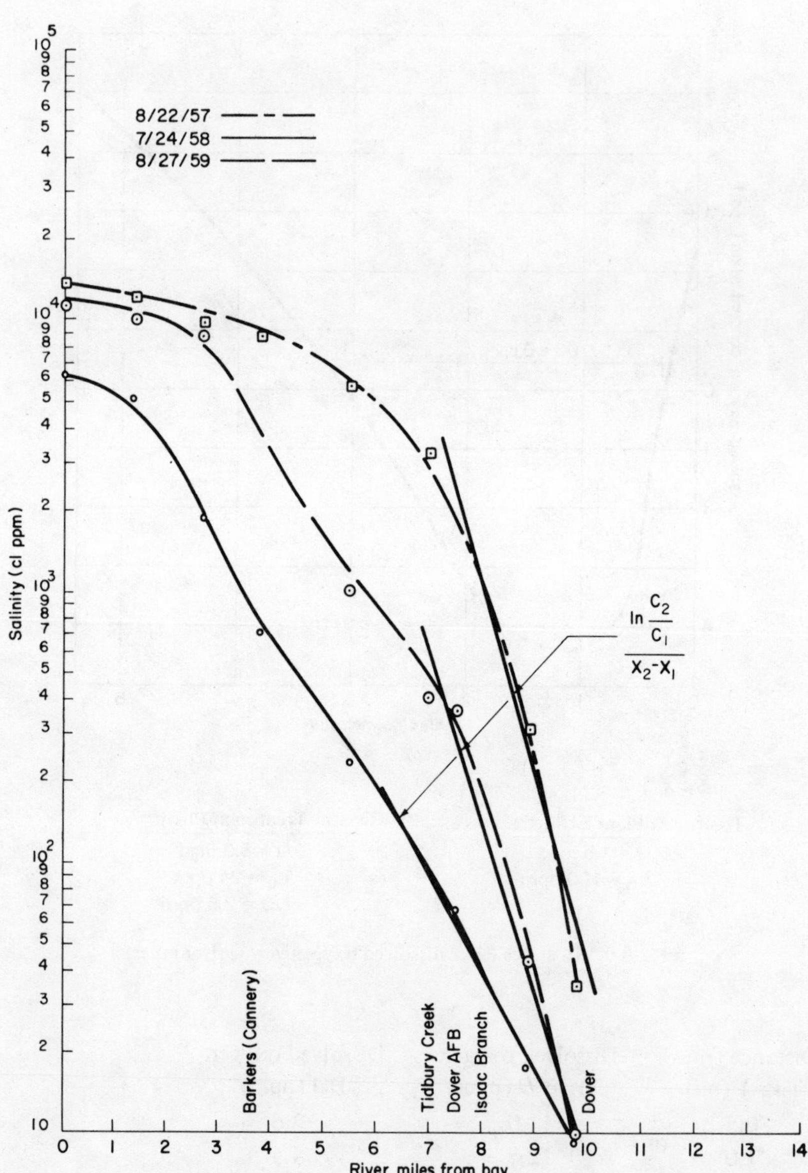

FIG. 8-10. St. Jones River L.W.S. salinity.

Calculation of Critical BOD Loading (L):

Initial Dissolved Oxygen Deficit (DO) = 8.4 - 2.5 = 5.9 ppm
Allowable Critical Deficit (D_c) = 8.4 - 5.0 = 3.4 ppm
Assume Initial River BOD Loading (L_o) = 10.0 ppm

$$\frac{1}{J_1 - J_2} = \frac{1}{\dfrac{-0.872}{\text{mi}} - \dfrac{0.465}{\text{mi}}} = -2.32$$

$$F = \frac{K_d}{K_2 - K_r} = \frac{0.402}{0.168 - 0.402} = 1.72$$

Critical Distance Downstream from Dover Treatment Plant (X_c) at L_o = 10.0 ppm:

$$X_c = \frac{1}{J_1 - J_2} \ln \left[\frac{J_2}{J_1} \left(1 - \frac{D_o}{L_o F}\right)\right]$$

$$X_c = -2.32 \ln \frac{-0.465}{-0.872}\, 1 - \frac{5.9}{(10)(1.72)}$$

X_c = 2.5 mi.

Dissolved Oxygen Deficit at X_c:

$$D = \frac{K_d L_o}{K_2 - K_d} (e^{-J_1 X_c} - e^{-J_2 X_c}) + D_o e^{-J_2 X_c}$$

$$D = \frac{(0.402)10)}{0.168 - 0.402} (e^{+0.872(2.5)} - e^{+0.465(2.5)}$$

$$+ 5.9 e^{+0.465(2.5)}$$

D = 6.43 ppm

Tabulation of Assumed L_o

L_o (ppm)	X_c (mi)	D (ppm)
5.0	4.2	1.86
7.0	3.1	3.48 ← D_c = 3.4
10.0	2.5	6.43

At L_o (ultimate) = 7.0 ppm
L_5 = 0.75 (ultimate) = 5.3 ppm
If Upstream BOD_5 = 1.0 ppm

$$\text{Allowable Plant BOD}_5 = 5.3 - 1.0 = 4.3 \text{ ppm @ 8.3 mgd}$$
$$= 23.0 \text{ ppm @ 1.5 mgd}$$

$$\text{Allowable Plant BOD}_5 \text{ at present avg. flow} = 23.0 \text{ ppm}$$

QUESTIONS

1. How would you define an estuarine water?
2. Describe one method of estuary analysis used by your author.
3. How do the chloride and dissolved oxygen concentrations change from low tide to high tide and back to low tide again?
4. Under what tidal conditions do critical oxygen concentrations occur and why?
5. How would extreme tidal contamination data be analyzed in the system proposed by the author?
6. Describe in general how the modified Streeter-Phelps technique is used in estuarine analysis for computing oxygen deficits.

REFERENCES

Edde, Howard: An Introduction to Determination of Estuarine Assimilative Capacity, *Natl. Council for Air and Stream Improvement Tech. Bull. No. 236*, 103 Park Avenue, New York, April 1970.

Barnes, R. S. K., and J. Green, eds.: *Estuarine Environment*, Essex, England, Applied Science Publishers, Ltd., 1972, 133 pp.

Janson, Donald: Delaware Bars Heavy Industry from Coast to Curb Pollution, *New York Times*, June 29, 1971, p. 1.

O'Connor, D. J.: "Stream and Estuarine Analysis," Chap. 6. Lecture Notes (Mimeographed), Manhattan College, New York, 1968.

Report of Kent County Comprehensive Sewage Study, O'Brien and Gere Engineers, Syracuse, N.Y., January 1967.

Singer, S. Fred: Federal Interest in Estuarine Zones Builds, *J. of Environ. Sci. and Technol.*, 3(2):124, 1969.

9
RIVER BASIN ADMINISTRATION

Since the passage of the Federal Water Pollution Control Act of 1956 (1956) the United States has experienced a deluge of water pollution control legislation and subsequent regulations, originating predominantly at the federal and state levels. During this same time period, however, there has been little evidence to suggest that increased legislation has resulted in decreased degradation of water resources. Water quality problems continue to besiege us at an alarming rate, despite the expenditure of large amounts of tax dollars. There are many examples that could be cited, but one is sufficient to demonstrate the point. Wicker (1968) reports on a stretch of an unidentified interstate river where $7.7 million had been spent on municipal sewage disposal plants since 1957; these facilities had reduced total pollution of the river by only 3 percent, while the amount of industrial waste discharged into the same river over the same time period had increased by 350 percent. This is an opportune time to bring out the fallacy of describing our accomplishments in pollution abatement by dollars spent or the number of treatment facilities available.

To report that recent legislation (prior to 1971) had done little to discourage overall water pollution would be closer to the truth. In 1971 the rediscovery and use of the Refuse Act of 1899 and the court's permission to use Citizens Suits have resulted in an enhanced public awareness and certain industrial policy changes. More recently, restrictions and qualifications have limited the use of these suits.

Our current water pollution problems are reflections of an expanding economy and population, increased labor and construction costs, inadequate private awareness and social consciousness, insufficient acceptance and use of basic economic principles, and incomplete assessments of benefits to be derived from water pollution abatement. Fox (1966), who for a long time has been writing about our dwindling natural resources, points out that our water problems occur in three major areas: (1) the urgent need for maintaining a continuously advancing science and technology, especially in water quality management; (2) balancing costs and returns from water development and use (It has proven to be uncommonly difficult to achieve a consensus of costs and returns, which, in turn, has often made it difficult to proceed with water programs.); (3) the inadequacy of water management institutions; laws, policies and organizational arrangements have not kept abreast of the requirements of a rising demand and an advanced technology. Legislation prior to 1971 had a sound basis and was well conceived, but its nonenforceability led to its ineffectiveness.

Before individual communities became aware that their neighbors were having the same problems and attempting the same solutions as themselves, the state and federal governments had moved in rapidly and established themselves as the "powers that be." Both have strengthened their positions to the point where communities are now almost completely dependent upon them for legislation, regulation, and development. Although local governments usually plan and finance local projects through bond issues, much of the revenues and the backing of these bonds is either state or federal. A consequence of this evolution is that local problems and desires are often overlooked in favor of the larger-scoped efforts.

The debate concerning regional versus state or federal control will undoubtedly continue for many years, due to the various types of programs applicable to both. The author believes that all these governments will and should play a significant role in the administration of river basin resources. However, regional control should predominate, despite the fact that very strong federal legislation and control are currently being considered by Congress. Keenan (1970) concludes that the river basin authority approach is the most efficient mechanism for the management of water resources. In Section 9-6 of this chapter the author proposes a method for implementation of the regional basin approach and describes some of the basic problems of this type of administration.

9-1. FEDERAL CONTROL

Although the Refuse Act of 1899 was the first basic federal legislation, it wasn't until 1948, when federal laws specifically concerning pollution were enacted by Congress, that the federal government became involved in abatement of stream pollution. Federal participation has been mainly concentrated in the areas of:

1. Direct grants for sewage treatment facilities–increasing amounts since 1956
2. Enforcement of stream standards (Water Quality Act of 1965)
3. Tax incentives (§169 Internal Revenue Code)

4. Special penalties, such as fines for oil pollution (Water Quality Improvement Act, 1970, §11)
5. Federal Permit Program
6. Encouragement of interstate compacts
7. Pilot projects–such as dealing with mine wastes, Great Lakes pollution (Water Quality Improvement Act, 1970, §14 and 15)

Since the Federal Water Pollution Control Act (1966) was passed, there has been a series of administrative changes at the Federal level. First the act was administered by the Federal Water Pollution Control Administration (originally part of the Public Health Service) in the Health, Education and Welfare Department, then by the Department of Interior, and then by an entirely separate agency, the Environmental Protection Agency. The new agency (1971) is attempting to maintain strong regional support at the federal level, largely through cooperative enforcement programs and industrial waste reference guides for reasonable treatment.

9-2. STATE AGENCY CONTROL

King (1958) concluded that the State "can protect present property rights and yet redistrict (reallocate) identical future ones." [p. 275] He also reports that "the Supreme Court observed that a State may consider the relation of rights and accomodate their coexistence, and in the interest of the community limit one that others may be enjoyed." [p. 286] So a State, for the purpose of conserving its natural resources, may regulate the production and use of these resources so long as the regulation is reasonably calculated to accomplish that end. Maas (1962) substantiates and clarifies the position of the State in water resources by maintaining that "the State is to serve the community in order to make more effective the community's function relating to common standards. In doing so, the State must understand and in some ways interpret the principles which govern the common life, but never seek to prescribe them."

The State has historically been the "final word" in dictating levels of waste treatment within its boundaries. It approves plans, holds hearings, and analyzes receiving streams to assess actual compliance as well as initial level of treatment required. There has been a lack of consistency between States, and therefore some inequities on a national scale have occurred. The States guard with jealous fervor their prerogative of "controlling" water quality within their governmental jurisdiction. However, they desperately need assistance in the actual improvement of water quality.

The Senate Bill known as the Muskie Bill of 1972 greatly diminished State control over pollution abatement, since it relied almost entirely upon effluent standards administered by a Federal rather than State agencies.

9-3. INTERSTATE OR REGIONAL BASIN CONTROL

Interstate Commissions attempt to utilize broader, more comprehensive planning of water resources than States. They prefer to exert their influence or carry out their desires through State governments as often as possible. For example, the Ohio River Sanitation Commission will usually encourage pollution abatement of a municipality in Ohio by cooperating with the Ohio State Health Department. A similar arrangement exists between the Delaware Water Resources Commission and its constituents, which also include the federal government as a partner. This commission has extremely broad powers for the direct implementation of measures affecting the quantity and quality of both surface waters and groundwaters. Regional, large-basin control is useful where large rivers flow through several states. During its path the large river is normally used and misused by many consumers. Only by interstate cooperation can pollution be controlled in these cases. These commissions, however, are regarded by your author as too large and unwieldy to solve the smaller, more complicated local water quality problems.

9-4. LOCAL DRAINAGE BASIN BOARDS

Local drainage basin agency experience is practically nil in the United States. The arguments for regional control have admittedly been underplayed, not due to the lack of potential merit, but to the lack of sufficient experiences and evaluations. The author believes that the method of marketing stream pollution capacity resources as presented in Section 9-5 of this chapter is better suited for local or small regional boards than higher authorities. This, of course, is unsubstantiated (and, in fact, even disputed by Martin [1970] and others). It is, however, considered by this author to be a reasonable conclusion for the following reasons:

1. The beneficiaries are easier to identify.
2. Regional or local knowledge and understanding of local problems and desires surpasses that of higher authorities.
3. The disparity of activity, costs, and benefits from one region to another would make it very difficult and time-consuming for a higher authority to apply the principles suggested.
4. Application of a marketing system by higher authorities would tend to favor political rather than hydrological boundaries, especially in basins crossing state lines.
5. Regional applications would present only minor bookkeeping problems, since all expenditures and revenues would be internal (much like a completely new political entity), whereas application by higher authorities would result in massive bookkeeping problems, require increased numbers of personnel, and

allow for the continuance of external dis-economies (a cost to people outside the region).

One must then face the paradox that local environmental management of water resources appears to be ideally suited for efficient and equitable pollution abatement, but also, at the same time, is very difficult to initiate. For example, Martin (1970) said that "it is safe to say that much of what is called local government is irrelevant to present-day needs through deficiencies in scale and resources" (p. 187). Further, "they [local governments] are all legally dependent upon the States, not only for the powers they exercise but indeed for their very existence" (p. 187). Martin holds little hope for local government–as we know it–to undertake the pollution problem. This is even more true in 1984 when economic conditions are even more critical. For they are "governments of limited power" . . . and "the geographical area over which they have jurisdiction is carefully circumscribed by law" (p. 187). Local government rules and procedures for operation are very inflexible after years of being in existence. Their resources, both monetary and human, are extremely limited.

What the author envisions, however, is a small basinwide board that possesses no conventional governmental jurisdiction. All persons within such a basin share in the common problem of stream water quality. All affect its quality and some intend to consume more of the basin resources than others. A banding together of these people would allow for sharing of a common problem. Unfortunately the legal powers to come into existence would have to be obtained from the states and/or the federal government. In the next section the author discusses how such a local basinwide water resources board could market pollution capacity resources. Boards could either be formed initially by state or federal legislation, or both. Once again, however, the incentives for forming a local board could originate from the local level of government.

One must agree, however, that local politics are allied closely with budgetary problems. Local voters tend to favor political machines that do not consume great quantities of money; therefore, it is important that these local boards be self-supporting and not require any direct local financing.

9-5. MARKETING STREAM RESOURCES

Every body of water has an assimilative capacity for any type of waste discharge. In the past the debate has centered about a growing federal position that the assimilative capacity of a stream should not be a major factor in determining waste treatment required. Rather, the federal governmental agencies have maintained that polluters should treat their wastes as highly as possible and at least to the secondary effluent level, regardless of the receiving water quality. Many state agencies have supported this federal mandate, and in some instances have required additional treatment to suit local water quality criteria levels. Although more complete treatment may become necessary, it is the author's contention that as potential polluters become larger as well as more numerous, and as monetary considerations become increasingly important in waste treatment decisions, the limited assimilative water resource capacity will play an intrinsic part in technical solutions.

Since the assimilative capacity of a water resource at any one given time is finite and limited, it should be marketable for all potential users. A dollar value can be placed upon the use of the resource for all purposes, and its change in dollar value computed as the water quality varies due to increasing or decreasing contamination. Therefore, knowing both the number of units of resource available for sale at each water quality and the unit value of these resources, an administrative body can begin to market them to all potential users.

Although the method that the author proposes to apply could be used for any nonconservative contaminant (those constantly diminishing with distance from introduction into a stream) of a waste, only oxygen-consuming contaminants (primarily biodegradable organic matter) are used as an illustration. This has been done to make the method as easily understood as possible, because dissolved oxygen is traditionally the major indicator of pollution, and because it represents one of the major problems in the water resource chosen for this illustration.

A large metropolitan area is selected for the example. The people and industries of the area discharge directly or indirectly (through tributaries) into a relatively large lake. Over the years of development in the metropolitan area the lake has been degraded from one of a fishing and recreational quality to a contaminated effluent receptacle for the municipality and its industries. The author proposes that a regional board be empowered to sell the assimilative capacity of the lake to those discharging into it. The operation and management of this board will be described and discussed in Section 9-6 of this chapter. Before the board could sell these resources it would require at least the following information:

1. identity of all discharges;
2. quantities of discharge;
3. existing and desired Pollution Index (a measure of water quality described in Chapter 7, Section 5);
4. benefits of waste treatment; and
5. assimilative capacity.

The identities of and loadings from discharges to the lake are presented in Table 9-1.

Using the method developed by Nemerow and Sumitomo (1970) and described in Chapter 7, Section 5, it has been determined that the present Pollu-

TABLE 9-1. Lake Discharges

Discharger	Average flow (mgd)	Organic matter as BOD without any treatment (lb/day)
A	129	43,500
B	40	74,000
C	75	300
D	2	100
E	14	100
F	5	5,000
Total:	265 mgd	123,000 lb/day

tion Index for the lake is 5.4, whereas for water-contact recreation (a goal of the board for a desirable water quality), the Pollution Index should be equal to 1.0.

Using updated data, with the dollar benefit calculations provided by Faro and Nemerow (1969) it has been determined that the Lake is presently "worth" $3.2 million less per year than if it were of a quality suitable for water-contact recreation. Thus, the loss of benefits amounts to about $8,800 per day. The itemized annual benefits derived by Faro and Nemerow are shown in Table 9-2.

At this point it is necessary to relate the available data in order to establish a methodology for fulfilling the objectives as listed by the author (Nemerow, 1966). The input data used in establishing a relationship are summarized below:

1. 123,000 pounds of oxygen-consuming material discharged per day.
2. $8,800 of benefits lost daily at present lake quality.
3. Average dissolved oxygen content of the lake presently is equal to 1.5 ppm.

TABLE 9-2. Itemized Annual Lake Benefits.
[*After Nemerow, 1966. By permission of publisher.*]

Water use	Present quality (*PI* 5.4)	Improved quality (*PI* 1.0)
Recreation	$1,479,416	$5,194,856
Land use	−595,291	1,266,952
Wastewater disposal	3,831,000	0
In-stream uses	80,931	0
Withdrawal uses	300,000	1,095,000
Total	$4,334,194	$7,556,808

4. Average daily streamflow into the lake is equal to 179 million gallons per day.
5. Lake volume equal to 37,078 million gallons.
6. A direct relationship exists between oxygen-consuming material and dissolved oxygen concentration of the lake.
7. No loss of benefits exists when the discharge of oxygen-consuming material is zero.
8. Pollution Index equals 0 when the discharge of oxygen-consuming material is 0.
9. Dissolved oxygen of the lake equals 8 ppm when the discharge of oxygen-consuming material is 0.
10. A minimum of 4 ppm of dissolved oxygen is required in the lake (lowest acceptable for fish propagation).
11. The lake is completely mixed.
12. Pollution Index equals 5.4 at present quality.
13. Pollution Index of 1.0 is the objective for water contact recreation.
14. 470,000 pounds of dissolved oxygen are presently available in the lake.
15. 2.5 million pounds of dissolved oxygen are available for sale daily under conditions of zero (0) discharge of oxygen-consuming material.

In Fig. 9-1 (Benefits of Water Quality Enhancement, 1970) the author shows a possible relationship among these data above (Table 9-1). It is important to note that Fig. 9-1 may not be an explicit relationship of the four parameters. It is known that there exists a relationship between dissolved oxygen and amount of oxygen-consuming material being discharged; between Pollution Index and oxygen-consuming material; and between Pollution Index and benefits lost. It is not unreasonable to assume, therefore, that there exists a relationship among all four parameters. The exact and precise graphical relationship may be difficult to show, but Fig. 9-1 can be substantiated theoretically and is presented here as one attempt at a relationship.

The two endpoints of the curve in Fig. 9-1 are known to be essentially factual; $8,800 of benefits lost daily at the present discharge, and zero loss in daily benefits when there is no discharge of waste materials. A third point on the curve was selected as the intersection of the two water quality objectives: Pollution Index equals 1.0 and dissolved oxygen concentration equal to 4.0 ppm. The curve shown was then drawn through the three points. The assumption of a direct relationship between dissolved oxygen and BOD results in 1.25 million pounds of dissolved oxygen being available each day for a discharge of 76,000 pounds of BOD per day and still meets the quality objectives of the lake. The remaining BOD (123,000 - 76,000) must be eliminated by individual polluters by means other than that of discharge into the lake.

Once the amount of resource available for sale is established, the board then

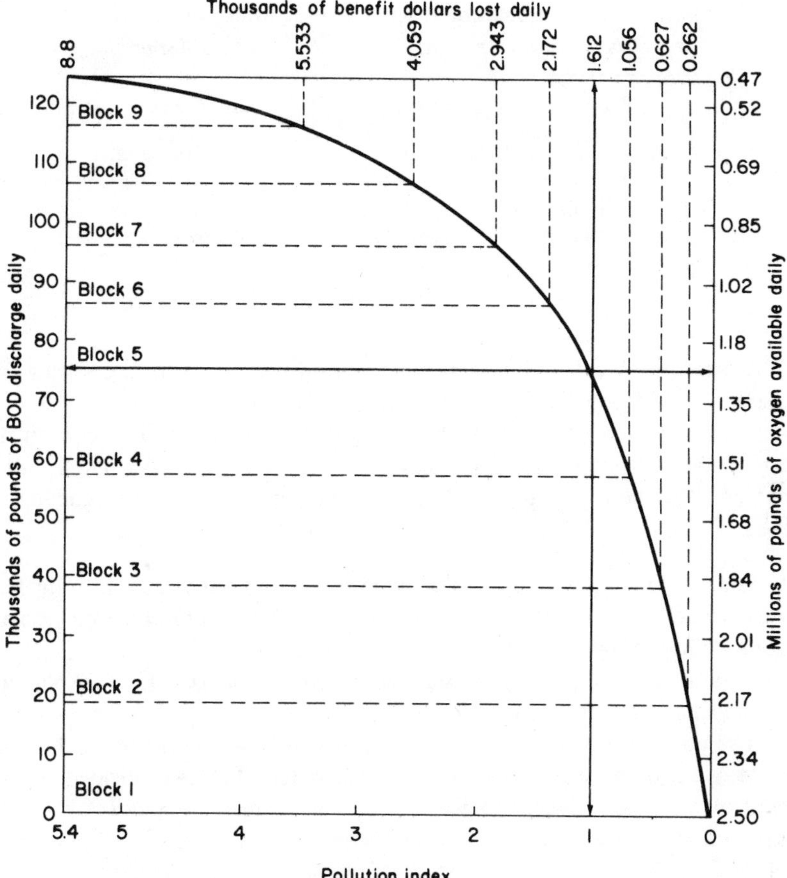

FIG. 9-1. Unit price calculation. (*After Benefits of Water Quality Enhancement. By permission of publisher.*)

has considerable flexibility in setting the price of the resource. For example, in a recent air pollution study (Newman, 1971) an arbitrary value of $.01 per pound of sulfur is being considered as a start, although the unit price may have to be raised as experience is gained. The stated objective in the air pollution study is to start with a unit price level just high enough to encourage a cheaper alternative such as sulfur removal by treatment. This same procedure could be used in our lake example. Much more work and/or data collecting is necessary to establish a "true worth" unit price. The author believes that the purchaser would appreciate and understand paying a true benefit unit charge rather than an arbitrary price designed solely to discourage any purchase. Furthermore, it is our objective to

TABLE 9-3. Calculation of Unit Price. [*After Benefits of Water Quality Enhancement. By permission of publishers.*]

Block	Pounds BOD sold per day	Incremental benefit lost ($)	Unit price per pound (¢)
1	19,000	262–0	262/19,000 = 1.4
2	19,000	627–262	365/19,000 = 1.9
3	19,000	1056–627	429/19,000 = 2.3
4	19,000	1612–1056	556/19,000 = 2.9

use the principle of supply and demand and to discourage only excessive waste discharges in order to maintain the desired water quality.

Let us assume—as a beginning—that the board is organized for nonprofit, its sole monetary objective being to recover the dollar benefits lost by allocating resources. It is now possible to devise the sliding scale payment schedule as follows:

1. 76,000 pounds of BOD will be permitted to be discharged into our lake in order to achieve our water quality objective of 4.0 ppm of dissolved oxygen, and/or a Pollution Index of 1.0.
2. Assume the board decides to sell the discharge privileges[1] (76,000 pounds) in four blocks of 19,000 each.
3. The unit price of each block is then calculated as the incremental loss of benefits accrued by selling discharge privileges (see Fig. 9-1) divided by the amount of discharge privileges being sold. In this manner we may compute the data shown in Table 9-3.

The amount of discharge permitted each customer could be a decision for the board, depending upon the local situation. Four possible methods of allocating purchase quota are suggested by the author:

1. Percentage of oxygen consuming waste.
2. Percentage of flow.
3. Relative value (taxes) to the community.
4. Relative need or desirability to the community of a particular customer (discharger) determined by some objective system.

While any of the above methods may be used, the author has selected the first method only to illustrate the use of the system.

[1] A similar procedure could be used with other contaminants such as mercury or heat (temperature).

TABLE 9-4. Customer (Discharger) Costs

Discharger	% of organic load	Block size (#)	Allowable block purchase (#)	Total purchase four blocks (#)	Total daily costs	Total annual cost
A	35.29	X 19,000	6,700	26,800	$569.50*	$207,867
B	60.18	X 19,000	11,440	45,760	969.97	354,039
C	.26	X 19,000	50	200	4.24	1,547
D	.08	X 19,000	15	60	1.27	464
E	.08	X 19,000	15	60	1.27	464
F	4.11	X 19,000	780	3,120	66.07	24,115
Totals	100.00		19,000	76,000	$1,612.32	$588,496

*6700# @ 1.4¢ = $93.80 and 6700# @ 1.9¢ = $127.30, etc.

Table 9-4 represents an allocation of the oxygen resources, assuming that all customers elect to purchase as much BOD stream resources as possible rather than treat their wastes.

Having purchased all they are allocated or allowed, the customers must then eliminate from the discharge the remaining oxygen-consuming matter. Discharger A may decide that, rather than provide treatment to remove his remaining 16,700 pounds of BOD and pay $207,867 annually for discharging 26,800 pounds of BOD, it is more economical for him to provide more waste treatment and purchase less pounds of stream resources. This freedom of choice is more consistent with our democratic principles and also makes more resources available to other customers should they decide to install waste treatment devices.

The board is in a position to sell resources once the unit prices have been determined. In any real situation the benefits foregone would be attributed to more than just oxygen-consuming wastes. A resource allocation for each contaminating element deemed to be degrading the aquatic environment would be needed. The foregone benefits would have to be prorated so as to establish unit prices for each contaminant of a waste. The board then would have achieved the primary objective of maintaining some minimum acceptable water quality. It has encouraged waste treatment by establishing a "pollution price" based not upon cost of waste treatment only (as some have advocated) but upon the benefits foregone, and has collected revenue that will be disbursed primarily for all means of maintaining the desired water quality.

9-6. RIVER BASIN MANAGEMENT

The proposed method of allocating water pollution capacity resources is perhaps best managed at the regional level. It may be a river basin or portion thereof.

This is to suggest that federal and state governments provide for implementation of their policies by smaller units capable of making better evaluations of regional needs and desires, as well as benefits of pollution abatement. The higher jurisdictions may still determine the minimum water quality upon leaving the basin, but the manner in which that goal is achieved and the water quality within the basin could be decided upon based on regional desires. One board may wish to collaborate with another adjacent board in order to set water quality standards acceptable to both waters. Representatives of the board, ideally, should include one spokesman from every type of major water use within the jurisdiction. As far as numbers are concerned, it can only be stated now that there should be sufficient members to present as many opinions as may be necessary for equitable decisions. These should not be so many as to tie up proceedings in lengthy and perhaps useless debate. It is proposed that board members be salaried. The board would need a full-time staff of engineer(s), and legal and financial advisors as well as secretarial support. One method of selecting members would be for the individual interest groups to select their own representatives for some staggered term of office.

The Sale of Resources

With the various problems of initiating a resource sales board solved, the daily operations and actual sales should amount to a relatively minor problem. The first order of business, if it had not already been done, would be a survey of the basin, or subbasin, to determine waste discharge sources and available pollution-carrying capacity resources. Coincidental to such a survey would be feasibility studies for various alternatives, at different objectives, and the benefits to be derived at each objective. Through a series of informational hearings, or perhaps even a ballot, the local beneficiaries would be informed of the alternatives, the costs and benefits, and an objective would then be established by the board in behalf of the citizens.

The board, with all interest groups represented, would then construct a payment schedule, as suggested earlier in this chapter, and distribute to the dischargers a summary sheet stating the allowable purchases and unit costs within each block of resources to be sold. After allowing sufficient time for the dischargers to evaluate their alternatives, a sale date would be set and all dischargers summoned. The actual sale would then proceed as discussed earlier.

The board may wish, for the first year of operation, to sell at extremely low prices in order to evaluate how the program will function. At each successive sale the unit price could be increased until the foregone benefits are recovered. The frequency of sales would be a matter for the board to decide. They might wish to hold a sale annually, semiannually, during low and high resource seasons, etc. This decision would have to be made known to the purchasers in advance. Utilizing the high and low resource seasons would permit flexibility of treatment plant operations, and take advantage of a natural situation. Such an option is not now presently available. At present, the engineer must design for the "critical" summer low-flow conditions only.

The installations of new waste-producing industries has always posed problems to any community. It is not anticipated that the resource sales concept will in any way alleviate such problems, but environmental control should be made easier. This can be illustrated by a comparison with current methods. If uniform treatment is required in a basin, all new industries in the basin will be made to provide the same degree of treatment. After a period of time, the receiving waters are again overloaded and the uniform treatment requirement then upgraded. In the meantime, however, the problem will persist.

A resource sales board would first have the opportunity to decide whether the new industry was even acceptable and/or desirable. If not, it might deny the right to purchase, or allocate only very little. If the industry is acceptable and/or desirable, the interest groups (board members) would have participated jointly in this decision, and would therefore have made some prior agreement regarding cost-sharing of the increased waste discharge. In any event, the regional beneficiaries, through the board, would have maintained complete control over utilizing the resources available.

The operations do not end with the sale of resources. The water quality must be monitored, as well as the benefit evaluations rechecked. The overall picture of benefits versus quality will constantly be changing, resulting in a varying price to purchasers at each sale. It will also be necessary to maintain surveillance over the dischargers. The payment of discharge fees will not in itself assume compliance, any more than present methods of waste treatment are assumed satisfactory.

It can probably be seen by now that the actual routine is very similar to that of existing state control agencies, with the following major exceptions:

1. Planning, development, and management of regional resources are carried out entirely at the regional level.
2. All users are represented in resource quantity, price, and allocation decisions.
3. Arbitrary standards are not established, but only minimum water quality standards are adhered to.
4. Dischargers pay for resources utilized.
5. The unit price for the use of the resource increases with its depletion.

Budgetary Considerations

The proposed board will not exist entirely upon funds collected through taxation. It will function essentially as a private corporation or public utility, and be paid by users of its product. It will differ from governmental agencies, private corporations, and/or public utilities in the following respects:

1. It will not have to play the annual guessing game of preparing a budget request for the coming year, hoping to get the funds, and having to spend them.
2. It will not necessarily have to grow continually in size in order to provide for a growing community.
3. Due to product limitations it will not have to expand continually in order to fulfill demands.
4. It will not be interested in showing a profit at the end of the year.

The first year of operation may require catalyst or "seed" funds from private grants, or state and federal aid. Beyond this point the need for such assistance should either diminish or disappear. Funds are then secured through resource sales, part of which will come from municipalities using the resource. It may appear that this would impose an additional tax upon the citizens, but this is not necessarily true. Localizing the management program would, of necessity, entail a decrease in the operations and sizes of state and federal agencies, although they would still provide a valuable overseeing and advisory service. Accordingly, there *should* be a decrease in the amount of tax dollars required. It is difficult to determine whether or not the taxes required for payment to the board would equal or exceed the anticipated decrease in state and federal taxes, but it appears to the author that there would be a greater return for each dollar spent.

The revenue collected should theoretically amount to no more than the benefit dollars that had been lost annually. From this revenue must come the necessary funds for salaries and operations of the board. Any remaining funds may be utilized in any or all of the following ways:

1. Research pertaining to local problems such as shoaling, treatment studies, storm water, etc.
2. Development of recreational areas, marinas, etc., so as to enhance the stated water quality level objective.
3. Subsidizing of desirable but needy installations; grants-in-aid, revenue reduction, and debt retirement.
4. Scholarships.

Although the board must make considerable effort, in essence all that is being done is accomplishing at the regional level, with regional funds, what is being attempted by the state and federal governments for larger areas.

QUESTIONS

1. What are the advantages and disadvantages of federal, state, and regional control of pollution?
2. What is meant by "marketing of stream resources"?
3. What objectives would a Water Resources Board have in the marketing system?
4. How was the data used to plot Fig. 9-1 obtained?
5. How is the amount of resource available to each consumer determined?
6. Who is best fitted to manage the board?
7. What budgetary considerations are involved in this system of pollution prevention?
8. For what will the board's revenue be used?

REFERENCES

Benefits of Water Quality Enhancement, Environmental Protection Agency, Water Quality Office Water Pollution Control Research Series 16110 DAJ 12/70, 1970.

Faro, R. C., and N. L. Nemerow: "Measurement of the Total Dollar Benefit of Water Pollution Control," *2nd National Symposium of Sanitary Engineering Research, Development and Design*, Cornell University, Ithaca, N.Y., July 16, 1969.

Federal Water Pollution Control Act of 1956 (as amended) Public Law 660, 84th Congress of United States, Sess., 1956.

Fox, Irving K.: We Can Solve Our Water Problems, *Water Resources Research*, 2(4):617, 4th qtr., 1966.

Keenan, John: *Intrastate Arrangements for Water Resources Planning and Development, Part 1, Overview*, Syracuse University Maxwell School, Syracuse, N.Y., February 1970.

King, Dominic B.: Regulation of Water Rights Under the Police Power, in *Water Resources and the Law*, University of Michigan, Ann Arbor, Mich., 1958.

Maas, A., et al.: *Design of Water Resources Systems*, Harvard University Press, Cambridge, Mass., 1962.

Martin, Roscoe C.: Environmental Management and Local Government, Section VIII Appendix p. 187 in *Benefits of Water Quality Enhancement*, Environmental Protection Agency, Water Quality Office, Water Pollution Control Research Series 16110 DAJ 12/70, 1970.

Nemerow, N. L.: *Water Pollution Capacity Resources Allocation*, Syracuse University Dept. of Civil Engineering Research Report No. 9, 1966.

Nemerow, N. L., and H. Sumitomo: Calculations of a Pollution Index, Federal Water Quality Office, U.S. Environmental Protection Agency Report, Part A, Section V, p. 13 *Benefits of Water Quality Enhancement*, EPA 16110 DAJ 12/70, 1970.

Newman, Barry: Paying to Pollute–Nixon Proposal to Tax Befoulers of the Air Could Bring New Type of Pollution Control, *Wall Street Journal*, p. 24, July 12, 1971.

Wicker, A. R.: Newspaper article appearing in *Syracuse Herald Journal*, November 28, 1968.

10 STREAM STUDIES—ILLUSTRATIVE COMPLETE CASE

No textbook on the subject of stream pollution would be complete without some actual practical illustrations of stream pollution capacity in receiving waters. Most "case" studies are frustrating to students, because in their application, they usually deviate considerably from theory. The most common deviation is that of streamflow; generally actual stream sampling occurs at two or three times the critical low flow. Only once has this author been able to conduct a stream study during the design critical flow (Nemerow, 1971). Often an industry is not operating normally or not operating at all during the stream study. At other times the study follows a heavy rainstorm or unusual discharge of waste. The stream analyst must demonstrate great patience with the acts of nature and those of pure chance which interfere with the ideal stream study. In fact, my students are taught to design for studies during industrial and municipal normal operation and critical streamflows, but to plan for the occurrence of the "unusual normalities." They are further encouraged to project their results to those that would have occurred under ideal study conditions. This represents a difficult and sometimes impossible task. But to expect and plan for the unusual event is to avoid frustration when it occurs.

This chapter presents an actual stream study which demonstrates the above comments. It does not represent an "ideal" study, and points out the occurrence of the unusual normal event. The stream was studied several times in fruitless attempts to overcome this phenomenon.

Although ancillary material is given to assist the reader in understanding all the many problems involved in stream studies, none of the data are superfluous. Some of the information may seem extraneous, but most of it has significant bearing on some aspect of the study.

CATTARAUGUS CREEK STUDY

Cattaraugus Creek arises in Wyoming County of Western New York State and flows westerly to Lake Erie. It runs along the border of Erie and Cattaraugus counties most of the way. Major sources of contamination include the village of Gowanda and two industrial plants: Moench Tannery and the Peter Cooper animal glue manufacturing operations. They are rather closely grouped plants, as shown in Fig. 10-1, a map prepared for an April 24, 1969 survey. The village of Gowanda and its associated industries are faced with orders from New York State to upgrade and improve their waste treatment plants. The degree of treat-

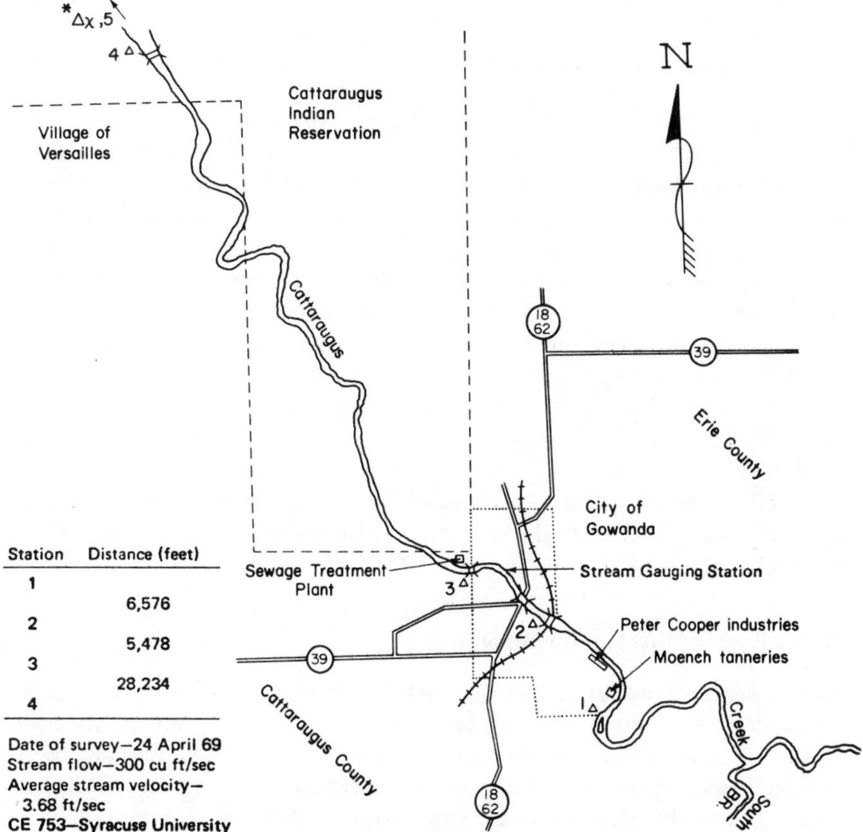

FIG. 10-1. Map showing stream survey area of Cattaraugus Creek in the vicinity of Gowanda, N.Y. Scale: 1 inch = 1 mile (*From U.S. Army Corps of Engineers maps of Cattaraugus and North Collins Quadrangles.*) $^*\Delta\chi$ is located at a point on the creek two and one-half ($2\frac{1}{2}$) miles downstream from Δ4. It is also referred to as Δ5.

ment to be provided was based upon some 1960 state studies and antiquated data on quantity of discharge and waste quality. The state's study was later updated somewhat by "time of flow" studies made in 1964. The author was employed by the major industries and began studying the stream and the industries in 1967. Three major stream surveys have been planned and completed since early 1967: (1) October 16–18, 1967; (2) June 25, 1968; and (3) May 28, 1969. The objective here is to predict how the Creek will react to the wasteload at various degrees of organic matter removal under critical streamflow conditions. New York State defines critical receiving water flow as the minimum seven-day low flow likely to recur once every 10 years. Our first objective, then, is to determine the design or critical flow.

Minimum Creek Flow Determination

The frequency series method for finding the recurrence interval was used. Actual data recorded at the gauging station (Fig. 10-1) were used in the computation. The recurrence interval is equal to $T = (N + 1)/M$, where N = number of consecutive years of record and M = the order of magnitude in decreasing order of severity of low flow. As shown in Table 10-1, the seven-day minimum flows were arranged in order of increasing magnitudes (decreasing intensity) with the lowest drought flow as $M = 1$. The T values were then computed for each flow recorded, and a plot of flow versus time of occurrence was drawn (Fig. 10-2). From this curve the once in 1-, 10-, 25-, and 50-year low flows were determined. These data are shown in Table 10-2. Since New York State uses the one in 10-year, seven-day value, we must design a system to perform satisfactorily at 61 cubic feet per second of creek flow.

It is now important for us to consider the character, quantities, and general diurnal, weekly, and seasonal distribution of wastes entering Cattaraugus Creek in the Gowanda area.

Wastes Entering Creek in Gowanda Area

The village of Gowanda (3,000 persons) is served by a well-operated primary sewage treatment system with combined sewers. The plant is quite old and operates at or above rated capacity many times each year. It is located just below Station 3 (see Fig. 10-1), and its effluent is discharged into the creek, which already contains the effluents from the two major industrial discharges.

Description of wastes from the Peter Cooper Corporation and the Moench Tanning Co. The glue that is produced at Peter Cooper comes from hide stock and from fleshings. Approximately 65,000 pounds of glue are manufactured per day. The treatment of the limed stock and the fleshings to produce the glue yields different types of liquid wastes. The composite liquid waste consists of

TABLE 10-1. Annual Drought Data for Cattaraugus Creek at Gowanda, New York

Year (Oct.–Sept.)	7-Day min. flow (cfs)	M (Magnitude of drought flow)	Time of recurrence T, in Years
1954–55	56	1	28.0
1944–45	57	2	14.0
1940–41	60	3	9.4
1951–52	70	4	7.0
1961–62	71	5	5.6
1963–64	76	6	4.7
1948–49	76	7	4.0
1952–53	77	8	3.5
1959–60	77.3	9	3.1
1964–65	77.4	10	2.8
1965–66	79	11	2.5
1958–59	80	12	2.3
1960–61	82	13	2.1
1950–51	85	14	2.0
1953–54	86	15	1.86
1945–46	91	16	1.75
1962–63	94	17	1.65
1941–42	94	18	1.55
1949–50	94	19	1.47
1946–47	96	20	1.40
1943–44	99	21	1.33
1957–58	101	22	1.27
1956–57	103	23	1.22
1947–48	112	24	1.16
1942–43	117	25	1.12
1955–56	140	26	1.08
1939–40	142	27	1.04

$$T = \frac{N+1}{M}$$

NOTE–Data taken from actual streamflow records where N = number of observations (27).

washwater, acid wastes, and liquid from the evaporators, which dry and concentrate the liquid glue.

The waste from the evaporators does not usually constitute the major pollutional load of the industry. The wastes from the fleshings constitute over 50 percent of the total pollutional load of the plant. Currently, the glue factory is settling its waste and discharging the effluent into Cattaraugus Creek. The solids are sold as fertilizer to nearby farmers. Under a proposed treatment process, fleshings would no longer be used, and a flocculation and coagulation system would

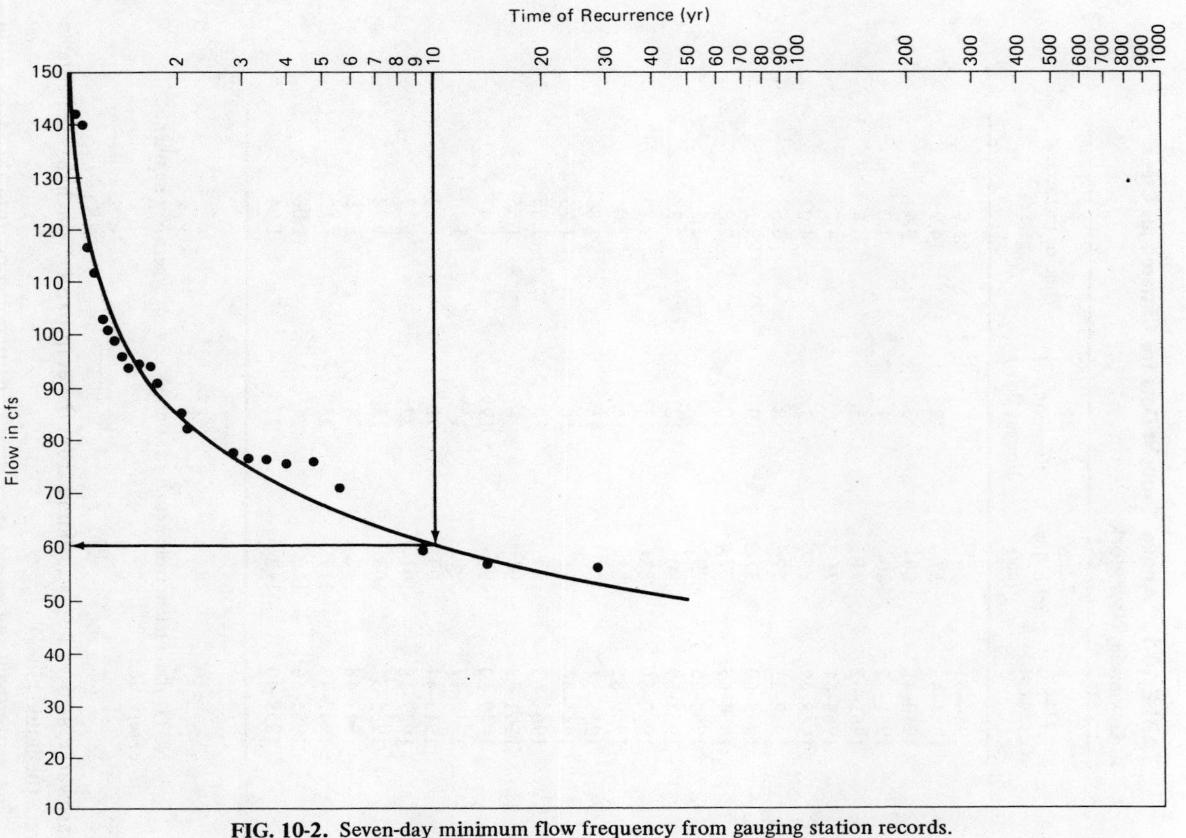

FIG. 10-2. Seven-day minimum flow frequency from gauging station records.

TABLE 10-2. Low Flows Expected (From Fig. 10-2)

Minimum creek flow expected once every	Flow (cfs)
1 year	150
10 years	61
25 years	54
50 years	50

be installed. From the results of laboratory studies, these treatments achieved about an 86 percent BOD reduction. Additional biological treatment may also be required to reduce the organic load on the creek further. The total waste flow from the glue plant is about 3 mgd.

The Moench Tanning Company produces chrome-tanned, upper-shoe leather. The main process steps involved are soaking, liming, dehairing, and defleshing before the actual tanning. Screens separate the hair particles, and then wastewater is discharged to two settling tanks. The material that has settled is washed into the creek during high flow periods. Results of pilot plant studies indicate that 90 percent BOD reduction can be achieved by a modified activated sludge process: high solids–high pH biological aeration.

The following table indicates what the current pollutional load is and what it would be if the wastes were first treated:

Waste Loadings

	Waste flow (mgd)	Primary treated waste (# BOD_5/day)	Treated waste (# BOD_5/day)
Glue plant	3.0	25,800	3,620
Tannery	0.25	2,900*	1,243
Total:	3.25	28,700	4,863

*Untreated in reality

The graphs constituting Figs. 10-3, 10-4, and 10-5 indicate the least-squares, best-fit curves for treated and untreated tannery wastes and animal glue primary treated waste, assuming a monomolecular first-order reaction.

Specific Creek Surveys

The first survey was carried out in the normal drought-flow period of the first year, on October 16-18, 1967. Samples were collected from seven stations from above the tannery nearly to the mouth of Lake Erie, 18 miles northwest. Sam-

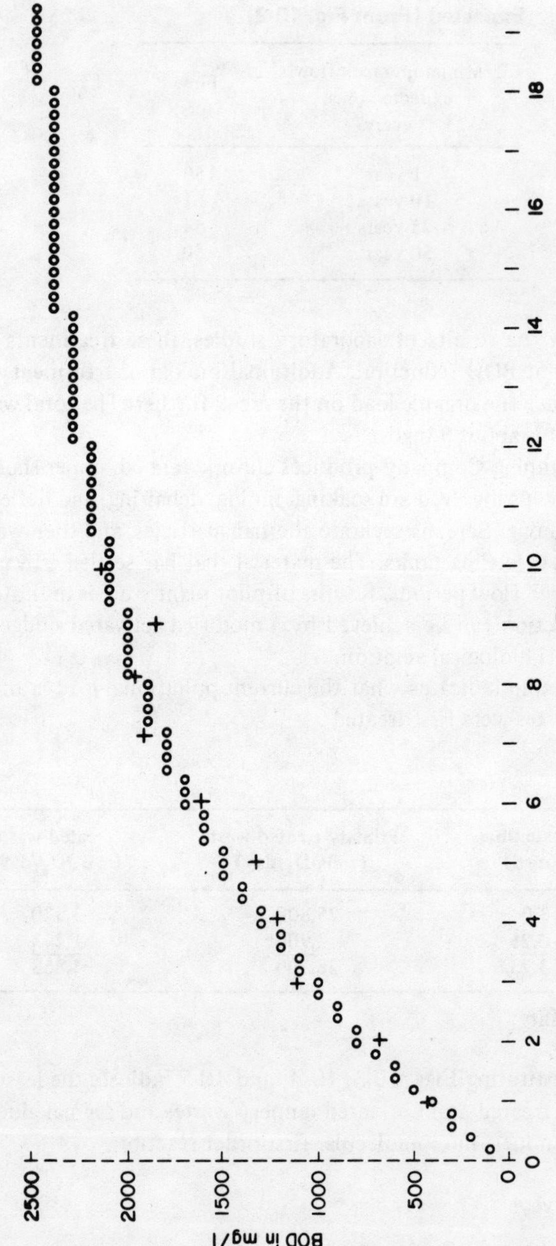

FIG. 10-3. Least-squares best-fit long-term BOD of equalized raw tannery waste (Ultimate BOD = 2,558 mg/l; K_1 = 0.076; Data points = +).

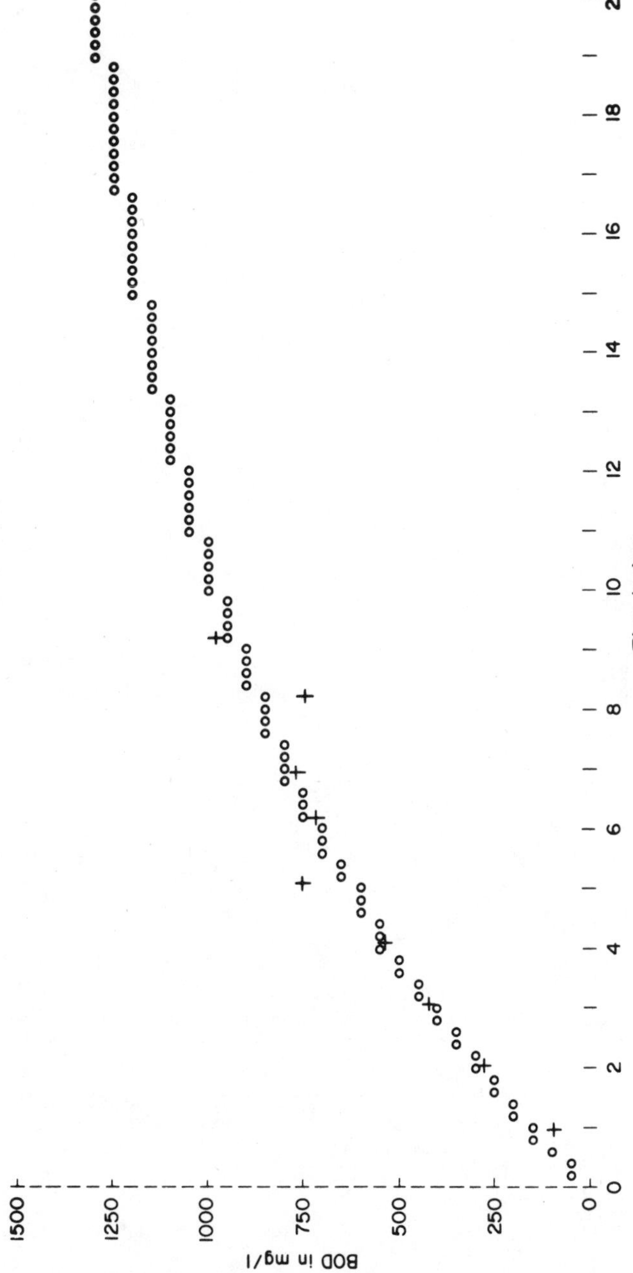

FIG. 10-4. Least-squares best-fit long-term BOD of poorly treated equalized tannery waste (Ultimate BOD = 1,146 mg/l; K_1 = 0.05; Data points = +).

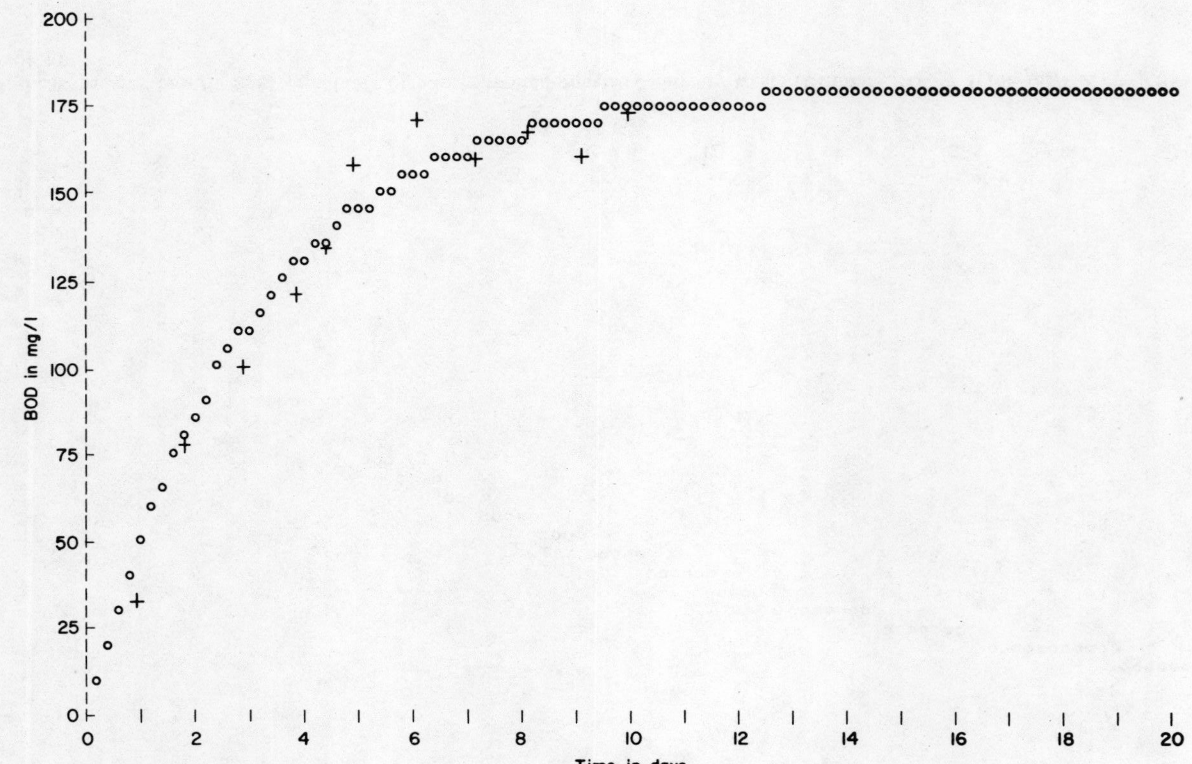

FIG. 10-5. Least-squares best-fit long-term BOD of treated glue plant waste (Ultimate BOD = 180 mg/l; K_1 = 0.14; Data points = +).

pling was concentrated at mile 18.2 (just above animal glue plant but below tannery) and mile 11.6 (about 6 miles below animal glue plant and 5 miles below the village treatment plant). Mile 18.2 was the upstream mix point for the tannery waste, and mile 11.6 was found to be the bottom of the dissolved oxygen sag curve. The stream gauge was read when each sample was collected. On the 16th and 17th, the location and extent of the oxygen sag were determined. On the 18th, four samples were collected at each of the two points of major concern to the tannery. The results are shown in Table 10-3.

Since the October 1967 sampling was not carried out during the critical flow period (353 cfs vs. 61 cfs), the results represent only an approximation of creek conditions that are expected to occur during the critical drought flow period. Preliminary results do indicate, however, that a 60-percent BOD reduction at Station 4 would result in sufficient dissolved oxygen (5.18 ppm) at Station 5 during any normal, yearly, low-flow period at near maximum stream temperatures.

The calculations from the data found in this survey are shown below. Stream "time of flow" was determined by making a series of cross-sectional velocity measurements. In this survey all factors except streamflow were as planned. Drought flows simply did not materialize as expected for this time of year.

Now, faced with the problem of waiting one whole year until the usual drought flows occur again, the author made a statistical study of what flows might be expected the following spring. This would save six months of waiting. The records of previous and recent monthly minimum creek flows were obtained from the U.S. Geological Survey records and are shown in Table 10-4.

The second survey was carried out on June 25, 1968. It consisted of sampling at only three locations: above the tannery (Station 1), which was considered a clean-water station; Station 4 at Aldrich Street, below both industrial discharges but just above the sewage plant effluent, and assumed to be the bottom of the sag curve; and Station 5 at Versailles Road, which is presumed to be the creek recovery point. Although the streamflow was less than the previous fall, it was still about $3\frac{1}{2}$ times that of the critical drought flow.

The results of the June 25 survey and calculations of the data are shown in the following pages. Several unexpected phenomena were encountered in this survey. First, the clean-water station averaged 9.7 ppm BOD during the entire day. The presence of some pollution in the creek prior to reaching the Gowanda area was sufficient to cast some doubt on the validity of this survey. Second, the omission of sampling just below the animal glue plant proved crucial, and questionable projections back to that point were necessary. And lastly, streamflow was still too high to predict with any real assurance the reaction of the creek to these wastes under critical flow conditions.

During the six years of record, mean monthly flows have been recorded less than 238 cfs during June in three years, or 50 percent of the years. Although

TABLE 10-3. Cattaraugus Creek Survey, October 1967

*Sta. No.	Location	Sampled		Flow at stream gauge		Temp. (°C)	DO (ppm)	BOD 5 day (ppm)
		Date	Time	Height	Flow (CFS)			
1	Mile 19.2 Upstream Moench	10-16-67	2:40 p.m.	2.325′	355	16	11.0	0.6
	Tannery	10-17-67	1:15 p.m.	2.32′	353	16	11.2	0.0
2	Mile 18.2	10-16-67	3:00 p.m.	2.325′	355	16	10.3	4.5
	Peter Cooper Intake-	10-17-67	1:30 p.m.	2.32′	353	16	10.5	3.3
(Tannery	Below Tannery	10-18-67	7:30 a.m.	2.32	353	14.5	9.4	7.1
Mix		10-18-67	10:30 a.m.	2.40	385	14.5	9.4	7.9
Point)		10-18-67	1:00 p.m.	2.45	405	14.5	10.0	8.0
		10-18-67	5:10 p.m.	2.63	509	14.5	9.9	8.3
3	Mile 17.7 Railroad Bridge	10-16-67	3:15 p.m.	2.325′	355	16	10.1	>16.6
	Downstream from Peter Cooper	10-17-67	1:50 p.m.	2.32′	353	16	10.2	>16.6
3A	Mile 17.2 Old Highway Bridge	10-16-67	3:30 p.m.	2.325′	355	16	9.6	16.6
	in Gowanda Main Street	10-17-67	2:00 p.m.	2.32′	353	17	9.1	>16.6
4	Mile 16.7	10-16-67	3:40 p.m.	2.325′	355	16	9.0	15.0
	New (1960) Bridge in Gowanda just above S.T.P. Aldrich St.	10-17-67	2:35 p.m.	2.32′	353	17	8.5	>16.6
5	Mile 11.6	10-16-67	4:20 p.m.	2.325′	355	17	8.0	12.3
(Sag	Versailles-Iroquois Road	10-17-67	3:00 p.m.	2.32′	353	17	7.5	15.2
Point)	Bridge West Channel	10-18-67	8:30 a.m.	2.32′	353	14.5	8.7	6.7
		10-18-67	11:35 a.m.	2.40	385	14.5	8.5	5.7
		10-18-67	1:30 p.m.	2.45′	405	14.5	8.2	16.4
		10-18-67	6:50 p.m.	2.63′	509	14.5	7.8	>16.6
6	Mile 1.5 Near Thruway on Rt. 5-20 Bridge	10-17-67	3:30 p.m.	2.32′	353	15	7.7	8.8

*See Fig. 10-1, p. 361.

Preliminary Analysis of Cattaraugus Creek During Oct. 1967 Survey

(Stream Velocity = 0.75 mi/2 hr or .375 mi/hr determined by surface floats)

velocity = 0.375 mi/hr

18.2 Station 2
11.6 Station 5
6.6 mi

$$\therefore \Delta t = \frac{6.6 \text{ mi}}{.375 \text{ mi/hr}} = 17.6 \text{ hr} \quad \text{or} \quad \frac{17.6}{24} = 0.735 \text{ days}$$

$$K_1 = \frac{1}{\Delta T} \log \frac{L_A}{L_B} = \frac{1}{.735} \log \frac{{}^*6.5 \times 1.46}{{}^*12.1 \times 1.46}$$

gives negative K_1 because of increased contamination entering creek between Station 2 and the Sag Point at Station 5.

Only way to get a positive K_1 is to use Station 4 as upstream point and Station 5 as downstream.

	Station 4	*Station 5*
10-16-67	15.0	12.3
10-17-67	†>16.6	15.2
Avg.	17.5	13.8

$$K_2 = K_1 \frac{\bar{L}}{\bar{D}} - \frac{\Delta D}{2.3\,\Delta t \bar{D}}$$

$$\bar{L} = \frac{17.5 + 13.8}{2} = \frac{31.3}{2} = 15.65$$

at 17°C · $\bar{D} = \frac{9.0 + 8.5}{2} = \frac{17.5}{2} = 8.75$ oxygen upstream ‡$9.8_{\text{Sat}} - 8.75 = 1.05$

$= \frac{8.0 + 7.5}{2} = \frac{15.5}{2} = 7.75$ oxygen downstream ‡$9.8_{\text{Sat}} - 7.75 = 2.05$

$$\bar{D} = \frac{1.05 + 2.05}{2} = \frac{3.10}{2} = 1.55$$

$$\Delta D = 2.05 - 1.05 = 1.0$$

$$K_2 = .189 \frac{15.65}{1.55} - \frac{1.0}{2.3(.555)(1.55)}$$

$$= 1.91 - \frac{1.0}{1.98} = 1.91 - .501 = 1.409$$

---

*Average values over the day

†Assume BOD = 20 ppm. Distance between Station 4 and 5 = 16.7 − 11.7 = 5.0 mi

$$\therefore \Delta t = \frac{5}{.375} = 13.3 \text{ hr} = .555 \text{ days}$$

$$= \frac{1}{\Delta t} \log \frac{17.5 \times 1.46}{13.8 \times 1.46}$$

$$= \frac{1}{.555} \log \frac{25.6}{20.1} = 1.8 \log 1.27$$

$$= 1.8 \times .105 = .189$$

‡Saturation level of dissolved oxygen at 17°C (See Appendix C)

Preliminary Analysis of Cattaraugus Creek During Oct. 1967 Survey (*Continued*)

Using Thomas' Nomogram

$K_2 t = 1.409(.555) = .783$ $\qquad \frac{D}{L_A} = .08$

$\frac{D_A}{L_A} = \frac{1.05}{25.6} = .041$ $\qquad D_t = 25.6(.08) = 2.048$ Nomograph

$D_t = 2.05$ act. sample

$\frac{K_2}{K_1} = \frac{1.409}{.189} = 7.45$

with 60 percent L_A reduction $L_A = .4 \times 25.6 = 10.24$

$\frac{D_A}{L_A} = .102$ and $\frac{D}{L_A} = .10 = .1 \times 10.24 = 1.024$

at 1/3 flow and 22°C the D could be $3 \times 1.024 \times \frac{1.06}{.9} = 3.62$ ppm

at 22°C Sat = 8.8 − 3.62 = 5.18 satisfactory

Cattaraugus Creek Survey 6/25/'68 (*BOD Values; 5 day, 20°C ppm*)

Station No. 1		4		5	
Above Tannery		Aldrich Street		Versailles Rd. Bridge	
time	ppm	time	ppm	time	ppm
All day		10:10 a.m.	38.1		
Composite	*9.7	11:10 a.m.	39.0	11:50 a.m.	21
		12:10 p.m.	66.5		
		1:10 p.m.	63.0		
		2:10 p.m.	75.0	2:00 p.m.	28.8
		3:10 p.m.	105.0	3:00 p.m.	19.8
		4:10 p.m.	111.0	4:00 p.m.	15.6

Stream vel. at Station 4 = 1 ft/sec
Stream vel. at Station 5 = 1.75 ft/sec

$$\text{avg } \frac{2.75}{2} = 1.375 \text{ ft/sec}$$

distance = 16.7 − 11.6 (Versailles) = 5.1 × 5,280 = 528.0
(Aldrich St.)

$$\begin{array}{r} 528.0 \\ 26{,}400 \\ \hline 26{,}928 \text{ ft} \end{array}$$

$$\frac{26{,}928}{1.375} = \frac{19{,}600 \text{ seconds}}{60 \times 60} = 5.45 \text{ hr}$$

Time of flow by floats 9:55 a.m. to 4:00 p.m. = <6 hr 5 min

∴ BOD *Station 1 = 9.7
(5 day, Station 4 = †38.5 [avg. 10:10 and 11:10 a.m.]
20°C) Station 5 = 17.7 (avg 3 and 4 p.m.) (ppm)

*This is an unusual, perhaps abnormal, high BOD concentration for this clean water station. It is the first time that a concentration above 1 ppm has been found here.

†To allow for the 5.45 hours of creek travel time between Stations 4 and 5.

For Stream Analysis for 6/25/'68

Station No. 1		Station No. 4		Station No. 5	
Avg BOD (5 day)	9.7	Avg 10, 11 a.m.	38.5	Avg 3, 4 p.m.	17.7 ppm
Avg DO (ppm)	8.63	Avg 10, 11 a.m.	4.45	Avg 3, 4 p.m.	8.65
Avg Temp.	21.63°C	Temp. 10, 11 a.m.	22.5°C	Temp. 3, 4 p.m.	23.5°C

Flow Time Sta. 4-Sta. 5–5.45 Hours
Streamflow–212 cfs

$$K_{1(4-5)} = \frac{1}{t} \log \frac{L_4}{L_5}$$

$$t = \frac{5.45}{24} = .2275$$

$$L_4 = 38.5 \times 1.46 = 56.2$$

$$L_5 = 17.7 \times 1.46 = 25.9$$

$$K_1 = \frac{1}{.2275} \times \log \frac{56.6}{25.9}$$

$$K_1 = 4.4 \times \log 2.18$$

$$K_1 = 4.4 \times .340$$

$$K_1 = 1.50$$

$$\therefore K_1 (\text{Str. } 20°\text{C}) = 1.50$$

$$K_{2(4-5)} = K_1 \frac{\bar{L}}{\bar{D}} - \frac{\Delta D}{2.3\, \Delta t \bar{D}}$$

$$\bar{L} = \frac{56.2 + 25.9}{2} = \frac{82.1}{2} = 41.05$$

$$\bar{D} = \frac{D_4 + D_5}{2} = \frac{4.21}{2} - 2.1$$

$$D_4 = \text{Sat}_{22.5°\text{C}} = 8.66 - 4.45 = 4.21$$

$$D_5 = \text{Sat}_{23.5°\text{C}} = 8.50 - 8.65 \leqslant 0$$

$$\Delta D = D_5 - D_4 = 0 - 4.21 = -4.21$$

$$K_2 = 1.50 \times \frac{41.05}{2.10} - \left[\frac{-4.21}{2.3 \times 2.1 \times .2275}\right]$$

$$K_2 = 29.3 + 3.85$$

$$K_2 = 33.15$$

$$\therefore K_2 (\text{Str. } 20°\text{C.}) = 33.15$$

$$K_2 t = 33.15 \times .2275 = 7.55$$

∴ Thomas Nomogram cannot be used

These reaction rates, however, cannot be used above Station 4 to compute the sag curve analysis which begins at Station 3.

$$D_t = \frac{K_1 L_A}{K_2 - K_1} \left[10^{-K_1 t}\ 10^{-K_2 t}\right] + D_A \times 10^{-K_2 t}$$

Since it is obvious that the DO is on the increase from Station 4 to Station 5, any treatment provided by the polluters will only decrease the extent of the existing sag.

Streeter-Phelps Sag Curve Analysis is only meaningful between Station 3 and Station 4. Since Station 3 was not sampled in the 6/25/'68 survey, BODs must be extrapolated logarithmically from Station 4 back to Station 3, and DOs can be extrapolated from Station 1 to Station 3.

For Stream Analysis for 6/25/'68 (*Continued*)

	DO 10/1967 (ppm)		DO 1960 Survey (ppm)	BOD
Station 1	11.1		8.7	At Station 3 obtained by extending
Station 3	10.15		7.3	BOD curve upwards as shown in
Ratio DO	91.5		83.5	Fig. 10-7 from 38.5 to 42 ppm
Station 3/1		87.5%		(linear) but (considering log = 44.6)
Avg				as shown in Fig. 10-6

∴ Assuming K_1 and K_2 values of survey exist also between Stations 3 and 4 and velocity of flow is similar, or 1.375 ft/sec

	Station 3	*Station 4*
BOD	44.6	38.5
DO	8.63 X .875 = 7.52	4.45
T =	22°C.	22.5°C.

$$t = \frac{5{,}280 \text{ ft}}{1.375 \text{ ft/sec}} = \frac{3{,}825}{3600 \text{ sec/hr}} = 1.06 \text{ hr} \quad \text{or} \quad .0443 \text{ days}$$

$K_1 = 1.5$

$K_2 = 33.15$

Flow 212 cfs

$D_3 = 8.78 - 7.52 = 1.26$

However, we have no assurance that these reaction rates (from Station 4 to 5) prevail also between Stations 3 and 4. It therefore becomes mandatory to compute K_1 and K_2 values between Stations 3 and 4.

$$K_1 = \frac{1}{t} \log \frac{L_3}{L_4} = \frac{1}{.0443} \times \log \frac{44.6 \times 1.46}{38.5 \times 1.46} = 22.55 \log \frac{65.1}{56.2} = 22.55 \log 1.158$$

$K_1 = 22.55 \times .064 = 1.44$ (stream rate at 20°C)

$$K_2 = K_1 \frac{\bar{L}}{\bar{D}} - \frac{\Delta D}{2.3 \bar{D} t} = 1.44 \times \frac{60.65}{2.735} - \frac{2.95}{2.3 \times 2.735 \times .0443}$$

$K_2 = 31.9 - 10.6 = 21.3$ (stream rate at 20°C)

$$\bar{L} = \frac{65.1 + 56.2}{2} = 60.65 \quad \bar{D} = \frac{1.26 + 4.21}{2} = 2.735 \quad \Delta D = 4.21 - 1.26 = 2.95$$

However, K_1 and K_2 must be converted to stream temperatures to compare the calculated dissolved oxygen deficit at Station 4 with the observed dissolved oxygen deficit.

$K_{1_{22.5°C}} = K_{1_{20°C}} \times 1.122 = 1.44 \times 1.122 = 1.62$

$K_{2_{22.5°C}} = K_{2_{20°C}} \times 1.04 = 21.3 \times 1.04 = 22.2$

Now to compute the calculated oxygen deficit at Station 4

$$D_4 = \frac{1.62}{22.2 - 1.62} \times 65.1 [10^{-1.62 \times .0443} \; 10^{-22.2 \times .0443}] + 1.26 \times 10^{-22.2 \times .0443}$$

$$D_4 = 5.1 [10^{-.0717} \; 10^{-.9850}] + 1.26 \times 10^{-.9850}$$

For Stream Analysis for 6/25/'68 (*Continued*)

D_4 = 5.1[.8478 − .1035] + 1.26 × .1035

D_4 = 5.1[.7443] + .130

D_4 = 3.80 + .130 = 3.93 ppm calculated

D_4 = 4.21 ppm observed

∴ reaction rates closely describe what is actually taking place between Stations 3 and 4. Now we can use these conditions to compute D_4 under various other conditions

K_1 = 1.62

K_2 = 22.2

t_{3-4} = .0443 days

D_3 = 1.26

Flow = 212 cfs

Temp = 22.5°C

No tannery or glue waste treatment

1. D_4 under above conditions = 3.93 ppm_{calc} and 4.21 ppm_{obs}
2. Assume that flow in stream was reduced to 1 in 10 year min. 7-day flow of 60 cfs and pollution from tannery and glue plant was reduced 60 percent. Same temp. and reaction rates

Calculate new D_3 as follows using the basic assumption that the pounds of oxygen deficit will be increased in direct proportion to the ratio of the original to the low flow.

1.26 × 8.4 × 212 = ppm × 8.4 × 60

$$D_3\text{, ppm,} = \frac{212}{60} \times 1.26 = 4.46$$

Calculate new L_3 as follows

L, ppm, = 9.7 + 65.1 × .40 = 9.7 + 26.04 = 35.74

Now calculate new D_4 as follows

$$D_4 = \frac{1.62}{22.2 - 1.62} \times 35.74[10^{-1.62 \times .0443}\ 10^{-22.2 \times .0443}] + 4.46 \times 10^{-22.2 \times .0443}$$

$$\frac{1.62}{20.58} \times 35.74[.7443] + 4.46 \times .1035$$

$$2.8[.7443] + .462 = 2.08 + .462$$

D_4 = 2.542 ppm

∴ No contravention of DO standards at Station 4 even at 1 in 10-year low flow, as long as total BOD is reduced by 60 percent.

TABLE 10-4. Mean Monthly Flow Cattaraugus Creek (cfs)

	Oct.	Nov.	Dec.	Jan.	Feb.	Mar.	Apr.	May	June	July	Aug.	Sept.
1960–1961	104	118	111	136	1,441	1,217	2,030	730	833	230	367	178
1961–1962	149	460	736	744	554	1,212	1,338	675	329	124	107	134
*1962–1963	294	463	380	366	222	2,740	1,346	376	179	145	160	114
*1963–1964	81.8	317	395	925	333	2,365	1,427	386	212	171	185	95.5
*1964–1965	114	332	876	747	1,434	890	1,391	465	238	161	107	93.3
1965–1966	119	609	997	512	1,212	1,912	1,112	843	337	151	126	158

*Minimum June daily flow: 102 cfs 1962–1963; 126 cfs 1963–1964; 152 cfs 1964–1965.

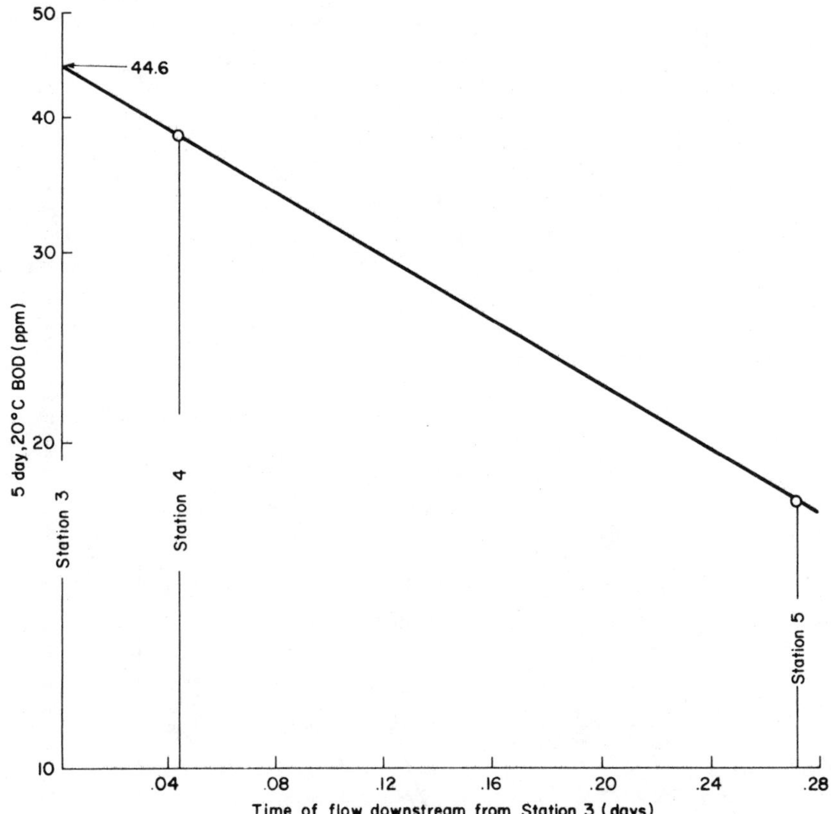

FIG. 10-6. BOD plot below major sources of pollutant entrance.

this flow is still above the minimum seven-day flow, with a return period of 10 years (60 cfs), it is considerably below the 335 cfs recorded during the 1967 creek sampling study. Furthermore, minimum daily flows during June of the above three years were less than 152 cfs.

With the above data in hand, it was decided to sample Cattaraugus Creek again in June 1968 to establish sufficient repetitive data to allow use of both the Streeter-Phelps and Churchill techniques in computing BOD reduction required to maintain the stream at Class C condition.

These results are in good agreement with previous stream studies carried out in 1967 in connection with Moench Tannery wastes. Generally, a 60 percent reduction in BOD of what is *currently* entering the creek should protect the creek from dipping below 4 ppm DO at Station 4, even during the period of extreme critical low flow (60 cfs).

The sag may continue a short distance below Station 4 before recovering, but

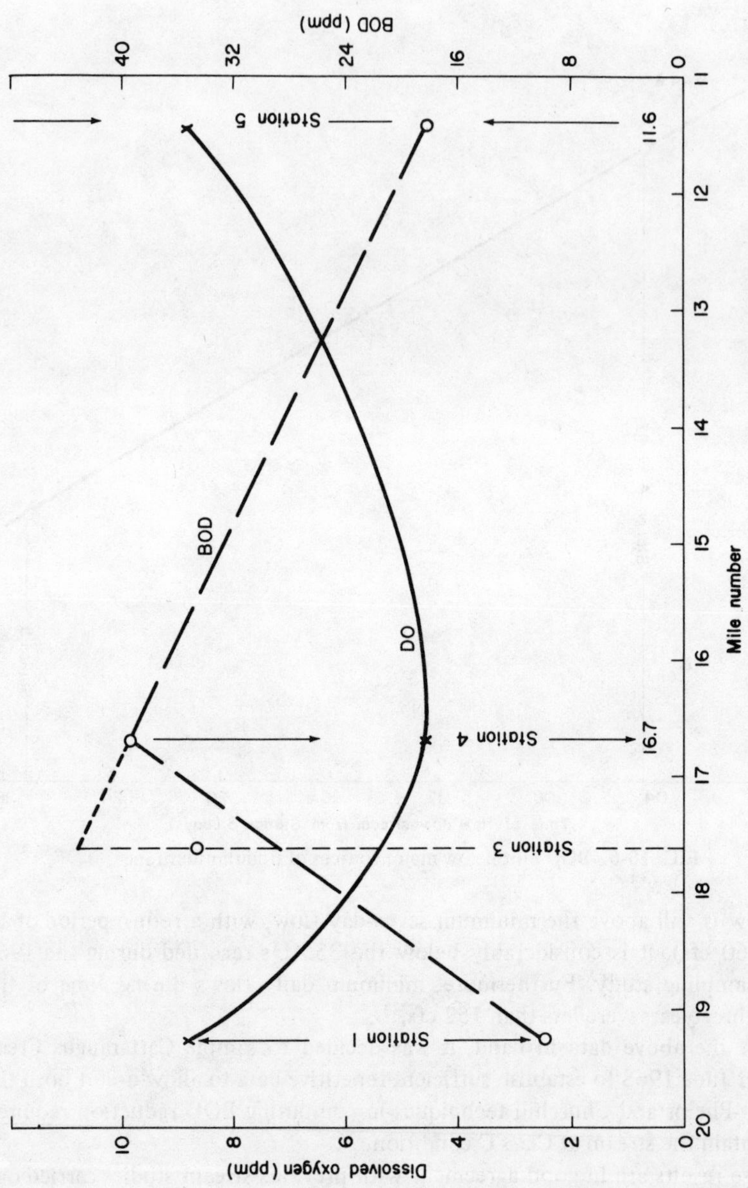

FIG. 10-7. DO and BOD plots of Cattaraugus Creek in vicinity of Gowanda, New York.

no use (or access) is made of the creek until Station 5 at Versailles Road Bridge. Protection of the creek at the point of leaving habitation (Station 4) with an extra part or two of dissolved oxygen to spare should offer ample protection to the creek between Station 4 and Station 5.

The third survey was carried out in the spring of 1969, when ample manpower was available once again to do a thorough sampling and analysis job. Unfortunately, the previous fall's flows were relatively high, and manpower was not available to assist in another survey. Therefore, one and one half years had elapsed between the first and third surveys. The third survey was on April 24, 1969—too soon for the normal spring low flow to occur. In fact, no usual low flows ever did occur during the spring of 1969.

Samples were collected from four stations, and one sample was collected about halfway into the Indian reservation. "Times of flow" were determined once again by taking cross-sectional velocities. Four "slugs" of streamwater were followed down the creek during the day at various times. These are shown in Table 10-5, and in Fig. 10-8, are plotted in running fashion from above the tannery to Versailles Road Bridge downstream. Much difficulty was encountered in attempting to use the Streeter-Phelps formulations in obtaining creek reaction rates k_1 and k_2. The determination of k_2 was especially difficult, because little dissolved oxygen sag occurred on this day as a result of the waste discharge. Perhaps this was caused by the relatively high streamflow of 300 cfs, almost five times the critical drought flow. To illustrate these problems—rather than to add new knowledge—the author includes the calculations attempted in trying to solve the Streeter-Phelps equations. Insufficient data are currently available to allow for the use of multiple regression techniques in developing a stream equation.

By the time the third survey was run, certain industrial operations had changed. For example, the tannery no longer used vegetable dyes in tanning sole leather, and the glue plant was "phasing out" its use of fleshings as a raw material. These materially lowered the organic loading on the creek and undoubtedly contributed considerably to the high oxygen levels found downstream. These changes occur quite often in the working life of the stream analyst. Change is the "name of the game," and one can expect poor correlation from one stream survey to another because of these "unusual normal" changes.

Streeter-Phelps Analysis of April 24, 1969, Creek Survey

The Streeter-Phelps analysis was performed on the data that was obtained from the survey and on data that was projected from low-flow and industrial records. The following are the equations used to determine the rates of deoxygenation and reaeration and the Streeter-Phelps equation.

TABLE 10-5. Original Data Tabulated from Cattaraugus Creek Survey, April 24, 1969 (*flow = 300 cfs; velocity = 3.68 fps*)

Station	Descriptions	Distance, ft*	Time, hr: min.*		BOD_5, mg/l	Temperature, °C	DO_{stream}, mg/l	$DO_{saturated}$, mg/l
1	below island	0	0:00	a, b, c, d	1.5	8.5	11.20	11.73
1′	just above tannery	1,320	0:06	a, b, c, d	1.5	8.5	11.20	11.73
2	railroad trestle–	5,256	0:24	a	41	8.5	10.95	11.73
	point of complete			b	30	9.5	10.70	11.46
	mixing of the			c	29	11.5	10.40	10.95
	two discharges			d	25	12.5	10.40	10.71
3	highway bridge–	5,476	0:25	a	36	12.5	10.50	10.71
	just above town's			b	21	12.5	10.40	10.71
	sewage treatment			c	20	12.5	10.40	10.71
	plant			d	20	12.5	10.40	10.71
4	"Versailles" bridge–	28,234	2:07	a	22	12.5	10.60	10.71
	just below Indian			b	20	12.5	10.50	10.71
	reservation			c	15	12.5	10.30	10.71
				d	12	12.5	10.10	10.71
X	$2\frac{1}{2}$ miles into reservation	13,200	0:50	a	18	a, b, c, d 12.5	9.80	10.71

*Indicates distances and times of flow between station and preceding station.

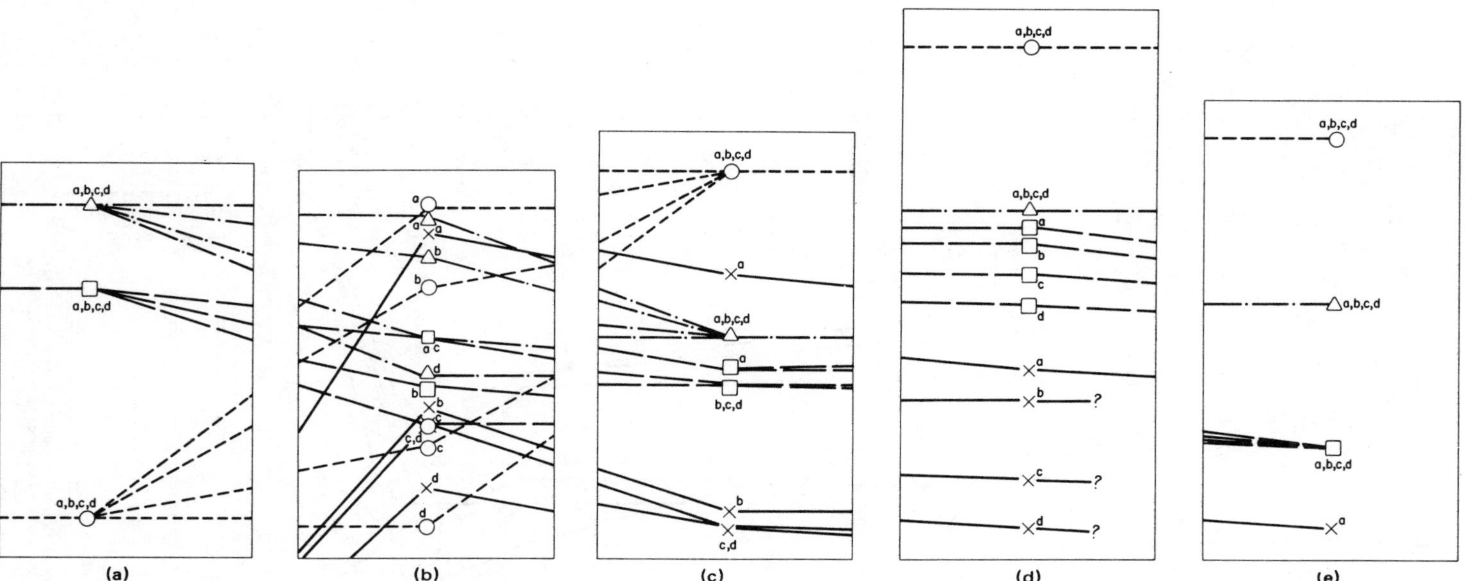

FIG. 10-8. Plot of four "slugs" of streamwater from Cattaraugus Creek from above Moench Tannery to Versailles Road Bridge downstream.

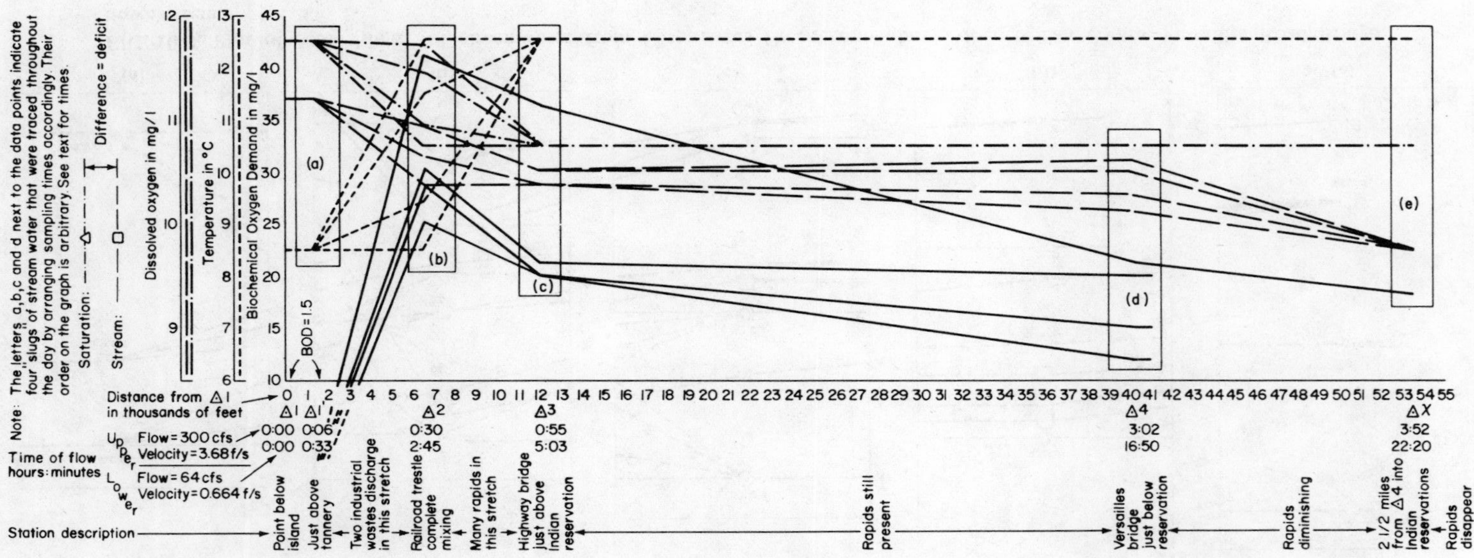

FIG. 10-8. *(Continued)*

K_1, deoxygenation rate for the stream conditions

$$K_1 = \frac{1}{T_b - T_a} \log \frac{La}{Lb}$$

where stream flows from a to b
T is time of sampling each point
L is ultimate first-stage BOD at each point

K_2, reaeration rate for stream conditions

$$K_2 = \frac{(K_1)(\bar{L})}{\bar{D}} - \frac{D_b - D_a}{(2.3)(\bar{D})(t)}$$

where stream flows from a to b
$\bar{L}$ is average value of BOD from a to b
$\bar{D}$ is average value of DO deficit
D is DO deficit at point on stream
t is time of flow from a to b

Streeter-Phelps Equation

$$D = \frac{K_1 L_0}{K_2 - K_1} (10^{-K_1 t} - 10^{-K_2 t}) + D_0(10^{-K_2 t})$$

where D is DO deficit at any time of flow t
L_0 is initial BOD load in stream
D_0 is initial DO deficit in stream
t is the time of flow

The Streeter-Phelps analysis described the effect of an organic load on a stream's dissolved oxygen content. The assumptions made to apply this analysis are that the stream's reaeration rate is constant; the organic loading (deoxygenation rate) is constant; the rates of reaeration and deoxygenation follow logarithmic paths; and the flow and flow time are constant across the reach of the stream that is being studied.

Computations for the stream reaction rates K_1, K_2. From the original data, Stations 2 and x were chosen to determine the deoxygenation rate K_1. This choice was made because between these two stations a fairly constant decrease in stream BOD was observed, and the longer time of flow yields more accurate results. In computing the reaeration rate K_2, the data from Stations 4 and x were used because of the lack of a constant value of oxygen deficit in the upper reaches.

$$K_1 \begin{cases} T_2 = 0{:}00 \text{ (hr)} \\ T_x = 3{:}54 \text{ (hr)} \\ L_2 = 30 \text{ mg/l} \\ L_x = 18 \text{ mg/l} \end{cases}$$

$$K_1 = \frac{1}{3.9 \text{ hr}} \times \log_{10} \frac{30}{18} \times \frac{24 \text{ hr}}{1 \text{ day}} = \frac{1.51}{\text{day}}$$

$$K_2 \begin{cases} \bar{L} = \dfrac{22 + 18}{2} = 20 \text{ mg/l} \\ \bar{D} = \dfrac{0.93 + 0.34}{2} = 0.64 \text{ mg/l} \\ D_x - D_4 = 0.94 - 0.34 = 0.60 \text{ mg/l} \\ t = 1 \text{ hour} \end{cases}$$

$$K_2 = \frac{(1.51/\text{day})(20 \text{ mg/l})}{0.64 \text{ mg/l}} - \frac{(0.60 \text{ mg/l})(24 \text{ hr/day})}{(2.30)(0.64 \text{ mg/l})(1 \text{ hr})}$$

$$K_2 = 37.41/\text{day}$$

It must be noted here that the values just computed for K_1 and K_2 are based on the base 10. The following are the constants K_1 and K_2 from above, converted to the Naperian base:

$$K_1' = 3.47/\text{day} \qquad K_2' = 98.50/\text{day}$$

Application of Streeter-Phelps. The oxygenation rates have now been determined from the original data. An organic loading of thirty milligrams per liter (30 mg/l) was chosen as representative from the four possibilities, because it is an average over the sampling period. A saturation oxygen concentration of 10.74 mg/l was used (see Appendices B1 and B2) at the ambient stream temperature of 12.5°C and, coupled with the measured DO at Station 2 of 10.40 mg/l, gave an initial deficit of 0.34 mg/l. These figures were applied to a computer program which solves the Streeter-Phelps equation for values of time, and plots the dissolved oxygen at these values of time. The results of this program-run appear as Printout 1.

Upon examination of Printout 1, it can be seen that an apparent sag of 9.79 mg/l appears after about 45 minutes of flow from Station 1. This result was considered to be false. Because of the extremely high rate of reaeration (K_2) in this reach, which is due to the presence of rapids, the oxygen is being supplied to the organic material as fast, or faster, than it is being taken up. There is, therefore, no sag as such. What is happening in the stream is that the dissolved oxygen value actually remains quite constant, varying to a slight degree above and below a mean value. The organic material, however, has not been completely degraded, and it continues to decompose at a fixed rate. When the stream passes point x, the rapids subside and the oxygen starts being used more rapidly than it can be supplied. This is the point where the Streeter-Phelps analysis can be more accu-

Printout 1. Streeter-Phelps Tabulative Program Using Naperian-based Values of *K*. *Q* = 300 cfs; *V* = 3.68 fps

```
READY

$SAVE

READY
$LIST
ØØØ1Ø PRINT "WHAT IS K1, K2";
ØØØ2Ø INPUT K1, K2
ØØØ25 PRINT
ØØØ3Ø PRINT "WHAT IS LØ,DØ,DS ?";
ØØØ4Ø INPUT LØ,DØ,D9
ØØØ45 PRINT
ØØØ5Ø PRINT "WHAT IS MIN., MAX., STEP,(DAYS)";
ØØØ6Ø INPUT S1,S2,S3
ØØØ7Ø PRINT
ØØØ71 LET A=1/(K2-K1)*LOG[(K2/K2)*(1-[DØ*(K2-K1)/(K1*LØ)]))]
ØØØ72 LET B=K1*LØ/(K2-K1)*(EXP(-K1*A)-EXP(-K2*A))
ØØØ73 LET C=B+DØ*EXP(-K2*A)
ØØØ74 LET C=D9-C
ØØØ75 PRINT "MAX SAG OF"C, "OCCURS AT TIME"A
ØØØ76 PRINT
ØØØ8Ø PRINT "TIME,DAYS", "        ", "D.O.,PPM"
ØØØ81 PRINT "       ";
ØØØ82 FOR I=Ø TO D9
ØØØ84 PRINT I;
ØØØ86 NEXT I
ØØØ9Ø FOR T=S1 TO S2 STEP S3
ØØ1ØØ LET R=K1*LØ/(K2-K1)*(EXP(-K1*T)-EXP(-K2*T))
ØØ11Ø LET D=R+DØ*EXP(-K2*T)
ØØ115 LET D=D9-D
ØØ123 PRINT
ØØ124 PRINT
ØØ125 PRINT T;TAB(5+6*D);D
ØØ13Ø NEXT T
ØØ138 PRINT
ØØ139 PRINT
ØØ14Ø PRINT "DO YOU WISH TO TRY ANOTHER SET OF TIME (1=YES,2=NO)";
ØØ15Ø INPUT Q
ØØ16Ø IF Q=1 THEN ØØØ5Ø
ØØ17Ø END
```

Printout 1. (*Continued*)

```
WHAT IS K1, K2?3.47,98.5

WHAT IS LØ,DØ,DS ??3Ø,.34,.74

WHAT IS MIN., MAX., STEP,(DAYS)?Ø,1,.Ø21

MAX SAG OF 9.79 ppm OCCURS AT TIME .031298 days   Day of Sampling

TIME,DAYS                               DO, ppm
              Ø     1     2     3     4     5     6     7     8     9     1Ø
0  Ø      0:00 Δ1                                                  1Ø.13
   Ø.Ø21  0:30 Δ2                                                  9.547Ø1
1  Ø.Ø42  0:55 Δ3                                                  9.53518
   Ø.Ø63                                                           9.59119
2  Ø.Ø84                                                           9.65172
   Ø.1Ø5                                                           9.7Ø9Ø8
3  Ø.126  3:05 Δ4                                                  9.76253
   Ø.147                                                           9.81225
4  Ø.168  3:52 ΔX                                                  9.85848
   Ø.189                                                           9.9Ø145
5  Ø.21                                                            9.94141
   Ø.231                                                           9.97856
6  Ø.252                                                           1Ø.Ø18
   Ø.273                                                           1Ø.Ø48
7  Ø.294                                                           1Ø.Ø78
   Ø.315                                                           1Ø.1Ø8
8  Ø.336                                                           1Ø.128
   Ø.357                                                           1Ø.158
Time of flow, hours

5/28/69, 15:11
```

*Add Ø.27 to all of the values in right-hand column.

rately applied. There are no data for determining the reaeration rate in this reach following point x. It is safe to assume that the deoxygenation rate remains constant, but a method for predicting K_2 must be used. G. M. Fair (Fair, Gordon, and Geyer, 1956, p. 846) has suggested the following table for predicting K_2 when K_1 and a general description of the stream are available.

Nature of receiving water	*f value*
Small ponds and backwaters	0.5–1.0
Sluggish streams and large lakes	1.0–1.5
Large streams of low velocity	1.5–2.0
Large streams of moderate velocity	2.0–3.0
Swift streams	3.0–5.0
Rapids and waterfalls	above 5.0

When an appropriate f value is chosen, to determine K_2 multiply the known value of K_1 by the f value; the resulting product can be used as the value of K_2.

The stream below Station x was judged to fall into the upper part of the fourth category, and a "Fair's f" value of 3 was used. The resulting value of K_2 was 10.42/day. Printout 2 contains the results of the second program run, using the above described data, and the deficit and saturation values appropriate to Station x.

The combination of Printouts 1 and 2 will give an approximation of conditions in the stream on the day of the survey, using the Streeter-Phelps method of analysis.

Projection of Data–Streeter-Phelps. The previous section revealed that the Streeter-Phelps analysis would hold true if considered in the reach following Station x, because of the rapids present in the first reach of the stream under conditions of high flow. The stream takes on different characteristics when the flow reaches a low value. It is during this period of low flow that the stream must be analyzed, for this is where the greatest amount of damage can occur when unabated pollution is discharged into the stream. As stated earlier, the low flow which is likely to occur on an average of once every 10 years is about 64 cubic feet per second (64 cfs).

The rates of deoxygenation for the two industrial waste discharges and the subsequent loadings on the stream must now be determined. This will give a picture of what will occur in the stream under the most adverse conditions: low flow and untreated industrial waste.

From Figs. 10-4 and 10-5 pages 321 and 322 it can be seen that the K_1 values for the glue factory and the tannery are 0.14/day and 0.05/day, respectively. Their corresponding hydraulic loadings are 3.0 mgd and 0.25 mgd, respectively. The stream low-flow value, in million gallons per day, is 40.7. Assuming no treatment, since the streamflow is one-fifth as great, the concentration of the organic

Printout 2. Streeter-Phelps Analysis Using Fair's $f = 3$ from Station $\chi \rightarrow$ Downstream
$Q = 300$ cfs; $V = 3.68$ fps

```
          Ø5/28/69          15:16

WHAT IS K1, K2?3.47,1Ø.42

WHAT IS LØ,DØ,DS ??17,.94,1Ø.74

WHAT IS MIN., MAX., STEP,(DAYS)?Ø,1.5,.Ø21

MAX SAG OF 7.27314 ppm   OCCURS AT TIME Ø.141323 days   Day of Sampling
```

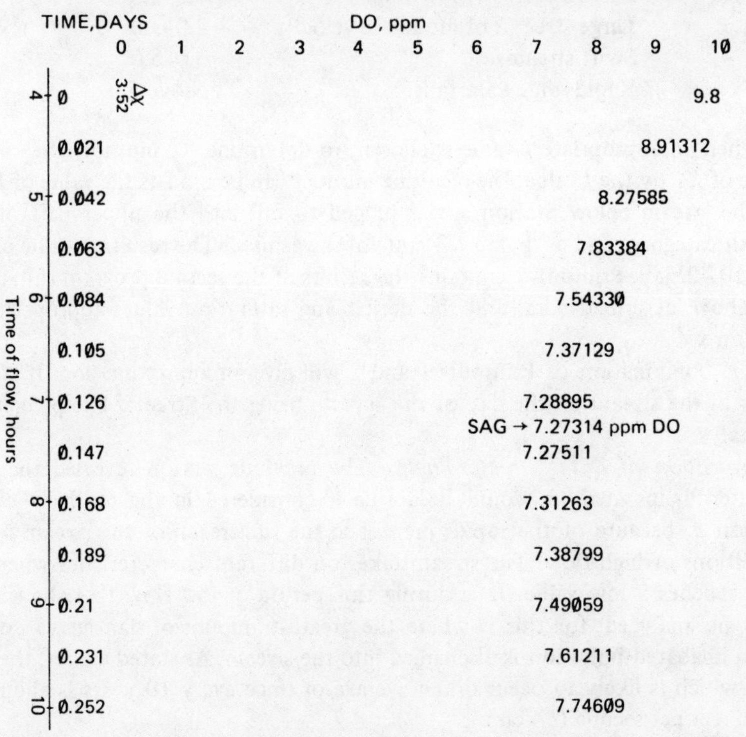

load is taken to be five times the original, or 150 milligrams per liter (150 mg/l). The waste flow from the glue factory is 12 times that of the tannery; therefore it is assumed that the effect of the glue factory waste is 12 times that of the tannery's. The single value of K_1 was chosen by an arithmetic average to be 0.133/day. Printout 3 contains the computer run of these values, using the saturation of 20°C because the low flow occurs during the summer, when temperatures are high. The original deficit of 0.34 mg/l was retained and used. An examination of Printout 3 shows the effect. After about half a day's flow, the stream goes

Printout 3. Streeter-Phelps Analysis: Q = 64 cfs; V = .664 fps; Low Flow, Untreated

```
$OLD JAB1

READY
$BASIC

              Ø5/27/69                  16:26

WHAT IS K1, K2?.133,.2

WHAT IS LØ,DØ,DS ??15Ø,.34,9.17

WHAT IS MIN., MAX., STEP,(DAYS)?Ø,3Ø,1

MAX SAG OF -35.3121 ppm     OCCURS AT TIME 6.Ø72Ø3 days
                ANAEROBIC

TIME,DAYS                      DO, ppm
        Ø     1     2     3     4     5     6     7     8     9

 Ø                                                         8.83

 1      -8.ØØ165

 2      -19.6784

 3      -27.3963

 4      -32.1Ø36

 5      -34.5456

 6      -35.3Ø9Ø

 7      -34.8522

 8      -33.531Ø

 9      -31.62ØØ

 1Ø     -29.3295

 11     -26.8186

 12     -24.2Ø65

 13     -21.58Ø7

 14     -19.ØØ45

 15     -16.5219

 16     -14.1624
```

Printout 3. (*Continued*)

```
17    -11.9446

18    -9.87819

19    -7.967Ø9

2Ø    -6.21Ø41

21    -4.6Ø4Ø7

22    -3.14175

23    -1.81568

24   -Ø.61719Ø
                ← stream regains oxygen after 25 days of flow
25     Ø.462788

26               1.43344

27                    2.3Ø383

28                         3.Ø8269

29                             3.77838

3Ø                                 4.89874

31                                    4.9511Ø

32                                       5.44225

33                                          5.87845

34                                           6.26541

35                                             6.6Ø834

36                                               6.91197

37                                                 7.18Ø57

38                                                  7.418Ø1

39                                                   7.62775

4Ø                                                    7.8129Ø

41                                                     7.97625

42                                                      8.12Ø28
```

Printout 3. *(Continued)*

43	8.24721
44	8.35902
45	8.45747
46	8.54413
47	8.62036
48	8.68742
49	8.74637
50	8.79819

anaerobic; that is, there is no oxygen to be had. Only after 25-plus days' flow does the stream recover oxygen. Printout 4 shows an expanded time scale of the first 14 hours of flow from "3." The velocity of the stream at low flow is 0.664 feet per second. The original stations can now be superimposed onto this time scale. After $11\frac{1}{2}$ hours of flow, which is found between Stations 3 and 4, the point of zero oxygen appears. Biological examination of this area (details follow) confirms this fact. The history of this reach is one of anaerobic conditions. It is felt, therefore, that the projections of the data have been accurate and that the assumptions made here have been valid. The mass discharge of untreated industrial wastes from the two factories has been going on for at least the past 60 years, and their effects on Cattaraugus Creek bear witness to this fact. Not only is the stream deprived of enough oxygen to function naturally; it has been changed into an almost sterile environment.

Results of industrial waste treatment–projected. The *treated organic load* due to the industries, as stated earlier, is

Glue plant	3,620
Tannery	1,243
Total	4,863 # BOD_5/day

At a low stream flow of 40.7 mgd, the concentration of organics as it mixes with the stream is

$$\frac{4{,}863\ \#/\text{day}}{40.7\ \text{mgd}} \times \frac{1\ \text{mg/l}}{8.34\ \#/\text{mg}} = 14.4\ \text{mg/l}$$

Referring back to the chart containing the values of "Fair's *f*," the stream at low flow is considered to be a large stream (64 cfs) of low velocity (0.664 fps). The corresponding "Fair's *f*" value of 1.5 was used. The Streeter-Phelps analysis

Printout 4. Streeter-Phelps Analysis: Q = 64 cfs; V = .664 fps; Low Flow, Untreated Effluents

```
$BASIC

          Ø5/27/69     16:19

WHAT IS K1, K2? .133, .2
WHAT IS LØ, DØ, DS? ? 15Ø, .34, 9.17
WHAT IS MIN., MAX., STEP, (DAYS)? Ø, 1, .Ø21
MAX SAG OF-35.3121 ppm        OCCURS AT TIME 6.Ø72Ø3 days
```

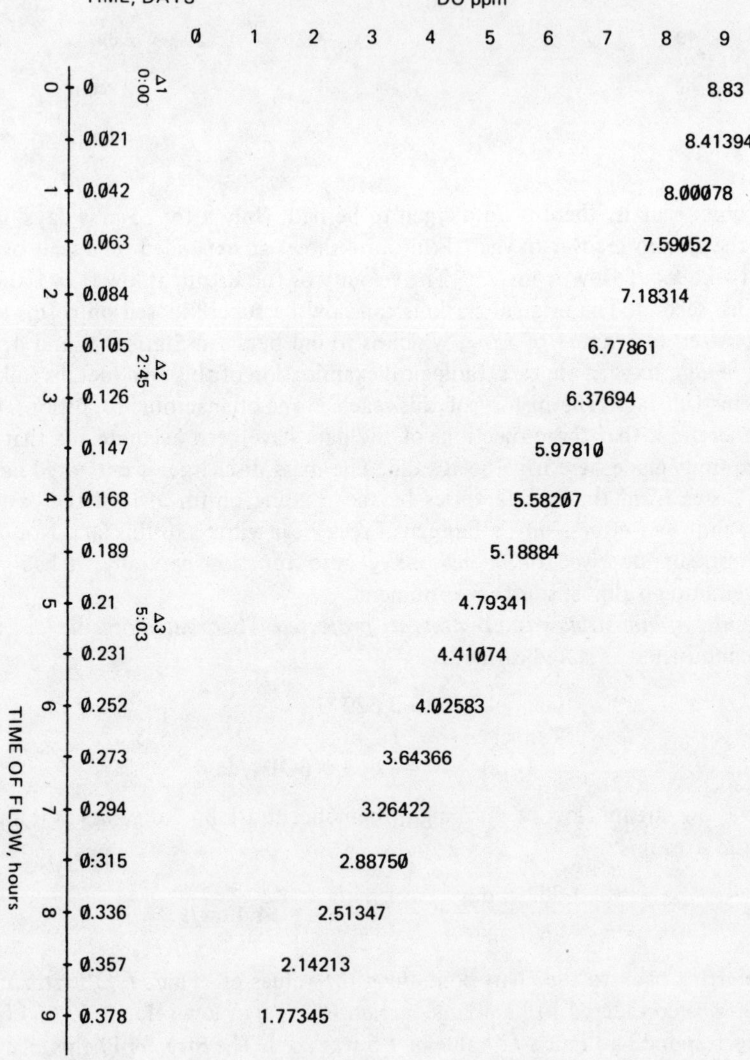

Printout 4. *(Continued)*

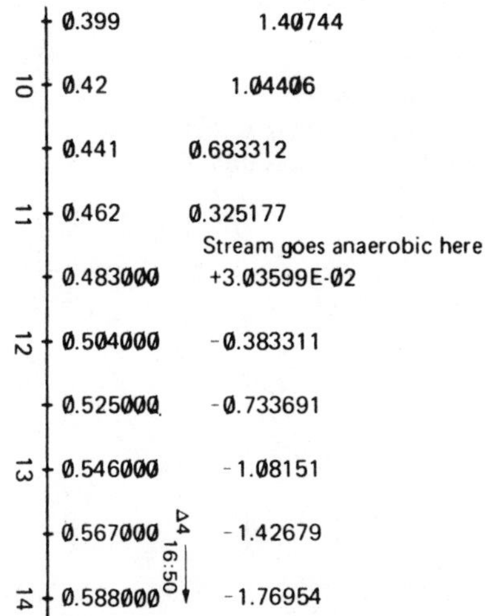

of the above data yields an oxygen sag curve as shown in Printout 5. The resulting prediction shows an abundance of oxygen throughout the entire area of the survey. After six days of flow, the dissolved oxygen concentration reaches a minimum of 4.81 mg/l. This is well above the minimum of 4 mg/l required for the best New York State stream classification *A*, which is waters used for fishing. It can be projected, therefore, that with both industries treating their wastes, resulting in a 90 percent reduction of BOD, the stream will be able to support high forms of fish and become a clean stream.

Thomas Method for Determining Sag. The Thomas method was applied to the original data and was found satisfactorily comparable. It was then applied to the projected data for low flow and treated industrial waste between Stations 2 and x.

$$K_1 = 0.133/\text{day}$$

$$K_2 = 0.200/\text{day}$$

$$t = 19.5 \text{ hours}$$

$$Da = 0.34 \text{ mg/l}$$

$$La = 14.4 \text{ mg/l}$$

$$K_2 \times t = (0.2/\text{day})(0.812 \text{ day}) = 0.162$$

$$Da = \frac{0.34}{14.4} = 0.0236$$

Printout 5. Streeter-Phelps Analysis: 90% Treated Effluents.
Q = 64 cfs; V = .664 fps; Low Flow

```
            Ø5/27/69      16:Ø2

WHAT IS K1, K2? .133, .2
WHAT IS LØ, DØ, DS? ? 14.4, .34, 9.17
WHAT IS MIN., MAX., STEP, (DAYS)? Ø, 7, .25
MAX SAG OF 4.8Ø698 ppm       OCCURS AT TIME 5.91Ø49 days
```

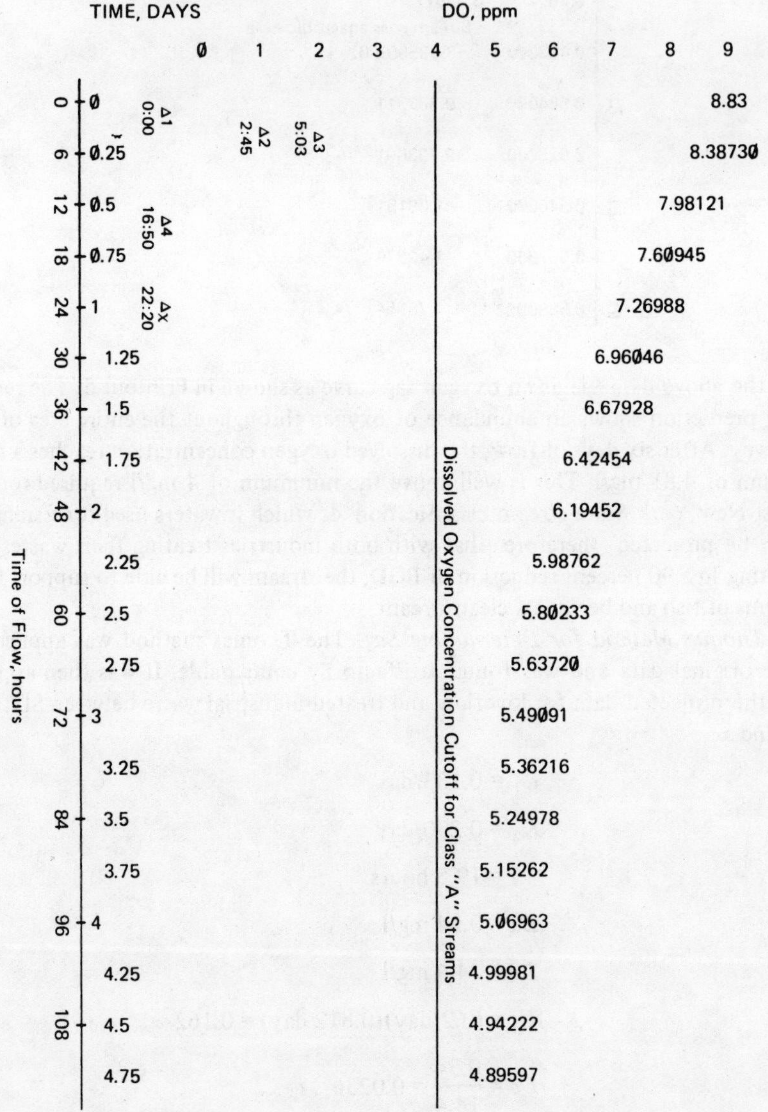

Printout 5. (*Continued*)

120	5	4.86024	
	5.25	4.83423	
132	5.5	4.81721	
	5.75	4.80850	
144	6	4.80744	
		SAG →	4.80698 ppm DO
	6.25	4.81342	
156	6.5	4.82587	
	6.75	4.84426	
168	7	4.86808	
	7.25	4.89685	
180	7.5	4.93014	
	7.75	4.96754	
192	8	5.00865	
	8.25	5.05312	
204	8.5	5.10060	
	8.75	5.15077	
216	9	5.20335	
	9.25	5.25806	
228	9.5	5.31463	
	9.75	5.37284	
240	10	5.43245	

$$\frac{K_2}{K_1} = \frac{0.2}{0.133} = 1.5$$

$$\frac{D}{La} = 0.23 \qquad \text{from nomograph}$$

$*Dt = 14.4 \times 0.23 = 3.32$ mg/l (for 0.8 day flow time)

*To obtain critical deficit, draw tangent to curve

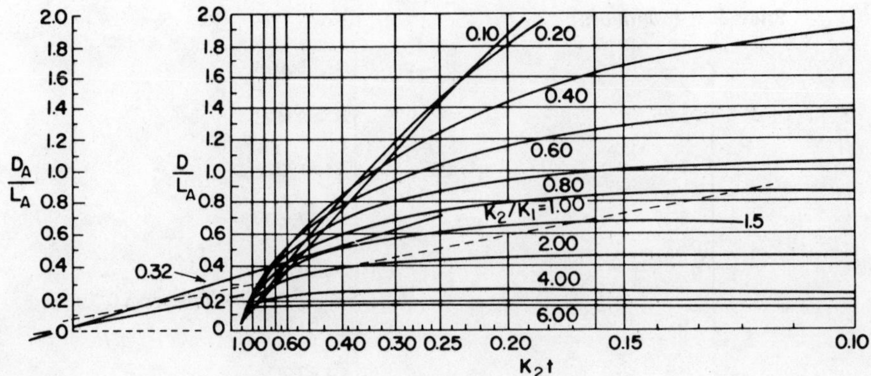

$$\frac{D}{La} = 0.23 \qquad \text{from nomograph}$$

$$Dt = 14.4 \times 0.32 = 4.60 \text{ mg/l} \qquad \text{critical deficit}$$

$$\text{DO in stream} = 9.20 - 4.60 = 4.60 \text{ mg/l (sag)}$$

(Compares with Streeter-Phelps value of 4.81 mg/l)

It must be obvious to the reader that the original K_1 and K_2 values of 3.47 and 98.50, respectively, were almost beyond reasonableness. Even the more logical values of 3.47 and 10.42, respectively, are open to question, since the latter value was not obtained by actual stream analysis. On the other hand, the use of true laboratory reaction rates of 0.133 and 0.20 is also unrealistic; stream conditions are bound to be very different from those found to exist in laboratory BOD bottles. Therefore, in this stream study the use of the Streeter-Phelps technique is, at best, a subjective estimate.

In an attempt to overcome the lack of a real significant oxygen sag (which was partially anticipated), a biological analysis of the creek was also made at the same time the stream was being surveyed for the third time. It was hoped that the biological analyses would at least corroborate the physicochemical analyses.

Biological investigations.[1] A discussion of the biotic characteristics of a stream should be included in all reports of the sanitary aspects of the stream. The basic assumption underlying biological surveying is acceptable. This assumption is that species differ in their ability to survive and reproduce in a given en-

---

[1]Largely a contribution by biology students enrolled in the Stream Sanitation Course at Syracuse University during 1969, especially John Keenan, currently Professor, University of Pennsylvania.

vironment, and that a knowledge of the inhabitants can help give a picture of the habitat. The biological survey was especially important in this Cattaraugus Creek study. The high-water conditions negated the value of typical sanitary measurements. Dissolved oxygen measurements obtained on the day of the survey implied that Cattaraugus Creek was indeed a healthy one. In this section the biological evidence is examined to see if that conclusion is supported.

Biological samples of two types were taken at each of the sampling points. The first consisted of a one-quart grab sample for plankton determinations and counts. These samples were concentrated with a clinical-type centrifuge to a final volume of 8 ml, including FAA preservative. The plankton counts were accomplished using the Lackey Drop-Sedimentation Method (APHA, 1965) and the results are summarized on the following sheet. Texts written by Smith (1950) and Palmer (1962) were used in making the identification as shown in Table 10-6. The numbers in parentheses refer to the percent of total numbers at that station, whereas the other figures represent cells per liter.

The dominant algae observed were all diatoms and were identified to the species level, where possible. *Navicula gracilis*, *N. graciloides* (about 75 percent of the *Navicula* forms were *N. gracilis*), *Meridion*, and *Cymbella cesati* are clean water types (Palmer, 1962). *Navicula* was the sole alga that exhibited a steady downstream decrease in numbers. However, all forms were observed at much higher levels at Station 1 than at those stations downstream. The data imply that, although there is a certain reduction in the total number of individuals as one travels downstream, there may be a trend towards a more balanced community. (*Gomphonema* may be exerting a disruptive influence on the quantitative representation of the data. This is because the numbers of this attached

TABLE 10-6. Plankton Count Above and Below Pollutant Entrance on Creek

Algae	Stations							
	1		2		3		4	
	Cells/ liter	% of total	Cells/ liter	% of total	Cells/ liter	% of total	Cells/ liter	% of total
Navicula (gracilis & graciloides)	1,960	(14.3)	1,960	(27.1)	1,180	(32.2)	1,000	(26.0)
Gomphonema (olivaceum)	1,570	(11.5)	338	(5.4)	850	(23.9)	170	(4.40)
Meridion	7,190	(52.4)	1,520	(24.3)	680	(19.1)	1,000	(26.0)
Fragilaria (capusina)	1,710	(12.5)	1,690	(27.1)	170	(4.8)	1,000	(26.0)
Cymbella (cesati)	720	(5.3)	–	–	–	–	–	–
Synedra	550	(4.0)	845	(13.5)	680	(19.1)	680	(17.7)
Tabellaria	–	–	169	(2.7)	–	–	–	–

form in a plankton sample may be primarily a function of local scouring.) The trend towards a more balanced community accompanies the decreased numbers of *Meridion.*

The saprobicity values for the diatoms used by Sladecek (1965) were employed to evaluate the condition of the stream. The procedure used was as follows: for each species the product of the abundancy (cells/l) times the saprobic valency. The decimal fraction of each group (xeno-, oligo-, β-meso, α-meso, and poly-) was determined and multiplied by (1, 2, 3, 4, 5) and the sum determined. The following values are obtained (where xeno = 1, oligo = 2, β-meso = 3, α-meso = 4, poly = 5)

Station 1 = 2.014
2 = 2.45
3 = 2.57
4 = 2.46

This implies oligosaprobic conditions upstream and approaching β-mesosaprobicity downstream.

The second group of samples was collected to examine the bottom fauna. For this purpose, Cattaraugus Creek is considered to be a fast-moving, rocky-bottomed stream. One sample was taken at each of the sampling points. The samples were collected with a D-net, and immediately preserved in 70% ethanol. Identifications were made using texts written by Pennak (1953) and Usinger (1963).

The examination of the macrofauna of a stream is valuable for several reasons. These organisms are less subject to drift than, for example, are the plankton. A knowledge of the organisms's respiratory mechanism is helpful to the investigator in determining the history of dissolved oxygen in the locale. For instance, a large population of stone-flies, which require a high level of dissolved oxygen, implies that the water has been well oxygenated for some time.

One of the best ways to describe the biota of a stream is to do so in terms of a density index expressing the number of individuals/family/sampling station. This is done since a polluted environment may support a large population consisting of only a few types, whereas an unpolluted ecosystem supports a diverse population of many groups (see Patrick, 1949).

However, the samples taken from Cattaraugus Creek are qualitative in nature, since only one collection was obtained at each station. Usinger (1963) reports on a study by Needham and Usinger (1956) in which they showed that 13 samples (using a Surber square-foot sampler) are needed to ensure (at the 95 percent confidence level) that all families are represented, and that 194 samples are required to give significant results on wet weights. The number of samples depends upon the number of species present. Hence, such quantitative measures as a diversity

TABLE 10-7. Bottom Organisms Found Above and Below Pollutant Entrance on Creek

Insects	Nos. of specific bottom organisms				
	Stations				
	1	2	3	4	5
Plecoptera Peltoperlidae	1				
Ephemeroptera Baetidae	4	2	3	2	
Trichoptera Hydropsychidae				1	
Annelid worms					
Oligochaeta Tubificidae					15
Hirudinea				1	

index had to be omitted from this study. Table 10-7 lists the numbers representing each of the listed families found at each sampling station.

The presence of the *Peltoperla* (*Insecta, Plecoptera, Peltoperlidae*), stone-fly, naiad, and the four mayfly naiads (*Insecta Ephemeroptera Baetidae*) at Station 1 is quite significant. These forms are characteristic of well-oxygenated waters. *Peltoperla* is a carnivore, whereas the mayfly naiads are herbivores. The further occurrence of Baetidaean naiads at Station 2 implies that the stream is still well aerated at this point. These forms continue down as far as Station 4. In addition, a caddis-fly larva (*Insecta, Trichoptera, Hydropsychidae*) and a leech (*Annelida, Hirudinea*) were collected at Station 4.

Leeches respire by means of exchange through the epidermis, which is lined with a rich network of capillaries. Ephemeropteran naiads respire with large paired tracheal gills on the lateral surface of most abdominal segments. Stone-fly naiads utilize gill filaments. These forms are all dependent upon dissolved oxygen, and require fairly high concentrations of it. Hence, it can be concluded that the waters of Cattaraugus Creek have had a past history of being well oxygenated down to Station 4.

At Station 5 the most significant results were observed. Tubificid worms (*Annelida, Oligochaeta, Plesiopors, Tubificidae*) were collected at this point. In these forms respiration occurs through the general body surface. Although these forms require dissolved oxygen, tubificids are able to withstand relatively long periods of low oxygen tension. Pennak (1953) reports that some workers have found that small percentages of populations of *Tubifex* itself can survive anaerobic conditions for 48 to 120 days. The tubificids are most probably indicative of the sewage pollution introduced below Station 3 rather than the industrial pollution introduced below Station 1. Tubificids indicate organic pollution in general, and specifically indicate sewage pollution (Pennak, 1953). These organisms imply polysaprobic conditions (see Chapter 1).

It is frequently desirable to consider what was *not* observed as well as what *was* observed. Most notably unobserved were crustaceans. It was surprising to collect neither crayfish nor amphipods. The substratum may not be suitable for these types, being too rocky and barren of decaying material. In spite of this, the presence of the immature insect forms indicates that the crustaceans should be present.

Conclusions of creek biological anlaysis. It is difficult to generalize, based on one superficial set of samples. More data are required, especially in the upstream sites, to make an accurate assessment. This would give a better picture of the untainted stream. (Biological data from one sample, then, are no more revealing than a sample employing physical and chemical methods.)

Based on the samples collected during the Cattaraugus Creek survey, the following conclusions can be drawn.

1. No severe effect on the plankton community was observed.
2. Cattaraugus Creek appears to be well aerated, at least to the level of Station 4.
3. There is no biological evidence of extreme duress in the stream until Station 5, at which point the tubificid worms become common.
4. There is no evidence of toxic materials in the stream.
5. The stream conditions are:
   (*a*) oligosaprobic upstream (see Chapter 1)
   (*b*) mesosaprobic down to Station 4 (approaching but not actually meso-) (see Chapter 1)
   (*c*) presumably polysaprobic at Station 5

Results, calculated and observed, of April 1969 Creek Survey. Stream as sampled:

1. flow = 300 cfs
2. velocity = 3.68 fps
3. natural waters turbid due to silt
4. odors given off from waters
5. rapids present in sampling reach
6. temperature averaged 12.5°C
7. microflora and fauna present
8. dissolved oxygen concentration greater than 4 mg/l in all reaches
9. $K_1 = 3.47$/day
   $K_2 = 98.50$/day Station 1 to 4
   $K_2 = 10.42$/day Station 4 on ($f = 3$)
10. organic loading = 30 mg/l (BOD_5)

Predicted low-flow conditions

1. flow = 64 cfs (one-in-10 year, seven-day flow)
2. velocity = 0.664 fps
3. temperature = 20°C
4. dissolved oxygen goes to zero when untreated waste is discharged
5. K_1 = 0.133/day[2]
   K_2 = 0.20/day[2]
6. organic loading = 150 mg/l–untreated
   organic loading = 14.4 mg/l–treated (90 percent BOD reduction)
7. dissolved oxygen concentration greater than 4 mg/l when waste is treated (90 percent BOD reduction)
8. biological examination suggests anaerobic conditions as past history of low-flow periods from Station 4 downstream

General Conclusions of the April 1969 Survey:

1. The stream as sampled contained sufficient oxygen for trout fishing classification; however, the great amount of suspended material would rule out the abundance of fish.
2. The stream as sampled demonstrated conditions that were not typical, as is shown by sections of the stream that indicated an anaerobic history, based on observed biological and calculated data.
3. The load exerted on the stream by the industrial waste discharge is the main cause of oxygen depletion indicated above.
4. By attaining a 90 percent reduction in industrially discharged BOD, the stream has enough natural aeration capacity to maintain a minimum of four parts per million dissolved oxygen under critical low-flow conditions.

Additional Creek Studies Made by New York State

Prior to the three creek surveys made by the author and his students, New York State sampled Cattaraugus Creek twice for partial and specific information. In September 1960 the state sampled five stations on four different days. During this period the creek flow was very low (77–120 cfs) and the water temperature was relatively high (17°-27°C). The data are shown in Table 10-8.

Industrial and municipal conditions have changed very materially since this particular survey was made. Not only has the tannery ceased sole leather manufacturing and the glue plant stopped using fleshing as a raw material, but also the glue plant has installed primary treatment with excellent solids removal. The vil-

[2]Assuming laboratory quiescence will actually be obtained at extreme low flow.

TABLE 10-8. New York State Health Department Results of Stream Survey Made in 1960

Station No.	Location	Sampled	Flow (cfs)	Temp. (°C)	DO (ppm)	BOD 5 day (ppm)
1	Mile 19.2	9-7-60	77	25	8.8	–
		9-8-60	est. 77	21	8.6	–
						*1.8 (9-20-60)
		9-21-60	120.2	21	8.4	–
		9-26-60	79	16	8.1	1.0
2	Mile 18.2	9-7-60	77	25	7.2	– *16.2
		9-8-60	est. 77	22	7.2	–
						*7.8 (9-20-60)
		9-21-60	120.2	21	7.0	–
		9-29-60	79	18	7.8	210
3	Mile 17.7	9-7-60	77	26	6.8	– *55.1
		9-6-60	est. 77	23	7.0	–
		9-21-60	120.2	21	8.0	–
		9-29-60	79	18	7.4	>162
4	Mile 16.7	9-7-60	77	27	1.6	–
		9-8-60	est. 77	24	2.0	–
		9-21-60	120.2	21	5.4	–
		9-29-60	79	18	5.0	142
5	Mile 11.7	9-7-60	77	27	5.4	–
		9-8-60	est. 77	24	5.6	–
		9-21-60	120.2	22	5.2	–
		9-29-60	79	17	6.2	22.2

*Composite sample of 24 hours (hourly).

lage of Gowanda has also grown and is contributing a larger pollutional load today than 10 years ago. The bottom of the oxygen sag curve, however, still was observed at Station 4 just above the sewage treatment plant outfall. In fact, on the 7th and 8th of September in 1960 the load of oxygen sagged to 2 or less ppm. This information prompted the state to insist upon greater than 90 percent BOD reduction, even 10 years later, when conditions had definitely changed for the better.

In addition, no time-of-flow studies were made by the state in the 1960 survey, and thus, it was impossible to compute creek reaction rates. Once again we have a situation in which some factors were ideal and others—vital to a successful survey—were less than ideal.

Determination of Allowable Loading for Cattaraugus Creek

Basis: 1964 Survey by New York State Health Department

(1) Between M.P. 16.0 and 11.0

DIST. = 5.0 miles Vel = 1.37 hrs/mi

time = 5.0 X 1.37 = 6.85 hrs = 1.07 ft/sec

$t = 0.286$ days

$$k_r = -\frac{1}{.286}\log\frac{181}{250} = -\frac{1}{.286}\log 0.725$$

$$= -\frac{1}{.286}(-.1397) = 0.488\ \text{day}^{-1} = \underset{\text{base}_{10}}{k_r}$$

$$K_2\ \text{base}_e\ (\text{at } 115\ \text{cfs}) = \frac{(DU)^{1/2} \times 24}{H^{3/2}} \qquad D = 81 \times 10^{-6} @ 20°\text{C} \quad V = \text{Vel} = 3860\ \text{ft/hr}$$

$$K_2 = \frac{(81 \times 10^{-6} \times .386 \times 10^4)^{1/2} \times 24}{H^{3/2}}$$

$$= \frac{(31.2 \times 10^{-2})^{1/2} \times 24}{H^{3/2}} = \frac{(.312)^{1/2} \times 24}{H^{3/2}} = \frac{.559 \times 24}{H^{3/2}}$$

$$K_2 = \frac{13.4}{H^{3/2}} \qquad \boxed{k_2 = \frac{5.84}{H^{3/2}}}\ \text{base}_{10} @ 20°\text{C}$$

H(ft)	$H^{3/2}$	$k_2(\text{base}_{10})$	f
1.0	1.0	5.84	12.0
2.0	2.83	2.06	4.4
3.0	5.17	1.13	2.3

(2) K_2 at low flows Vel = 0.37 ft/sec = 1330 ft/hr

$$\underset{\text{base}_e}{K_2} = \frac{(81 \times 10^{-6} \times .133 \times 10^4)^{1/2} \times 24}{H^{3/2}}$$

$$= \frac{(10.8 \times 10^{-2})^{1/2} \times 24}{H^{3/2}} = \frac{(.108)^{1/2} \times 24}{H^{3/2}}$$

$$K_2 = \frac{.3286 \times 24}{H^{3/2}} = \frac{7.88}{H^{3/2}}$$

$$\boxed{\underset{@20°\text{C}}{\underset{\text{base}_{10}}{k_2}} = \frac{3.42}{H^{3/2}}}$$

H(ft)	$H^{3/2}$	$k_2(\text{base}_{10})$	f
1.0	1.0	3.42	7.0
2.0	2.83	1.19	2.44

Stream Parameters (day^{-1})-low flow

	at 20°C	*at 25°C*
k_r	0.488	0.610
k_2	1.19	1.330
f	2.44	2.18

Determination of Allowable Loading for Cattaraugus Creek (*Continued*)

A. Class "D" Section–Allowable Loading

$Dc = 8.4 - 3.0 = 5.4$ mg/l @ 25°C $C_s = 8.4$ mg/l

Background DO = 8.4 mg/l (100% Sat.)

$^*Qs^r = 60$ cfs $f = 2.5$
$^\dagger Qw_t = 7.78$ cfs $Dw = 6.4$ mg/l

$^\ddagger Qw = 7.78 - 0.90$ cfs = 6.88 cfs without Hospital and Collins Flow

$$Da = \frac{60(0) + 6.88(6.4)}{66.88} = \frac{44}{66.88} = 0.66 \text{ mg/l}$$

$f = 2.5$

$$\log_{10} La = \log Dc + \left[1 + \frac{1}{(f-1)}\left(1 - \frac{Da}{Dc}\right)^{.418}\right] \log f$$

$$= \log 5.4 + \left[1 + \frac{1}{(2.5-1)}\left(1 - \frac{.66}{5.4}\right)^{.418}\right] \log 2.5$$

$$= .732 + [1 + .667(1 - .122)^{.418}]\ .398$$

$$= .732 + [1 + .667(.878)^{.418}]\ .398$$

$$= .732 + [1 + .667(.947)]\ .398$$

$$= .732 + [(1.631)\ .398]$$

$$= .732 + .649$$

$\log La = 1.381$

$La = 24.4$ mg/l or 8,800 #/day in Stream

$$La = \frac{QsLs + Qw_1Lw_1 + Qw_2Lw_2 + Qw_3Lw_3}{Qs + Qw_1 + Qw_2 + Qw_3}$$

$Qs = 60$ $Qw_1 = 0.62$ $Qw_2 = 1.10$ $Qw_3 = 5.16$

$Ls = 3.0$ $Lw_1 = 368$ $Lw_2 = 48$ $Lw_3 = ?$

$$(66.88)24.4 = 60(3) + .62(368) + 1.10(48) + 5.16\,Lw_3$$

$$1630 = 180 + 228 + 53 + 5.16\,Lw_3$$

$$1630 - 461 = 5.16\,Lw_3$$

$$1169 = 5.16\,Lw_3$$

$$(\text{TOD})Lw_3 = 226 \text{ mg/l or } 6{,}300\ \#/\text{day}$$

B. Class "B" Section–Allowable Load

$Dc = 8.4 - 4.0 = 4.4$ mg/l $f = 2.5$

$$\log La = \log 4.4 + \left[1 + \left(\frac{1}{2.5-1}\right)\left(1 - \frac{.66}{4.4}\right)^{.418}\right] \log 2.5$$

$^*Qs^r$ = critical flow in creek
$^\dagger Qw_t$ = total wastewater flow
$^\ddagger Qw$ = wastewater flow (Moench & Peter Cooper) without other wastewaters

Determination of Allowable Loading for Cattaraugus Creek (*Continued*)

$$= .643 + [1 + .667(1 - .15)^{.418}].398$$

$$= .643 + [1 + .667(.85)^{.418}].398$$

$$= .643 + [1 + .667(.934)].398$$

$$= .643 + [1.623].398$$

$$= .643 + .645$$

$$\log La = 1.288$$

$La = 19.4$ mg/l or 7,000 #/day in Stream

$La = 19.4$ mg/l

$$19.4(66.88) = 180 + 228 + 53 + 5.16\,Lw_3$$

$$1300 - 461 = 5.16\,Lw_3$$

$$839 = 5.16\,Lw_3$$

$(TOD)Lw_3 = 163$ mg/l or 4525 #/day

Allowable Loading (TOD) at Gowanda

Source	Flow (cfs)	TOD (#/day) Class "D"	TOD (#/day) Class "B"
1. Cattaraugus Creek	60.0	970	970
2. Moench Tannery	0.62	1,225	1,225
3. Gowanda STP	1.10	285	285
4. Peter Cooper	5.16	6,300	4,525
	66.88	8,780	7,005
Allowable Load		8,800	7,000

Waste Loadings

Source	Qw mgd	Qw cfs	Waste Loading (TOD) (#/day)	Waste Loading (TOD) (mg/l)
1. Gowanda STP	0.71	1.10	285	48
2. Moench Tannery	0.40	0.62	1,225	368
3. Collins & Gowanda State Hospital	0.59	0.90	515	106
	1.70	2.62	2,025	
4. Peter Cooper	3.33	5.16		
	5.03	7.78		

NH_3 Loadings

Source	NH_3 (#/day)	NOD (#/day)
1. Moench Tannery	110	497
2. Collins & Gowanda State Hospital	60	270
3. Gowanda STP	15	68
	185	835

Stream

NH_3 = 2.0 × 67.88 × 5.4 = 680 #/day (3,060 #/day of NOD)

Background:

= 0.1 × 60.0 × 5.4 = 32.4 #/day (146 #/day NOD)

Peter Cooper = 680-32-185

Allowance = 680–217 = 463 #/day NH_3 or 2,080 #/day of NOD

In 1964 the New York State Health Department made another survey of the creek to establish times of flow at critical creek low flows and to compute creek reaction rates. Their computations are shown on the Determination of Allowable Loading for Cattaraugus Creek and Gowanda. The deoxygenation rate k_r was obtained by using the Streeter-Phelps formulation between Stations 4 and 5. The reaeration rate k_2 was obtained by using the O'Connor-Dobbins formula rather than the Streeter-Phelps. As shown in the calculations, the value of k_2 is highly dependent upon the depth of creek flow. At critical low flows the creek depth is lowered drastically. Although the creek was not sampled at critical low flow (it was sampled at 115 cfs), the state assumed a depth of 2.0 feet at 60 cfs. If, in fact, the depth had been closer to 1.0 feet, the real k_2 value would have been triple that used by the state. Thus, the state's assumption of creek depth at critical flow is just another example of the stream analyst's frustration at attempting to make precise predictions of stream reaction.

It must be said that the state's presumptions were not entirely unfounded. They developed curves for four known streamflows and flow times in order to obtain creek velocities (see Fig. 10-9 and Table 10-9).

The state study evolved values of k_1 of 0.488 and k_2 of 1.19 at 20°C and two feet of depth. This roughly yielded Fair's *f* value of

$$\frac{1.19}{0.488} = 2.5$$

By using 2.5 as the reaction rate ratio, a total oxygen demand of 7,000 pounds per day was found permissible in the creek in order to maintain a minimum dis-

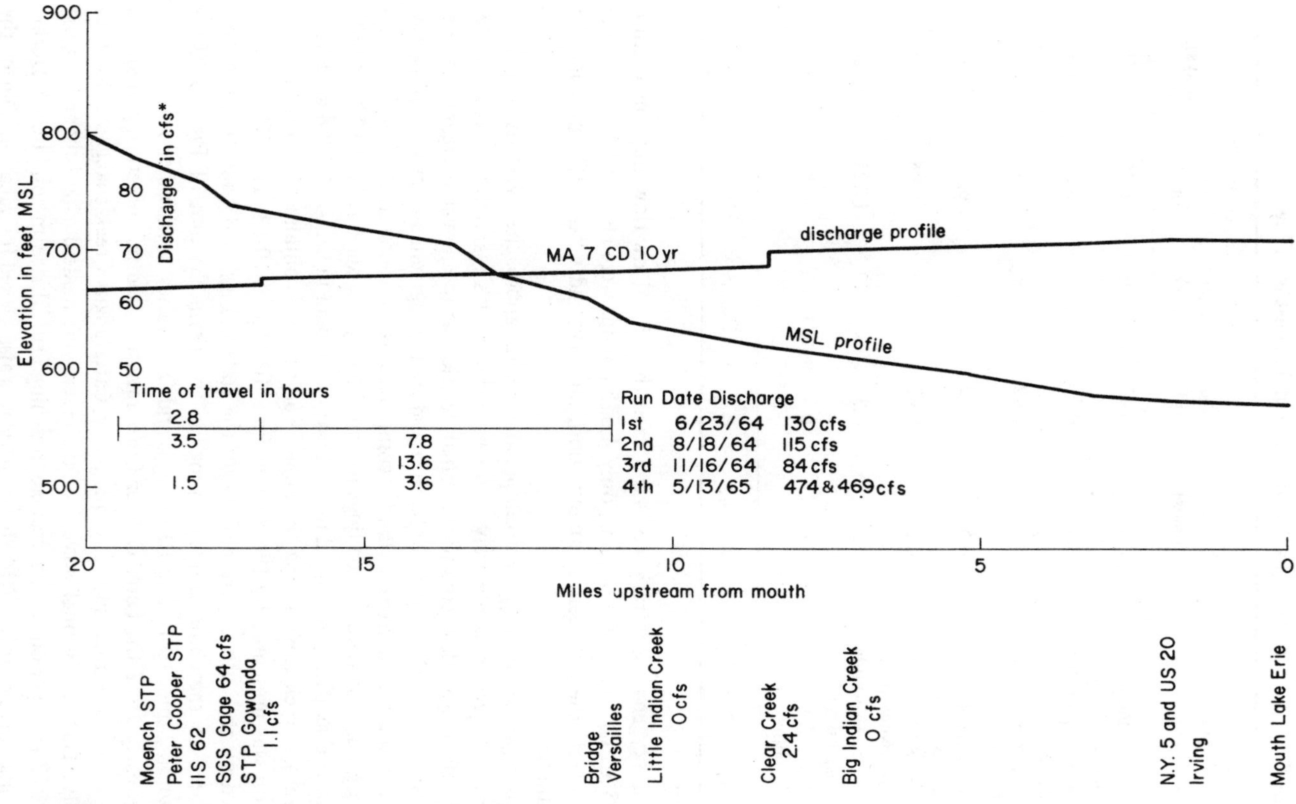

FIG. 10-9. Cattaraugus Creek.

TABLE 10-9. Cattaraugus Creek, Time of Travel-Low Flows

Sta. to Sta.	Distance (miles)	Vel (fps)	Time (hrs)	Accumulative Time (hrs)	(days)
(1) Village line to Moench	0.3	0.42	1.05	1.05	0.048
(2) Moench to Peter Cooper	0.7	0.42	2.45	3.50	0.146
(3) Peter Cooper to R.R. Bridge	0.3	0.42	1.05	4.55	0.190
(4) R.R. Bridge to USGS gauge	0.5	0.42	1.75	6.30	0.262
(5) Gauge to Gowanda STP (New Bridge)	0.5	0.42	1.75	8.05	0.336
(6) STP to Gowanda Hospital STP	2.3	0.37	9.20	17.25	0.720
(7) Hosp. STP to Versailles	3.4	0.37	13.60	30.85	1.290
	8.0 mi.		30.85	30.85	1.29

solved oxygen concentration of 4.0 ppm at the critical low flow and temperature of 25°C. Of this 7,000 pounds, only 4,525 pounds were allocated to the animal glue plant–the upstream contamination, tannery, and village effluent using the remainder.

Thus, we see how many attempts and several approaches were used in an attempt to ascertain as precisely as possible the creek's reaction to organic waste-loads at critical flow conditions. Final decision as to allowable organic oxygen demand was reached by a conference and mutual agreement between all parties concerned after all creek data and computations were presented. The final oxygen sag curve at reduced loadings was plotted by the state (Fig. 10-12) at both B and D Classifications, based upon the author's data in the June 1968 survey.

Actual creek flow during the June 1968 survey as obtained from the gauging station was presented by the state in Figs. 10-11 and 10-13. The BOD data obtained by the author were plotted by the state in Figs. 10-12 and 10-14. Actual data fed to the state computer program to obtain the plot of Fig. 10-10 are shown in Tables 10-10 and 10-11 for B and D classifications, respectively.

In early 1972 the Peter Cooper Corporation ceased manufacturing of animal glue at the Gowanda site. The quality of Cattaraugus Creek changed quite drastically. Three additional surveys of the creek were made during July and August of 1972 (the period of the year approaching critical stream conditions). During the first survey (July 27th) the tannery was also closed for vacation. During the

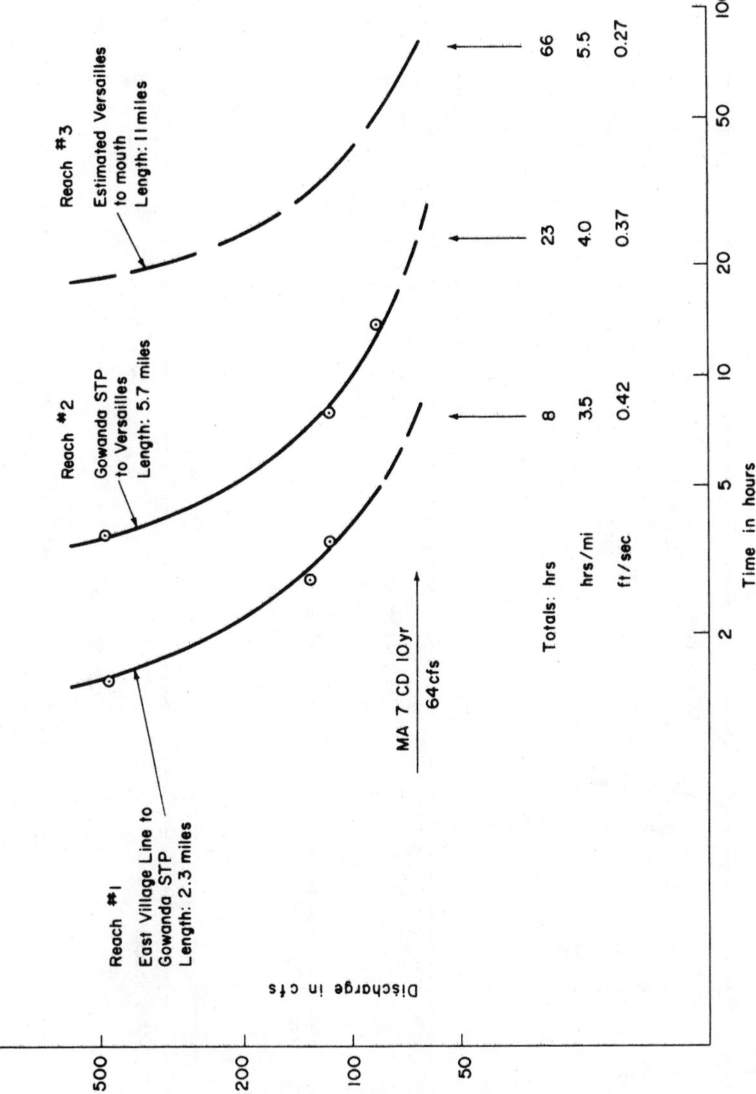

FIG. 10-10. Cattaraugus Creek. Relationship of discharge to time-of-travel.

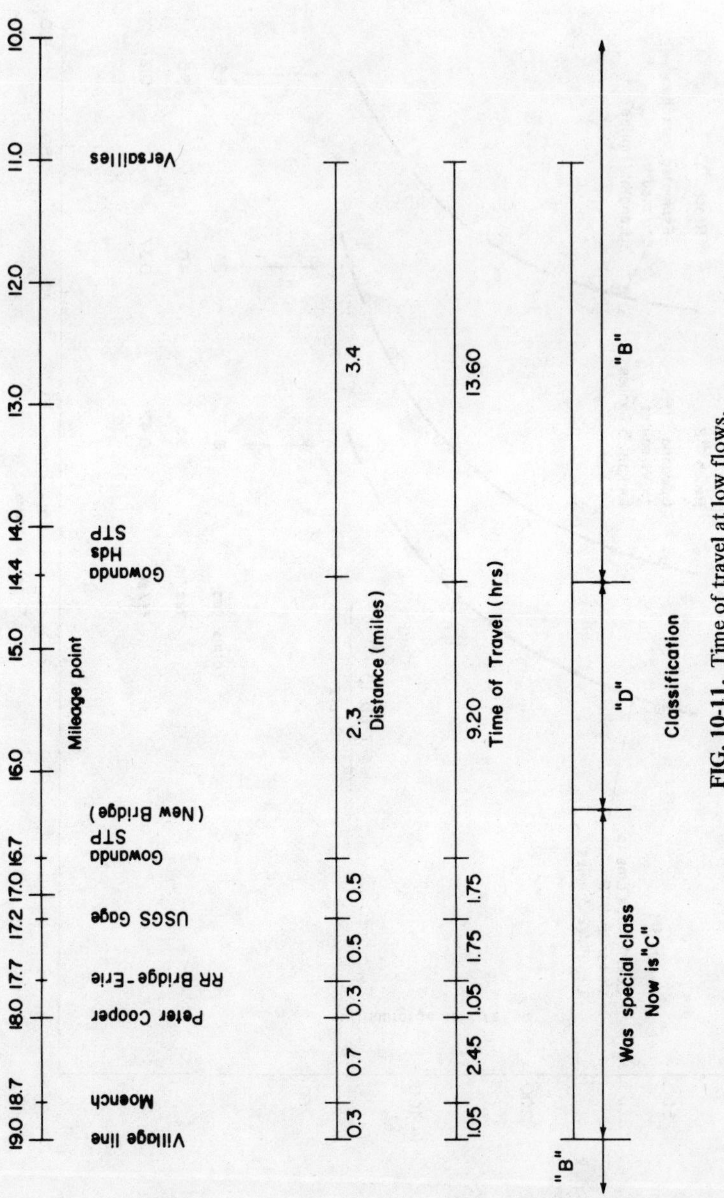

FIG. 10-11. Time of travel at low flows.

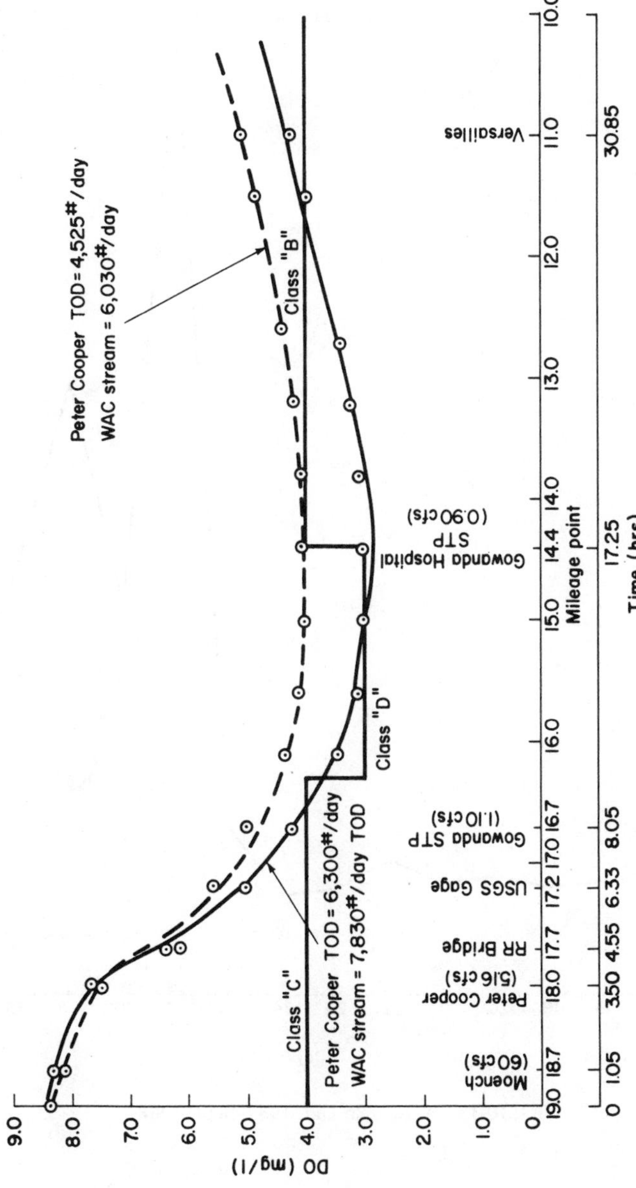

FIG. 10-12. DO vs. time and mileage point ($f = 2.5$; $k_1 = 0.488\ \text{day}^{-1}$ [base$_{10}$]; $k = 1.22\ \text{day}^{-1}$; $Q_{\text{stream}} = 60.0$ cfs; $C_S = 8.4$ mg/l).

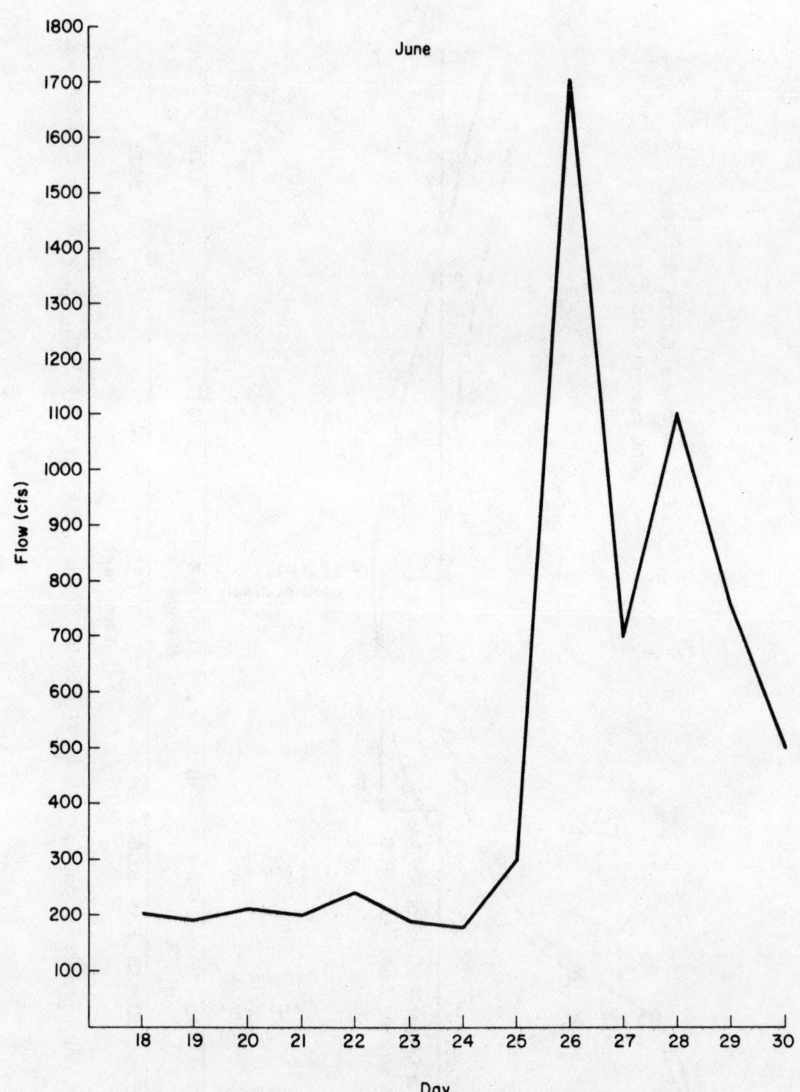

FIG. 10-13. Water year Oct. 1967–Sept. 1968.

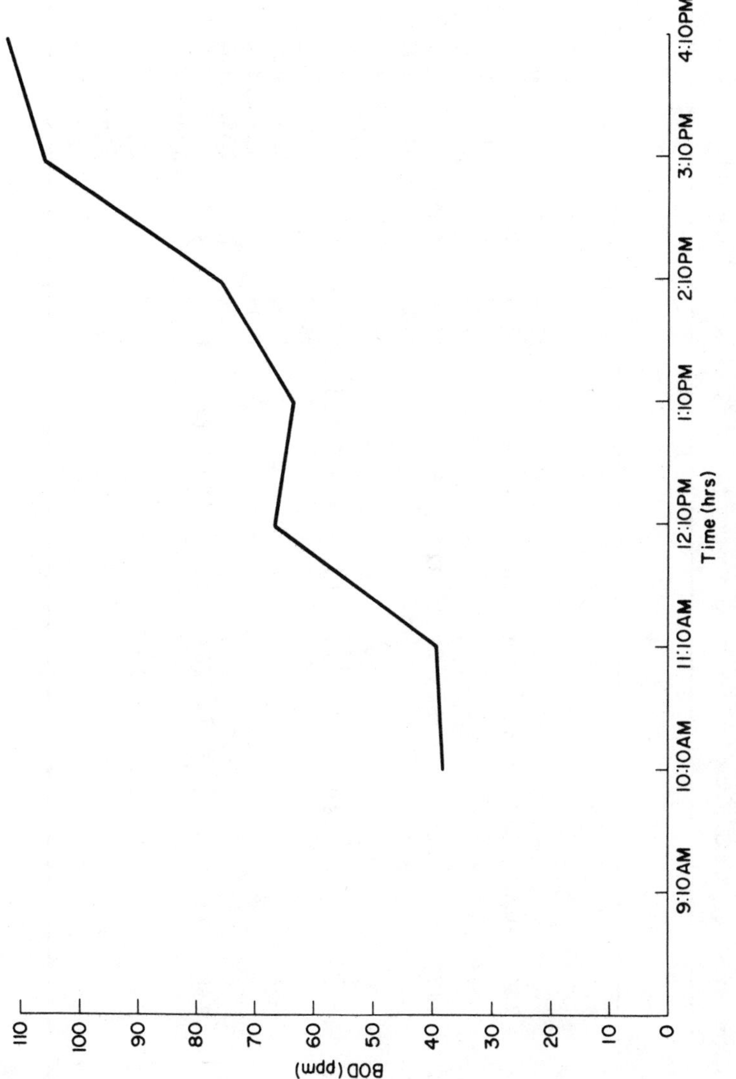

FIG. 10-14. Cattaraugus Creek survey 6/25/'68 (Aldrich Street).

TABLE 10-10. Computer Summary Sheet—WAC. *(Class "B" Section Governing [K_1 & K_2 are base "e" for program #14].)*

Source	Q_s	L_s	Q_w	L_w	D_s	D_w	K_1	K_2	C_s	L_a	D_a	T_c	D_t	Min. DO
1. Village line to Moench	60.0	3.0	0	0	0	0	1.12	2.80	8.40	3.00	0			8.40
										↓				
										2.84		0.05	0.15	8.25
2. Moench to Peter Cooper	60.0	2.84	0.62	368	0.15	6.4	1.12	2.80	8.40	6.57	0.21			8.19
										6.21		0.05	0.52	7.88
										5.877		0.10	0.76	7.64
3. Peter Cooper to Gowanda STP	60.62	5.877	5.16	163	0.76	6.4	1.12	2.80	8.40	18.20	1.20			7.20
										17.2		0.05	1.97	6.43
										15.91		0.12	2.80	5.60
										14.72		0.19	3.40	5.00
4. Mix Gowanda STP and then to Gowanda Hospital STP	65.78	14.72	1.10	48.0	3.40	6.4	1.12	2.80	8.40	15.27	3.45			4.95
										13.65		0.10	4.01	4.39
										12.49		0.18	4.26	4.14
										11.14		0.28	4.36	4.04
										9.98		0.38	4.33	4.07
5. Hospital STP to Versailles	66.88	9.98	0.90	106	4.33	6.4	1.12	2.80	8.40	11.25	4.36			4.04
										10.06		0.10	4.33	4.07
										9.00		0.20	4.20	4.20
										8.04		0.30	4.00	4.40
										6.58		0.48	3.57	4.83
										5.94		0.57	3.33	5.07

TABLE 10-11. Computer Summary Sheet—WAC. *(Class "D" Section Governing [K_1 & K_2 are base "e" for program #14].)*

Source	Q_s	L_s	Q_w	L_w	D_s	D_w	K_1	K_2	C_s	L_a	D_a	T_c	D_t	Min. DO
1. Village line	60.0	3.0	0	0	0	0	1.12	2.80	8.40	3.00	0			
to Moench										↓		0.05		
										2.84			0.15	8.25
2. Moench to	60.0	2.84	0.62	368	0.15	6.4	1.12	2.80	8.40	6.57	0.21			8.19
Peter Cooper										6.21		0.05	0.52	7.88
										5.877		0.10	0.76	7.64
3. Peter Cooper to	60.62	5.877	5.16	226	0.76	6.4	1.12	2.80	8.40	23.14	1.20			7.20
Gowanda STP										21.87		0.05	7.22	6.18
										20.23		0.12	3.33	5.07
										18.72		0.19	4.13	4.27
4. Mix Gowanda STP	65.78	18.72	1.10	48.0	4.13	6.4	1.12	2.80	8.40	19.20	4.17			4.23
and then to										17.16		0.10	4.92	3.48
Gowanda Hospital										15.71		0.18	5.26	3.14
STP										14.02		0.28	5.41	2.99
										12.56		0.38	5.39	3.01
5. Hospital STP to	66.88	12.56	0.90	106	5.39	6.4	1.12	2.80	8.40	13.80			5.40	3.00
Versailles										12.34		0.10	5.35	3.05
												0.20	5.19	3.21
										9.85		0.30	4.93	3.47
										8.07		0.48	4.39	4.01
										7.29		0.57	4.09	4.31

TABLE 10-12. Additional Analyses Cattaraugus Creek.* *(Made by John North during Summer 1972 after closing of Peter Cooper.)*

Survey 1—No Wasteload from Industry
Thursday, July 27

Stations Old	Stations New	Flow at GS	Time	DO	T°C	BOD ppm	pH	Suspended Solids ppm	Volatile Suspended Solids
1		190		8.9		0.2	8.5	67.5	30.5
2				9.0		0.6	8.6	13.0	8.5
3				9.1		NR	8.6	21.5	9.0
4	1		11:06 a.m.	9.1	20°	NR	8.6	20.0	13.0
5	2		11:37 a.m.	9.0	20°	0.5	8.4	14.5	11.0
6	3		12:13 p.m.	9.0	20°	2.3	8.5	9.0	4.0
7	3A		12:26 p.m.	9.0	20°	1.0	8.6	8.5	4.0
8	4		1:09 p.m.	9.1	20°	0.1	8.5	0.5	0
9				10.2		NR	8.5	0	0
10				9.1		0.2	8.6	0	0
11	5		5:35 p.m.	9.9	23°	5.2	9.0	0.5	0
12				9.6		0.3	9.0	0	0
13 (clear Ck.)				8.6		0.4	8.4	14.5	12.5
14				9.2		1.7	8.6	4.5	0
15	6			8.7		0.5	8.3	11.0	0

*Dyes were used to sample stations according to flow.

TABLE 10-12a. Additional Analyses Cattaraugus Creek.* *(Made by John North during Summer 1972 after closing of Peter Cooper.)*

Survey 2—Tannery Operating at High Efficiency
Monday, August 14

Stations Old	Stations New	Flow at GS	Time	DO	T°C	BOD ppm	pH	Suspended Solids ppm	Volatile Suspended Solids
1		140							
2									
3									
4	1		7:40 a.m.	8.5	20°	0.3	8.65	115/50	
5	2		8:21 a.m.	8.8	20°	0.4	8.65	60/58	
						Leach Aid			
6	3		8:54 a.m.	8.5	20°	14	8.65	240/125	
7	3A		9:06 a.m.	8.8	20°	1.0	8.7	55/40	
8	4		9:52 a.m.	8.5	21°	1.2	8.7	55/45	
9			(6)	8.5		0.8	8.5	225/85	
10									

*See footnote at end of table.

TABLE 10-12a. *(Continued)*

Survey 2–Tannery Operating at High Efficiency
Monday, August 14

Stations Old	Stations New	Flow at GS	Time	DO	T°C	BOD ppm	pH	Suspended Solids ppm	Volatile Suspended Solids
11	5		5:00 p.m.	9.9	22°	0.4	8.9	20/4	
12									
13 (clear Ck.)									
14									
15	6			7.6		1.8	8.35	205/65	

*Dyes were used to sample stations according to flow.

TABLE 10-12b. Additional Analyses Cattaraugus Creek.* *(Made by John North during Summer 1972 after closing of Peter Cooper.)*

Survey 3–Tannery Operating at Low Efficiency
Thursday, August 17

Stations Old	Stations New	Flow at GS	Time	DO	T°C	BOD ppm	pH	Suspended Solids ppm	Volatile Suspended Solids
1		180							
2									
3									
4	1		8:22 a.m.	8.75	19°	0.4	8.5	44/10	
5	2		8:55 a.m.	8.5	19°	1.0	8.65	40/8	
6	3		9:24 a.m.	8.5	19°	1.0	8.65	40/5	
7	3A		9:39 a.m.	8.55	19°	1.2	8.65	34/5	
8	4		10:21 a.m.	8.8	19.5°	0.1	8.55	36/12	
9			(6)	8.7		1.5	8.55	45/18	
10			(7)	9.0		0.6	8.60	30/6	
11	5		(8)	9.1		0.4	8.65	26/11	
12			4:40 p.m.	9.6	23°	0.1	8.85	6/5	
13 (clear Ck.)									
14									
15	6								

(6) Dump ground in Indian Reservation
(7) 1.5 mi. from Aldrich Street
(8) 2.0 mi. from Aldrich Street

TABLE 10-12, 10-12a, and 10-12b Tannery Data

	Flow (gpd)	BOD (ppm)	Lb/day	Lb/1,000 lb Hide	Lbs. Hide Processed	% BOD Reduction
Survey 1	0	0	0	0	0	
Survey 2	173,000	195	280	3.70	*74,000	88
Survey 3	348,000	387	1,130	15.30	†74,000	81

*No beaming–tanning only
†Beaming and tanning

TABLE 10-13. Surveys of Cattaraugus Creek *(July–August 1972)*

	DO Sag (%) Saturation		
Station	I	II	III
1	101	94.25	95.0
2	100	97.58	92.3
3	100	94.25	92.3
4	101	96.64	96.5
5	109	113.50	113.0

second survey (August 14th) the tannery was operating, and its waste treatment plant was performing at high efficiency. During the third survey (August 17th) the tannery was operating, but its waste treatment plant was performing at lower efficiency due to a higher loading. Creek samples were collected at several stations and at times which corresponded to the time of flow. These results are shown in Table 10-12. The excellent water quality of the creek, from a dissolved oxygen standpoint, even with the lower efficiency of treatment facilities at the tannery, is shown in Table 10-13.

These results show the recovery characteristics of a watercourse when one large industry ceases production and another smaller industry treats its wastewater adequately.

QUESTIONS

1. Why are long-term daily streamflow records vitally important in planning a comprehensive stream survey?

2. What problems develop for the stream analyst when he is forced to sample during mean or higher flows?
3. What alternatives are open to the stream analyst when he is unable to carry out a survey under critical low-flow conditions in the stream?
4. Of what importance to the stream analyst are industrial plant production records?
5. What are the dangers in evaluation of a stream over too long an extension of time? Over too short a period of time?
6. How does the stream analyst overcome the problem of the failure of easily accessible stream sampling points corresponding to desirable ideal sampling points?
7. Why do the geographical and the geological characteristics of the streambed play such an important role during surveys under critical conditions?
8. Can specific and significant conclusions about a stream be drawn from one biological analysis of the stations? Explain.
9. What are the difficulties of attempting to use results of surveys by others to substantiate, expand, or explain your findings?
10. What is the ultimate objective of the stream analyst in planning, undertaking and evaluating streams?
11. How many stream surveys were carried out on Cattaraugus Creek by the writer?
12. When was the first survey carried out? What was the limiting constraint in the first survey?
13. When was the second survey carried out? What were the three limiting constraints in the second survey?
14. When was the third survey carried out? What were the three limiting constraints in the third survey?
15. Were any other Cattaraugus Creek surveys run? If so, describe the constraint.
16. After New York State had set effluent standards, and conditions had reverted to a relatively stable condition, were any additional surveys made? Were they different? Did they produce the required results?

REFERENCES

Fair, Gordon, and John Geyer: *Water Supply and Waste Disposal*, second printing, John Wiley & Sons, New York, January 1956.

Needham, P. R., and R. L. Usinger: Variability in the Macrofauna of a Single Riffle in Prosser Creek, California as Indicated by the Surber Sampler, *Hilgardia*, **24**(14):383, 1956.

Nemerow, N. L.: *Liquid Wastes of Industries–Theories, Practices and Treatment*, Addison-Wesley Co., New York, 1971, Chap. 15.

Palmer, C. M.: Algae in Water Supplies, *U.S. Public Health Service Publ. No. 657*, U.S. Dept. HEW, Robert Taft San. Eng. Center, Cincinnati, Ohio, 1962.

Patrick, Ruth: A Proposed Biological Measure of Stream Conditions, Based on a Survey of the Conestoga Basin, Lancaster County, Pennsylvania, *Proc. Acad. Nat. Sci., Philadelphia*, **101**:277, 1949.

Pennak, R. W.: *Fresh Water Invertebrates of the United States*, Ronald Press, New York, 1953.

Sladecek, V.: The Future of the Saprobity System, *Hydrobiologia*, **25**, Dr. W. Junk Publishers, The Hague, 1965.

Smith, G. M.: *The Fresh-Water Algae of the United States*, McGraw-Hill Book Company, New York, 1950.

Usinger, R. L.: *Aquatic Insects of California*, University of California Press, Berkeley, Calif., 1963.

11
LAKES

11-1. GENERAL

Both natural and artificial lakes present special pollution problems for the environmental engineer. Although the number of natural lakes remains rather fixed because of geological conditions, the number of man-made lakes has increased dramatically in recent years. This increase can be correlated to the urgent demand for water by the geometrically increasing population for its many purposes. Lakes are multipurpose in use today. Some of these uses include recreation, such as swimming, boating, and fishing; the municipal and industrial water supply; power production; flood control; and agricultural irrigation. These uses are not necessarily compatible. The quantity and the quality of the water required at a given time can be different. In this chapter we will discuss some of these requirements as well as their dependence upon the physical, chemical, and biological factors of such lakes.

According to Worthington (1966) "large stretches of open water have great influence on all forms of plant, animal, and human life." He continues, "from the human point of view, the main effects are on climate, water supply, fisheries, forests, agriculture, transport, and health and the natural beauty of the countryside."

The main factors of the environment affected by lake formation are the geology, physiology, and climatology—the rocks, land surface, and the air above the lake.

Inundation of a land surface by impounded water will have a considerable effect on the groundwater table in the region of the lakes. It may even cause a recharging of subterranean aquifers and provide greater quantities of water over a broad geographical area. It is also possible for these man-made lakes to alter the climate. This is accomplished by daily lake breezes and nightly land breezes, with subsequent effect on rainfall and crop growth. The wind change is manifested by the fact that during the day, lake water will heat up slower than the surrounding land and cause a breeze to occur from the lake to the land. The opposite is true at nighttime, with the breeze directed from the land to the lake. The degree of difference in temperature between the two bodies (lake and land) determines the wind force. Wind will carry with it moisture which may be discharged as rain, if and when its dew point is reached as it moves. The rain subsequently results in increased crop growth in the area of the wind path. Vegetation created by the new lake alters the land use, value, and economy of the region.

The *qualitative* changes that occur as a result of lake creation are often more dramatic and important to man. Most of these are a result of providing time for the normally slow biological and chemical reactions to occur over a relatively small area. An example of this is the deoxygenation of the water that develops as a result of bacterial degradation of the organic matter. Some portion of this organic matter may have been contributed by the submerged land decomposing bottom vegetation. It is usually too difficult and costly to remove all of the land vegetation prior to flooding.

It is possible to grow fish and develop fisheries as a result of inundation. This will be a vast change from the original use of the land before flooding. Phytoplankton and floating vegetation will develop in the water near the surface. These will be influenced by the water circulation and distribution of nutrients in the lake. In fauna, birds and mammals—both terrestrial and aquatic—will also appear near the lake shores. When these are coupled with the natural vegetation, the potential for developing a wildlife conservation area becomes possible.

As far as man's needs and interests are concerned, a newly formed lake will provide opportunities for hunting, fishing, and nature loving and cultivating. Before the flooding for lake formation, the chances for all kinds of human activity are more limited.

For moderate temperature reservoirs, the great increase in biological productivity following inundation, and its return to a new steady-state level after a period of time, are well established. Impoundments also influence the flow of streams, evaporation, transportation, seepage, runoff, and water quality.

However, there are problems with the creation of lakes, that remain to be solved. The two major ones are eutrophication and oxygen depletion. These will be considered separately in the next sections.

Modelling of lakes to predict the effects of nutrient addition and/or removal can be found in Section 3 of Chapter 13. This section further explains the input and output parameters of such phenomena as eutrophication.

11-2. EUTROPHICATION

Eutrophication can be defined as a condition of lakes in which excess nutrients have caused an augmentation of algal production. This rapid growth of algae results in several undesirable effects on treatment of water for drinking and for fisheries, as well as for bathing and overall tourism types of recreation. The increase in algal growth is directly attributed to overfertilization of lakes by man's activities–usually by wastewater discharges.

Algae are best controlled by limiting the inflow of phosphates for the following reasons given by Thomas (1969):

1. Phosphate is present only in traces in oligotrophic lakes.
2. Natural tributaries running into these lakes contain very little phosphate as long as they are not subjected to pollution by the influence of man, but they contain large quantities of nitrates.
3. Fewer phosphates than nitrogenous compounds are washed out of agricultural land.
4. Rainwater often contains large quantities of nitrogenous compounds that can be utilized by plants.
5. Addition of phosphate alone to lake water is sufficient to increase the growth of bacteria and blue-green algae.
6. Bacteria and blue-green algae living in lake water are able to bind gaseous nitrogen organically, or to produce the growth factors for algae.
7. Some blue-green algae produce toxins that are very toxic to warm-blooded animals.
8. Out of sludge and putrified parts of organisms, nitrogenous compounds return in larger quantities than phosphate compounds in the biochemical cycle.
9. In eutrophic lakes, nitrates are eliminated from time to time by the process of denitrification.
10. It is cheap and easy to eliminate phosphates from sewage water (by $FeCl_3$ in the activated sludge process).

Thomas also divided Central European lakes (1969, pp. 34–35) into three groups: small lakes, medium-sized lakes, and large lakes. Their characteristics and some effects of their size are selectively presented here.

11-3. SMALL LAKES

Small lakes possess a surface area of less than 0.5 square hectometers and depths of less than 20 meters. These lakes receive a larger proportion of leaves, pollen, insects, and other organisms than the other bigger lakes. The dead organisms decompose in the lake water and set free all the nutrients for growth of planktonic and shore algae. In these little lakes, oxygen depletion as well as the appearance of much FeS, MnS, and H_2S occurs in the deep waters during the summer stagnation period. When depths are less than 10 meters, stormy summer weather frees these nutrients from the hypolimnion into the surface waters for use by the plankton. Small lakes are very sensitive to contaminants from sewage, and should be protected from them.

11-4. MEDIUM-SIZED LAKES

Medium-sized lakes are 20 to 50 meters deep and have a surface area of more than 50 hectares. Today, because these lakes do not have large volumes of water, they are sensitive to sewage inflow. These contaminants cause extreme development of algae, which destroys the balance of the food chain in the lake. The main effect of this is that the oxygen is used up in all areas below the depth of 5 to 10 meters. This destroys the salmonids and makes treatment for drinking water difficult and costly. On the shores of these lakes, masses of algae collect, decompose, and create trouble and complaints. Plankton algae develop in immense numbers in highly fertilized lakes of this size, and when they float on the surface they present an unpleasant appearance.

11-5. LARGE LAKES

Large lakes are more than 50 meters deep, and once were our truly oligotrophic lakes on which agricultural practices by man had no influence. Runoff nutrients have less effect here than in smaller lakes. During the summer, the phosphate content is relatively low, while nitrate ion content is high. Even in the surface water, phytoplankton development is small and there is little production of littoral plants. Because of the lack of phosphates, plants cannot fully utilize the nitrates. In clean, large lakes (which are rapidly decreasing in number) both $CaCO_3$ and oxygen concentration are uniform throughout the entire depth. The bottom fauna is well developed and dominated by salmonids.

Although eutrophication causes numerous kinds of nuisances and is mainly seen as irreversible, there is some evidence that by removing the nutrients the cycle of eutrophication can be halted and even reversed (Edmonson, 1969).

Edmonson reports (1969) that the problems of eutrophication in North America have been approached in many different ways, with no particular unity of approach. The nutrient budget system, phosphate loading in kilograms per hectare, as in agriculture, and the relevant bioassay techniques are being used to assess the degree and state of eutrophication of lakes.

Although treatment of sewage to remove phosphate as a nutrient is one method being used to control eutrophication, Nemerow and Rand (1968) found an indication that biological treatment of sewage renders the effluent more stimulating to a given culture for algal growth in a particular receiving water than does primary treatment.

11-6. OXYGEN DEPLETION NUISANCES

Oxygen depletion nuisances are caused by excessive nutrients which enhance eutrophication, and then finally, oxygen depletion. Oxygen depletion can arise from the primary effect of direct organic matter inputs to the lake, or from secondary effects of dying plankton and decaying algae. Lakes devoid of oxygen promote the early and sudden death of fishes and the release of obnoxious odors of H_2S, CH_4, and NH_3 gases. The effects of direct inputs of organic matter can be ameliorated somewhat (with certain restraints) (Nemerow and Rand, 1968) by treating these wastes before discharging them to the lakes, or by rerouting such contaminated wastes around lakes to flowing waters of greater dilution. The indirect effects of plant nutrients must be remedied by removal of nutrients from all point and nonpoint discharges into the lake. This can be accomplished primarily by tertiary methods of waste treatment, but also by diversion around the lake. Other methods, such as copper-sulfating the lake, adding clean dilution water, decreasing the lake water temperature, and increasing the pH are being used to some extent to control bloom nuisances.

Since warm water in lakes is less dense than cold water (down to 4°C), warmer water exists at the lake surface. When a lake starts to warm up in the spring season, it is difficult for the wind to mix the lake water completely. This is especially true in lakes greater in depth than about 15 feet (5 meters). In a well-stratified lake there may be three distinctly separated layers. The surface stratum, known as the epilimnion, is relatively warm, and usually well mixed by the wind action. The middle stratum, known as the thermocline section, separates the lighter top from the heavier bottom sections, and prevents them from mixing. In the thermocline, a rapid drop in temperature is observed as depth increases. The lowest stratum, known as the hypolimnion, is usually devoid of oxygen since it cannot receive it from above, and bacterial respiration soon uses all oxygen formerly present.

The release of this relatively cold hypolimnion layer in the outlet from a lake can cause nuisances downstream such as fish kills and odors.

Steichen et al. (1979) report the use of a pump capable of recirculating Ham's Lake surface waters near Stillwater, Oklahoma to the hypolimnion to effect lake water uniformity and thus overcome this problem. They used changes in various physical, chemical, and biological parameters to measure the pump's efficiency. They concluded that: (1) within two weeks the pump completely destratified the lake thermally; (2) a longer period of time was needed for destratification of dissolved oxygen than for thermal destratification; and (3) algae dominance shifted from blue-green before pumping, to green after pumping.

REFERENCES

Edmonson, W. T.: Eutrophication in North America, *Nat. Acad. of Science*, Washington, D.C., p. 127, 1969.

Man-made Lakes, *Institute of Biology and Academic Press.* Edited by R. H. Lowe-McConnell. Introductory Survey by E. B. Worthington, p. 3, 1966.

Nemerow, N. L., and M. C. Rand: Algal Growth Affected by Degree and Type of Wastewater Treatment. In *Algae, Man and the Environment*, edited by D. F. Jackson, Syracuse University Press, 1968, Chap. 18, p. 391.

Steichen, J. M., J. E. Gaston, and C. E. Rice: The Effect of Lake Destratification of Water Quality, *Journ. American Water Works Association*, p. 219, April 1979.

Thomas, E. A.: The Process of Eutrophication in Central European Lakes, *Nat. Acad. of Science*, Washington, D.C., p. 33, 1969.

12 COASTAL AREA CHARACTERISTICS AND POLLUTION

Oceans have better cleansing or assimilative processes than water resources on the land because of the following reasons (Osterberg, 1982):

1. Unlike land, the ocean is three dimensional, so the concentration of pollutants is on the surface, but is eventually mixed throughout the entire water depth.
2. Our contact with the ocean is minimal compared with our intimate contact with the land.
3. Many pollutants adhere to and settle to the bottom of the ocean. At that location these sediments are more isolated from man than on land.
4. The ocean is not pure water, but is a toxic mixture of practically every element known to man.
5. Marine food chains are generally longer than food chains on land. Therefore, man, at the top of the food chain, will get fewer contaminants from seafood than from a comparable amount of food from the land.

Coastal zones have been badly defined. Carmichael (1980) gives a definition of a coastal zone in its natural state as: the maximum distances to which the influence of certain selected ecological factors extend landward and seaward from the seashore. A landward factor could be salinity, while seaward could be the depth to which sufficient light penetrates to permit the growth of a notable cover of algae, sea plants, or coral reef.

Coastal ecosystems are influenced by land and air ecosystems, and also affect these systems. For example, mangrove forests on coasts may assist in protecting inland areas from hurricane induced tidal waves; at the same time these mangroves are known to have the ability to absorb certain potentially toxic metals.

For a detailed discussion of the various types of coastal zones, their influence on pollutants, and their reaction to the same contaminants, one can refer to the following excellent review by Carmichael (1980):

"Over half of the U. S. population lives in a 50-mile wide strip of land along the nation's coasts (Carmichael, 1980). Nearly one-half of the U. S. multibillion dollar fishing industry depends directly upon shore waters. The coast provides a home for heavy and light industry and a prosperous recreation industry as well. And billions of dollars worth of goods arrive and leave from our coastal ports daily.

"The coastal zones are under pressure. More than 50% of the Atlantic shellfish beds were closed in 1979 due to pesticides, oil, and sewage contamination. Over the years, man has altered two-thirds of our offshore barrier islands, in some cases destroying these natural storm buffers. And land available for recreation is at an all-time low."

12-1. CHARACTER OF SEAWATER

There is about 2700 times as much seawater on this earth as fresh water. The seawater properties are dissimilar to those of fresh water because they vary not only with temperature and pressure but with salt concentration. The temperature and oxygen concentration of seawater vary from time to time and place to place, primarily at the surface of the seawater where the water interacts with the atmosphere.

Oceans cover 70 percent of the earth's surface. Actually, our continents can be considered mere islands between our five great oceans. Each ocean is huge—especially when related to a sea, gulf, or bay, which may contain considerable coastal area. For example, the smallest ocean, the Arctic, is more than five times as large in area as the Mediterranean Sea (even when you include its arms of the Aegean, Adriatic, Black, and Tyrrhenian Seas) which itself covers an area of 1,145,000 square miles.

Density

The density of seawater naturally depends primarily on its salt concentration, s, but also on its temperature, t, and the atmospheric pressure, p, of the air-water vapor mixture above it. A mean density, ρ, of seawater can be taken as 1.025 grams per cubic centimeter. Most physical oceanographers refer to the density of seawater, ρ, s, t, p, as a value of sigma σ, s, t, p, because of the difficulty of measuring and expressing normal density changes in significant figures generally beyond the third decimal point. $\delta s, t, p$, is measured by

$$\delta s, t, p = (\rho s, t, p, -1)(1000)$$

This results in the mean $\delta s, t, p$ of 25.23, a more significant and easier number to work with.

It is important to understand that temperature decreases and pressure increases as oceans deepen. Some typical values are given.

Ocean Depth (ft)	**Average Temperature (°F)	Pressure ($tons/in^2$)
600	60.7	
1200	50.1	
1800	44.7	
2400	41.8	
3000	40.1	
3600	39.0	
5400	36.8	
6000	36.0	1
7200	34.8	
9000	35.3	
12000	35.2	2
3 mi		3
4 mi		4
5 mi		5
6 mi (*35,640 ft)		6

*Deepest known point in ocean.

**Variation in temperature of ocean diminishes with depth (for example the ocean surface varies about 56°F, while at 9,000 ft depth, the variation is only 7°F or less.)

In this subject, the most significant physical parts of the ocean to us are the continental shelf and the coastline itself. Shorelines are continuously changing their slopes. Lands tend both to sink and to rise. Evidence of their sinking are islands which once were hilltops of land. When lands rise, one can observe the shoreline caves and cliffs which were formed by years of wave action. The continental shelf is actually sunken land located offshore and often as shallow as 600 feet under water. The shelf is built constantly by rivers emptying their contents into the sea, and may extend hundreds of miles beyond the coastline. The change in the shape of the oceans and continents which is occurring all the time, makes environmental protection of the coastlines difficult. The Mediterranean Sea, for example, is known to be changing very rapidly, and thus effects a great number of earthquakes. Most of this change is known to be brought about by the constant rotation of the earth which in turn causes a slipping of rocks—a known precursor to earthquakes.

The ocean currents depict the actual movement of the waters. The chief currents in the ocean are a result of the prevailing winds as affected by the rotation of the earth. Therefore, the currents flow in a direction of about 45° to the right of the wind direction in the Northern Hemisphere, and to the left in the Southern Hemisphere. Minor currents may also be caused by differences in temperature and salt concentration.

Salinity

The salinity of seawater caused by the elements shown in Table 12-1 has been defined as "the total amount of solid material in grams contained in a kilogram of sea water when all the carbonate has been converted to oxide, the bromine and iodine replaced by chlorine, and organic matter completely oxidized" (Forsch, Knudson, and Sorensen, 1902). This yields a slightly lower value than that of the amount of dissolved solids in grams per kilogram. The chloridity of a sample seawater can usually be used to estimate the salinity of the ocean water by the following formulation:

$$\text{Salinity} = 0.03 + 1.805\ (\text{chloridity})$$

where the chloridity is in parts per thousand (0/00) and is determined by precipitation of silver chloride from a sample. Conductivity measurements can also be used. Concentration of salt also affects the density of the water and the temperature at which it freezes. For example, the freezing point depression, ΔT, is equal to 0.102710 Cl, where Cl is the chloridity in parts per thousand. Salt water has a freezing point temperature of maximum density of −1.332°C (as compared to +4°C for distilled water). Since the salt concentration in seawater is higher than that corresponding to its freezing point, seawater will generally freeze before it cools sufficiently to reach its maximum density.

Stratification

As we learned from our studies of lakes, cooler water becomes more dense and settles to the bottom leaving a stratified warmer layer on top. In coastal zones, this phenomenon is accented by the balance of opposing effects on density of

TABLE 12-1. Seawater Ions

Anions	Cations
Cl^-	Mg^{++}
Br^-	Ca^{++}
SO_4^{--}	Sr^{++}
HCO_3	K^+
F^-	Na^+
H_3BO_3	

with Na^+ and Cl^- the major ionic forms.

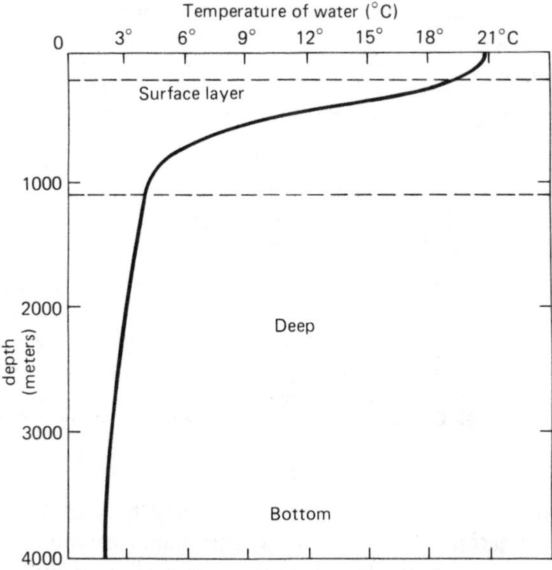

FIG. 12-1. Typical lake temperature variation with depth.

increasing salinity and temperature. In the oceans there usually exists a permanent thermocline which is deep enough to be relatively unaffected by seasonal cycles (Fig. 12-1). Often there comes and goes a shallower thermocline which steepens the vertical gradients of salinity and density.

12-2. POLLUTANTS OF CONCERN IN COASTAL WATERS

We are most interested in contaminants that resist chemical or biological change when discharged into rivers emptying into coastal waters. These contaminants are known as conservative material, and will tend to concentrate in the coastal zones. To complicate this phenomenon, we find that the uses of coastal areas may be quite different from the traditional and usually individual uses of flowing freshwater streams.

Heavy Metals

Because of the accelerated increases in industrial activity in the last 10 years, the amount of heavy metals being released to freshwater streams and reaching coastal waters is also increasing. These metals arise from condensation from the atmosphere and from mining activities, as well as from other industrial manufacturing. Some trace metals such as copper, zinc, and molybdenum are required in

TABLE 12-2. Some Heavy Metal Concentrations in the Oceans

Substance	μg/liter	Total in ocean (10^6 metric tons)
Copper	3	4×10^3
Cadmium	0.11	1.5×10^2
Chromium	0.05	7
Iron	10	14×10^3
Lead (estimated natural)	0.03	4
Mercury	0.03	4
Nickel	2	3×10^3
Vanadium	2	3×10^3
Zinc	10	14×10^3

Source: Hood, 1972. Pollution of the World's Oceans: Topics in Ocean Engineering, vol. 4. Edited by C. L. Britichmieder, Houston, Texas, Gulf Publishing Co.

low concentrations for biological growth, but may become toxic in high concentrations. Of greatest concern are lead, chromium, cadmium, mercury, selenium, arsenic, and antimony. The natural concentration of many heavy metals in oceans is given in Table 12-2. Heavy metals in river waters have resulted in reduction of fish stocks, such as salmon, and in an increase of metals within seafood, such as oysters. The net result of this is that society will have less fish food, and will have seafood with higher concentrations of metals. It is certainly possible, even highly probable, that metals released in estuaries and/or coastal areas will be complexed and find their way to the bed of the ocean where they will be picked up by organisms and seafood. In fact, Turekian (1971) has presented some evidence in this connection.

Plant Nutrients

The addition of plant nutrients to the ocean, especially phosphate and nitrate, has had considerable effect on the quality of local coastal zones. The nutrients arise mainly from domestic sewage effluents and nonpoint runoff from agricultural lands. The greatest impact of these nutrients in coastal waters is that of overabundant growth of plants that are unusual or nonindigenous to the area. This excessive plant growth usually means a reduction in diversity of species, and results in an imbalance of food-chain materials essential for intermediate organisms. In turn, *their* demise means an increase in BOD and turbidity of water. Its value for recreation and fishing is now reduced. Much of this type of pollutant originates many miles upstream from the coastal zone. It finds its way to the coastal area by cycling through several biotic species which then become larger species on their way to the open sea. There has been much documented data to show the concentration of nutrient elements in many coastal waters adjacent to urban centers. When the excessive nutrient in the form of higher plants arrives at the coastal zone, it must compete for oxygen resources with primary organic matter. It is also often inhibited from biodegradation by toxic chemicals such as the heavy metals discussed in the last section.

Several methods have been suggested to alleviate this concentration of pollution from the coastal zones (Coastal Zone Workshop, 1972). These have been summarized and reexpressed by the author below.

1. Make optimum use of tidal flushing, dispersion, and piping systems in the coastal zone.
2. Reduce all contaminants to stay within the assimilated capacity of the overall system. This involves better utilization of both mathematical and physical models.
3. Introduce treatment methods that greatly reduce important growth-stimulation nutrients.
4. Encourage closed-cycle systems in which the major concentrated effluents are disposed of on land or in the deep sea.
5. Balance the nutrient quality so that desirable seafood species are encouraged, enhancing productivity.
6. Spray land with nontoxic, dilute organic wastes to be utilized by land-based plants.

Organic Additions

Petroleum related. Most of the petroleum is produced in and/or transported through the coastal zones of the world. With the advent of offshore drilling and deep well oil injection, the amount of oil reaching the environment is well known. The National Academy of Science (1971) estimated the petroleum losses to the environment (Table 12-3). However, the quantities of oil reaching the environment are not as important as the particular location and the extent of petroleum emitted in each specific incident.

There is some evidence that petroleum components have already entered the marine food chain. The long-term effect of accumulated petroleum in marine life is not clear to date. Sources of oil in the marine environment include accidental types, such as spill, and purposeful types from sewage, storm sewers, filling station washdowns, and other municipal and industrial point sources, as well as diffused landfill operations. There is also some evidence that the major source of oil in marine coastal waters is chronic small discharges during shipment of oil products, rather than losses of crude oil.

Halogenated hydrocarbons. These compounds, which are rare in nature, are relatively stable when they enter marine water because of resistance to microbial

TABLE 12-3. The Involvement of Petroleum with the Marine Environment (millions tons/year). [*After NAS, 1971*]

World oil production (1969)	1820.
Oil transport by tanker (1969)	1180.
Injections into marine environment through man's activities	2.6
Offshore oil production (seepage from wells)	0.1
Tanker operations	0.5
Other ship operation	0.5
Accidental spills	0.2
Deliberate dumping	0.5
Refinery operations	0.35
Industrial and automotive wastes	0.45
Torrey Canyon discharge	0.117
Santa Barbara blowout	0.003–0.011
Atmospheric input from continents through vaporization of petroleum products	90.
Natural seepage into marine environment	< 0.1

action of the chlorine-to-carbon bond. Because of this they are used extensively in agriculture and in certain industrial products. Typical and recently highly publicized compounds include, DDT, PCB, and PBB. DDT compounds arise both from manufacturing plant spill and from runoff from agricultural use as an insecticide. Polychlorinated biphenyls (PCB) are used as dielectric fluids in capacitors and transformers, from which very small amounts of PCBs have been known to escape and find their way into the marine environment. Polybrominated biphenyls (PBBs) are used as fire retardants, and aren't as likely to be found in the environment. Other chlorinated hydrocarbon insecticides of a relatively stable, highly toxic nature, such as dieldrin and endrin, have also been detected in coastal ecosystems. Materials that are manufactured by industry in relatively large quantities, such as polyvinyl chloride, which are relatively insoluble in water but soluble in fats and oils and resistant to chemical and microbial breakdowns, tend to be a major problem in coastal areas since they are capable of moving great distances in these forms. The concentrations of PCB and DDT in coastal waters of the world were reported as early as 1972, and some are summarized in Table 12-4.

Plastics and Other Synthetic Organics

Because of the depletion and costliness of natural metallic materials, the production of plastics and other synthetics has increased geometrically in the last few

TABLE 12-4. DDT and PCB Concentrations in Fish (IDOE, 1972)*

Region	Material	Concentration[a] PCB	Concentration[a] DDT	Data source[b]
Open North Atlantic	Pelagic fish			
	muscle	1-10	0.6-3	Harvey, 1972
	liver	1000-6000	95-4800	
	Midwater fish and crustacea	8-59	3-12	
Open South Atlantic	Midwater fish and crustacea	2-14	1-8	Harvey, 1972
Denmark Strait	Groundfish			
	muscle	2-360	3-30	Harvey, 1972
	liver	300-1000	9-260	
Northwest Atlantic shelf	Groundfish			
	muscle	37-187	3-74	Harvey, 1972
	liver	1870-21,800	390-2680	
Gulf of Mexico	Whole fish or muscle	<1-530	1-150	Giam, 1972
Northeast Pacific	Euphausiids	9.2 (mean)	2.7 (mean)	Claeys, 1972
	Pink shrimp	23 (mean)	2.5 (mean)	
	Flatfish	23 (mean)	10.8 (mean)	
Scottish west coast	Fish muscle	<100-1500	<30-480	Holden, 1972
	Fish liver	200-42,600	70-5800	
Baltic Sea	Herring	150-1500	100-1500	Jensen, 1972
	Cod	16-180	9-340	

[a] Expressed in micrograms/kilogram (ppb) wet weight; DDT values include all metabolites.

[b] All personal communications.

*IDOE, "1972 Baseline Studies of Pollutants in the Marine Environment and Research Recommendations," May 24-26, New York City, 1976, 54 pp.

decades. It is only reasonable to expect some of these materials in coastal waters since they also possess the characteristic of resistance to degradation.

Carpenter (Coastal Zone Workshop, 1972) detected plastics in the form of hard, polystyrene spheres (0.1-0.2 mm in diameters) in Narragansett and Buzzards Bays, and even more in Niamtic Bay. The spheres contained 5 ppm of PCBs, apparently absorbed from the seawater. The same plastics have been observed in the intestines of larval fishes, winter flounder, grubby sculpin, and young herring. Ingestion of the spheres may cause intestinal blockage and also may affect larval fish by causing abnormal fry. Carpenter and Smith (1972, *Science*

175:1240) also reported polyethylene and other plastics floating on the Sargasso Sea in concentrations of about 300 grams/km^2.

Solid Wastes

All types of potential contaminants reaching coastal waters originated in so-called refuse disposal landfills. When refuse is eroded from land or intentionally discharged into marine waters, it settles and covers the benthic life. Because of toxic matter present in many solid wastes, benthic deposits may be sterile—others lacking toxic matter may exhibit teeming biological life. Some contaminants besides toxic metals include petrochemicals, nutrients, and pathogenic sewage-originated matter, as well as various other industry-originated contaminants.

Most of the solid wastes reaching coastal waters were considered by their depositors as being fixed stationary in their landfills, or at least "out of sight" and "causing no problems" in the bottom of the sea. Whether the coastal waters recover from this obvious misconception depends upon the availability of oxygen for biological assimilation of the organic matter, and the mixing taking place between the bottom water layers and the overlaying epilimnion. Stabilization may also take place in the bottom of the organic deposits which are covered with sand and silt eroded from effluent streams.

We can conclude then, that the prime problems caused by solid wastes are (1) increase in turbidity from suspended matter, (2) destruction of benthic life by obiliteration of the bottom, and (3) toxic substance addition.

Radioactivity

As the use of nuclear energy for power expands and surface nuclear explosions continue, especially from countries more recently entering the field—we are bound to discover an increase in radionuclides in coastal waters. We must have confidence that world government powers will agree to limit or even completely eliminate surface nuclear explosive testing. However, wastes from nuclear powered electric energy generating plants reaching the coastal environment are increasing, and probably will continue to do so. These wastes come from lakes, cleanups, and from fuel fabrication and reprocessing plants. A typical peaceful nuclear power plant is about 1000 megawatts in electrical capacity. The National Academy of Science (1971) gives the release of radionuclides from a typical reactor in Table 12-5.

The most important contaminants are the long-lived biologically dangerous plutonium wastes, of which containment for 100,000 years will be required. Accidental environmental catastrophes in marine areas, such as ship collisions and aircraft dropping or crashes, are also a possible source of coastal pollution.

TABLE 12-5. Annual Release of Radionuclides Estimated for a Pressurized-Water Power Reactor of 1050 megawatts Electric Capacity

Isotope	Half-Life	Microcuries/yr	Isotope	Half-Life	Microcuries/yr
Liquid Wastes					
^{3}H	12.26 yr	4×10^{9}	^{131}I	8 days	6.61×10^{3}
^{54}Mn	314 days	9.7×10^{-1}	^{132}Te	78 hr	6.99×10^{2}
^{55}Mn	2.58 hr	2.64×10^{1}	^{133}I	2.3 hr	2.8×10^{2}
^{55}Co	71 days	2.95×10^{1}	^{133}I	21 hr	5.13×10^{3}
^{80}Co	5.26 hr	3.48	^{134}I	53 min	2.16×10^{1}
^{82}Sr	50.4 days	9.1	^{134}I	6.7 hr	2.6×10^{3}
^{90}Sr	28 yr	5.76	^{134}Cs	2.1 yr	8.69×10^{2}
^{90}Y	64 hr	1.06	^{136}Cs	13 days	8.36×10^{1}
^{91}Sr	9.7 hr	2.49	^{137}Cs	30 yr	4.58×10^{3}
^{91}Y	59 days	2.11×10^{1}	^{140}Ba	12.8 days	2.28
^{92}Y	3.5 hr	5.13	^{140}La	40.2 hr	2.35
^{99}Mo	66 hr	1.25×10^{4}	^{144}Ce	285 days	7.82
Gaseous Wastes					
^{85}Kr	10.4 yr	5.62×10^{3}			
^{133}Xe	5.27 days	1.58×10^{3}			

Source: Preliminary Facility Description and Safety Analysis Report, Salem Nuclear Generating Station, Burlington Co., N.J. Docket No. 50–272.

Since most power plants require great quantities of water (about 1.5 cubic feet per second per megawatt of electrical energy generated), the marine zone is a prime location for these plants. The genetic effect of the low level of radiation on coastal zone organisms is not well established.

Pathogens

The discharge of human pathogens–mostly viruses–into coastal zone waters is proceeding relatively unabated. We are concerned about this fact for two reasons: (1) human residences and development in coastal areas are increasing, and (2) humans use these areas for recreation. The National Academy of Sciences (1971) provides a tabulation of both enteric bacteria and viruses of major importance (Table 12-6).

Much study is currently necessary to assess the mode and extent of disease transmission through the use of coastal waters for both recreation and food sources.

Thermal and Other Power Plant Pollution

As we mentioned previously, with the increased nuclear generating plants by 1980, it is estimated that 32 percent of all power plants will be located in coastal

TABLE 12-6. Pollutants, Actual or Potential, of Microbiological Nature, Occurring in Coastal Waters as a Consequence of Contributions of Human or Animal Waste Products

Enteric viruses		Enteric bacteria, human or animal origin
Human origin	Animal origin	
Polioviruses	Simian	Salmonellae
Coxsackie-viruses A	Porcine	Shigellae
Coxsackie-viruses B	Bovine	Vibrio parahemolyticus
Echoviruses	Foot and Mouth	
Reoviruses	Vesicular Exanthema of Swine	
Adenoviruses		
Hepatitis viruses		

areas. Besides the thermal problem—the cooling water intake velocity must be less than 1 ft/sec to prevent fish impingement on intake screens. This generally means a relatively large intake cross-sectional area in order to provide the volume of cooling water required. The heat problem can only be solved by (1) dissipation, or (2) adequate dilution water. Dissipation is provided by either cooling towers or ground and water cooling. Deep water below the thermocline—generally used for cooling purposes in the warmer seasons—may contain low oxygen, which might be reduced even further by its increased use and by a raise in temperature.

Other coastal power plant contaminants include (1) chlorine residual used to control cooling water slime microorganisms, (2) trace heavy metals from boiler compounds or corrosion product, (3) turbulent flow from plant, causing erosion of sediment and disturbance of benthic organisms, (4) by disturbing the constant natural seasonal temperature cycle marine organisms will be affected, usually adversely, and (5) leachate from boiler ash storage areas.

Hedgpeth and Gonor (1969) found that the rate of temperature change may be very important for bottom organisms in tidal waters. A one or two degree rise in midsummer ambient temperature can cause marine microorganism deaths in southern latitudes, while little or no effect may be observed with the same thermal change in northern latitudes. Time of exposure to any temperature is important to the survival of coastal biological life.

Dredging, Filling, and Mining

These operations, which are being actively pursued today (primarily for coastal renovation) upset the relatively stable condition existing at the benthos water. The sludge solids are moved to new sites, where siltation and habitat destruction develop. Dredging releases sulfides, causing the dual effects of unpleasant odors and lowered oxygen, and thereby species are therefore endangered. Heavy metal and petroleum resuspensions may increase the relative toxicity in the coastal area.

Dredging of coastal waters is often practiced to develop more beach area for bathers. Dredged sand also includes chunks of coral—some as large as footballs—which makes walking on the beach difficult. After the beach is cleaned, periodically, the rocks disappear and the beach is cleaner and wider; but the sediment and silt stirred up by the giant dredge lingers on the sea. The layer of silt kills the algae and marine life on which lobsters, sea urchins, and coral crabs thrive.

McPherson of the Miami Herald (p. 8D, Oct. 12, 1983) found that dredging apparently increased the salinity of the Loxahatchee River in Florida, upstream of the dredging. Salt water at this location has killed or damaged hundreds of towering cypress trees which line the river. Future, more ambitious dredging will push salt water upstream into the scenic headwaters. He concluded that development of the river (including the dredging) "could have both beneficial and deleterious effects depending upon what people want that estuary to be." Evidently, lower stretches would be improved by clear and sparkling water with sea grass beds and plenty of fish and mangroves lining the shore. But, upstream the "salt-water intrusion" could damage the winding stream.

12-3. OVERALL USE AND ABUSE OF COASTAL AREAS

The recent Coastal Zone Workshop (1972) concluded that:

(*a*) The projections for the use of coastal zone waters for once-through cooling at electric generating stations point to a significant deterioration of water quality due to heating and associated effects. General degradation of the environment through release of toxic substances has occurred in heavily populated areas and is still on the increase, although notable remedial efforts are being asserted in certain localities.

(*b*) *The aquatic ecosystem of the coastal zone has a finite assimilative capacity for a particular contaminant without significant deleterious effects.* The assimilative capacity of any particular part of the coastal zone is determined by physical processes such as currents mixing, geomorphology, types of sediments, types of water chemistry, and biology.

Each region of the coastal zone should be considered on its own merit; its own uses and characteristics, and the contaminant load should be determined on a performance basis—an approach that requires a much better understanding of the aquatic ecosystem than exists today.

Ocean Outfalls

In 1977, the United States Congress granted the EPA the authority to issue to publicly owned treatment works, permits that allowed wastewater that has not received secondary treatment to be discharged through ocean outfalls to specified marine waters. "We have already known that some nearshore coastal environments were very sensitive and delicately balanced so that a high degree of treatment (more extensive than secondary) would be required if outfalls were to be allowed, while others were more resilient, so less than conventional secondary treatment would be adequate. But, not enough was known about the specific relationships between sewage quality, degree of treatment, and marine ecosystem effects."[1]

Offshore Drilling

When an offshore well is being drilled for oil, a special mixture of clay, water, and chemicals is essential to the operation. As this mud is pumped down into the drilled hole, it cools the rapidly rotating bit, lubricates the drilling string as it turns in the wellbore, and carries rock cutting to the surface. The mud also serves as a plaster to prevent the surrounding rock from crumbling or collapsing into the wellbore, and provides the hydrostatic head necessary to control downhole pressures and to keep extraneous fluids (such as seawater) from entering the wellbore.

The chemicals are added to the drilling mud to enhance the mud's ability to perform its many tasks. These chemicals can range from bactericides, calcium removers, corrosion inhibitors, defoamers, and emulsifiers to filtrate reducers, shale-control inhibitors, thinners, dispersants, and weighting agents.

Some reported findings and the literature have shown that[2]

1. drilling fluid is 10 times more toxic than industrial effluents such as untreated wastes from oil refineries or pulp mills;
2. carcinogens are discharged during drilling operations;
3. drilling compounds are actively taken up by marine organisms, and at the same time, accumulated in them;
4. chemicals discharged during drilling persist for years in the sea bottom sediments;
5. a wide variety of organisms that normally live on the sea floor *cannot* grow on sediments contaminated by drilling fluids; and
6. effects of chemicals on coral may be delayed for a year before they can be observed.

---

[1]EPA Research Highlights 1979, EPA 600/9-80-005 Research and Development, Washington, D. C., Jan. 1980, p. 72.

[2]Ibid., p. 74.

Abuses of Coastal Zone

Balancing immediate human needs with long-term ecological effects in the coastal zone is a delicate and critical problem. Population increases can put too much pressure on peninsular and island aquifers, with resulting saltwater contamination of freshwater supplies.

Nutrients circulated and distributed in a coastal salt marsh by tidal action enrich plant roots and increase overall plant productivity. Toxic pollutants, such as oil discharges and chemical wastes, move in the same pattern, but with potentially adverse effects on the ecosystem.

Examples of Estuary or Coastal Zone Degradation

1. Occasionally, eelgrass, a plant of significance in the marine food chain, inexplicably browns, withers, and dies. Labyrinthula, slime mold that lives on eelgress, may be responsible.
2. In 1972, when red tides first appeared in Massachusetts waters, public officials closed all shellfish areas to avert widespread outbreaks of paralytic shellfish poisoning. Industry losses were severe. Florida has long suffered this same problem.
3. In Nahant Bay, decaying accumulations of another alga, pilayella littoralis, send out noxious fumes that drive away bathers from densely populated coastal areas.

12-4. ASSESSING MUNICIPAL, INDUSTRIAL, AND AGRICULTURAL POLLUTANT WASTEWATER LOADS ON THE COASTAL AREA

As part of the picture of coastal zone pollution we must determine the origin of the predominant contaminants that influence it. This is a relatively difficult and complicated undertaking when compared to that associated with rivers, or even lakes. The original source of these contaminants may be many kilometers from the actual shoreline. On the other hand, some sources may exist directly on or very near the coastline, but are difficult to measure or even detect, because of the tremendous dilution afforded by the seawater. All major contaminants result from either municipal, industrial, or agricultural activities. The assessment of each may be carried out quite differently. Before considering the separate approaches for each major source, we should try to visualize a coastal area as depicted in Fig. 12-2.

You can see from this conceptual illustration that contaminants originate from sea-based as well as land-based sources; that some are perennial while others have been recently discharged; and that some originate from distant

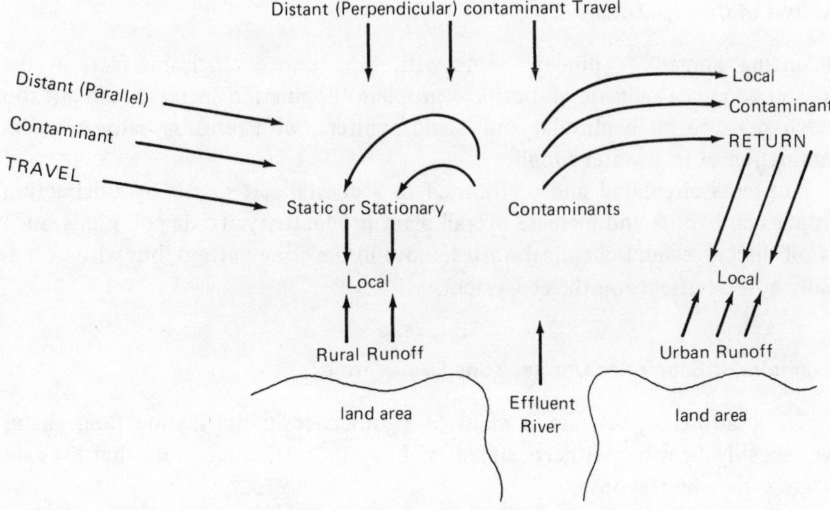

FIG. 12-2. Contaminant contributions to coastal zone pollution.

sources and at times that are predictable only with poor reliability. In an attempt to unravel the complex origin and age of contaminants in the coastal zone we must first consider the land-based types as separate from the sea-based ones.

Land-Based Contaminants

The municipal, industrial, and agricultural land-based contaminants originate either in the onshore local area under study or from distant shoreline sources (see Fig. 12-2). When making a coastal area survey, generally, we are seeking the local sources that influence the water quality. The distant sources are either presumed diluted sufficiently so as to be insignificant, or are seen as obligations and concerns of others. It should be pointed out here and emphasized that neither is necessarily a valid presumption. Distant-originated contaminants may, in fact, exert a significant effect on local coastline seawater quality. But their origin must be traced by either a larger, more comprehensive and inclusive sea study or by collaboration with another survey team studying the distant, land-based sources. In our analysis here we can only present an approach for evaluation of local land-based contaminants. However, the same sort of system analysis could be applied to the distant sources, and the two studies combined.

Municipal Local Land-Based Contaminants. These contaminants originate from human wastewaters, which include all of the normal household activities, but none of the outside work activities. The significant contaminants are mainly organic and suspended matter, and bacterial in nature. They generally reach the coastal zone either from sewers discharging directly into the sea, or indirectly into a flowing tributary of the sea. They may be treated or not, prior to being discharged. The quality and quantity of these contaminants may be obtained by any one or more of the following methods:

(*a*) analyses of sewage system effluents—which include average flows and chemical, physical, and biological analyses;

(*b*) statistical records which include up-to-date sewered population numbers and estimated contribution of each person per day to the contaminants in the sewer;

(*c*) surface involved and population density of areas contributing to the flow in the sewerage system. This data must be combined for actual use with data on the per capita contributions of contaminants.

(*d*) predicted urban population from previous census data, and professional estimation of percentage of this population contributing to the sewerage flow. This data must also be combined for actual use with the per capita contributions of contaminants.

The reliability of these methods in yielding accurate predictions of total contaminants decreases as one progresses from (*a*) to (*d*) above.

Industrial Local Land-Based Contaminants. These contaminants originate from the production of goods in manufacturing establishments. Information related to production activities may not be as easily obtained as that from strictly household ones. The reasons for this are partly political and partly economic. These difficulties in obtaining data must be overcome, or at least circumvented, by the sea pollution analyst. In the 1950s and 1960s another reason existed—that of not having ample and significant industrial data which would allow us to predict overall contaminant loads. However, this obstacle was overcome in the 1970s. We now have a way to convert industrial data to pollutant load with considerable assurance of obtaining accurate results. For the sake of simplicity we will call it the "production method."

The production method entails the use of the best information available from each industrial plant even though the basic data will not be the same from plant to plant. However, a priority of informational ranking is proposed consisting of: the *first*, that of having actual industrial wastewater analyses with accompanying flows; the *second*, information directly related to the daily or yearly production of goods; the *third*, obtaining data that yields—directly or indirectly—the consumption of water, also on either a daily or yearly basis; and the *fourth*, which is the number of employees actually working within a given industrial plant. All four methods contain potential errors, such as an error in reporting information, or a change in plant situation since the information was made public, or the reported production, water, or worker data not coinciding precisely with data on the production of the contaminants in the effluents. Using the results of actual wastewater analyses is subject to criticism. For example, the analyses may not truly represent industrial production practices. The primary advantage of the production method is that it yields a more accurate prediction of the contaminant loads from each plant. The main disadvantage is that the information may be different or even impossible to obtain. Questionnaires can be devised to yield one or more of the four types of production data. When the questionnaires are completed, they can be successful in providing proper production data. When they cannot be completed, for one of the previous reasons, alternative means of obtaining production data must be used.

The Environmental Protection Agency effluent guidelines can be used to convert production data to contaminant loads. We suggest using 30-day average EPA values to give a more general prediction of universal conditions. These guidelines pertain to effluents only. It is necessary to extrapolate raw waste information from effluent loading data. This is done with adequate knowledge, and experience as to what constitutes the best practical treatment utilized to obtain these effluent guideline levels.

REFERENCES

Carmichael, J.: *Coastal Ecosystems: a Review*, United Nations Environmental Program, Nairobi, Kenya, 1980.

Forsch, C., M. Knudson, and S. O. L. Sorenson: *K. dansldevidensk Selsk*, 6(12):151, 1902.

Hedgpeth and Gonor: Aspects of the Potential Effect of Thermal Alteration on Marine and Estuarine Benthic, in *Biological Aspects of Thermal Pollution*, Vanderbilt Univ. Press, 1969, pp. 80–118.

Coastal Zone Workshop, *The Water's Edge: Critical Problems of the Coastal Zone*, Ketchum, B. H., ed., Wood's Hole, Mass., 1972.

Marine Environmental Quality, *Ocean Sciences Comm., Nat. Acad. Science*, Washington, D.C., 1971, 107 pp.

Osterberg, C. L.: IAEA Bulletin 24(2):32, June 1982.

Radioactivity in the Marine Environment, *Nat. Acad. Science*, Washington, D.C., 1971.

Turekian: Rivers, Tributaries and Estuaries, in D. W. Hood, ed., *Infringement of Man on the Oceans*, NYC Interscience, 1971, pp. 9–73.

13
MODELS FOR RECEIVING WATER QUALITY ASSESSMENT

13-1. INTRODUCTION

In recent years, water resource agencies and consulting engineers have expressed a great need for predictive models for estimating water quality at a certain point and time in a water resource below potential input of contaminants. The development of such models has been continuous, but hampered somewhat by our inability to identify and quantify all the input parameters necessary for accurate predictions. Because of the rapid improvement of computer capabilities, the use of models–rightly or wrongly–has proceeded without much hesitation. As pointed out, however, by Hines et al., (1980) many decision makers have failed to accept models for river quality assessment, and many actually view models with considerable mistrust–apprehensive, mainly, of the following circumstances:

1. application of a model to a variable or process that is too complex for formulation of a practical, applied model;
2. application of a sophisticated, general-case model without adequate understanding of the particular river in question;
3. failure to recognize the importance of basin hydrology in defining the critical planning and management decision periods for model simulation;
4. misapplication of model calibration and verification procedures;
5. use of a poor data base for interpretation, calibration, and verification; and
6. failure to format results for ease of user understanding.

However, the authors do admit that with careful thought and interdisciplinary teamwork, these deficiencies can be corrected, and conceptual models transformed into practical, useful tools for river quality assessment.

13-2. MODELS FOR RIVERS

The simplest model is one that will allow the user to predict the dissolved oxygen at some critical point in the river. In this case dissolved oxygen is generally plotted graphically as a dependent variable on the Y-axis. Other significant parameters can be derived, but with greater difficulty and questionable accuracy. Hines et al., (1980) in a triangular representation, show some of these other parameters (Fig. 13-1).

A river quality model must be developed from a well-established and proven theoretical relationship. This necessitates a concept including mathematical solution techniques and a readily adaptable computer program. In the mathematical solution, models will include parameters to define reaction rates, proportionality constants, or other system characteristics. In the well-known Streeter-Phelps sag equation (see Chapter 6)

$$\frac{dD}{dt} = K_1 L - K_2 D$$

K_1 and K_2 are such reaction rates of river deoxygenation and reaeration, respectively.

The major problem of modeling arises from applying these reaction rates in

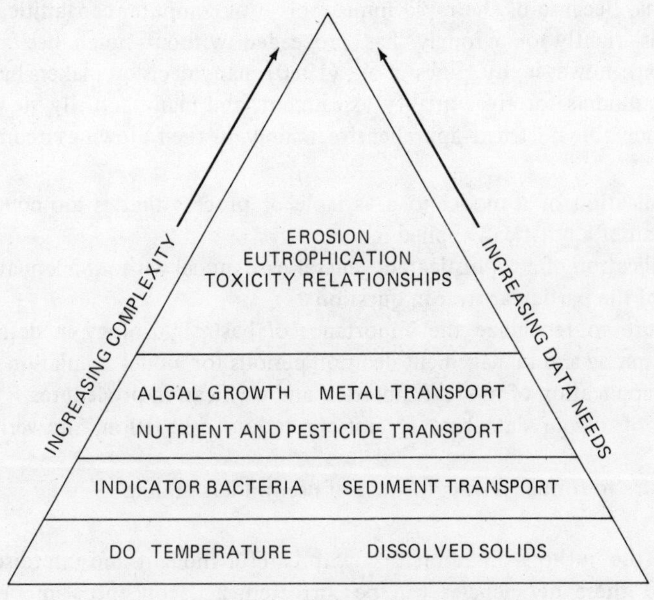

FIG. 13-1. (*After Hines et al., 1980*)

accurate values to the particular receiving water in question, at the precise location and time of its flow.

Other mathematical solution techniques, such as the use of Churchill's Least-Squares stream equation (Chapter 6) depend upon derivation of that equation during the exact environmental conditions under which one wishes to predict future effects. Since no future conditions are exactly like past ones, certain discrepancies in results can be expected. Hines et al., (1980) concluded that "all proposed river-quality models, whether conceptual or applied, should be accompanied by a statement of limitations, predictive accuracy, and suggested applications."

Weber, Kisiel, and Duchstein (1973) recommend that before selecting a particular model, one should make a short premodeling data collection and analysis program designed to determine such factors as mix- and waste-loading characteristics, streamflow patterns, expected ranges of model parameter values, and the presence and absence of algal growth or benthal deposits.

Velz collected a great deal of instream as well as offshore data on the Willamette River (1951). His findings were sufficient to formulate a model of simpler configuration than originally considered, and to recommend expending more investigative effort on factors having the largest effect on the dissolved oxygen system of the river. Velz found that it was important to recognize the overriding importance of river hydrology in controlling river quality. He recommends optimizing his data in two steps: first, a numerical range is established within each model parameter (can vary); and second, numerous computer runs are conducted, and values for each parameter are simultaneously varied within the range previously established. During the computer runs attempted, Velz uses a least-squares procedure to obtain a "best fit" of model outputs to a set of observed river quality data. However, he maintains, and is corroborated by Hines et al., (1980) that these fit derivations are merely little more than a computerized "curve fitting" process rather than what they need: "a fine tuning process governed by scientific understanding." They both believe that verification is the essential step in substantiating the predictive capability of a river. If the model predictions are what Hines et al. call "acceptably close" to the independent observations, the model is considered verified.

Data used in predicting the model should not be used in verifying the model. Once the model is deemed acceptable, it must be reverified when conditions in the stream change significantly. For example, if a river is dredged, or conversely, filled in part, to control flooding or promote wetland conservation, a new model should be developed.

Verifying data of the Willamette River was collected beginning in 1973, and eventually resulted in two "control curves" (Hines et al., 1980). They are shown here as examples of useful models (Figs. 13-2 and 13-3). In Fig. 13-2, the stream analyst can predict the dissolved oxygen level at a critical river location at any

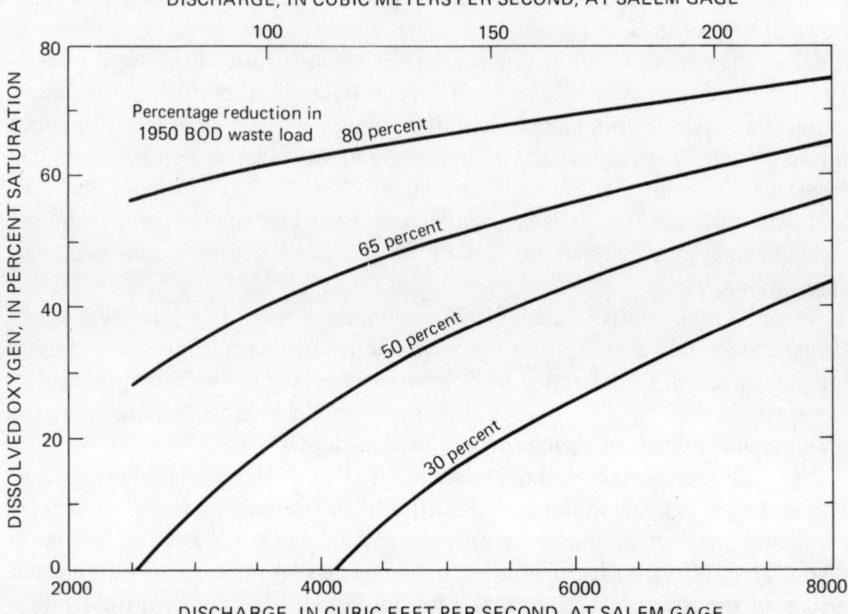

FIG. 13-2. Relation between DO and summer runoff for Willamette River, Oreg. (1950–51), at river mile 6. Curves represent DO-runoff relations at different BOD waste-loading conditions. (Adapted from Velz, 1951, p. 68a.)

river with a specific reduction in offshore BOD load. In Figure 13-2, the analyst can determine the corresponding dissolved oxygen level knowing the exact BOD load. The curve model then is extremely useful to the analyst. In Fig. 13-3, the analyst can determine the corresponding dissolved oxygen level knowing the exact BOD load in the river at various streamflows with various treatment reductions of BOD. The planner can then make administrative decisions to suit society's wishes.

In some cases of river analysis, growth of algae create depletions of oxygen during the dark hours and an increase of oxygen during the daylight hours. Unless this photosynthetic effect is accounted for, oxygen predictions downstream from sources of organic contaminant entrances will be in error. Only a few models allow for the accurate prediction of photosynthetic activity. We must generally presume that such activity occurs diurnally—that is, daily. A *diurnal* model developed by O'Connor and DiToro (1970) measures the diurnal variations of dissolved oxygen as a result of algal activity. This model has been preferred because of its ease of application. The model depends upon a correct assessment of a mass balance of all oxygen inputs and extractions from a flowing

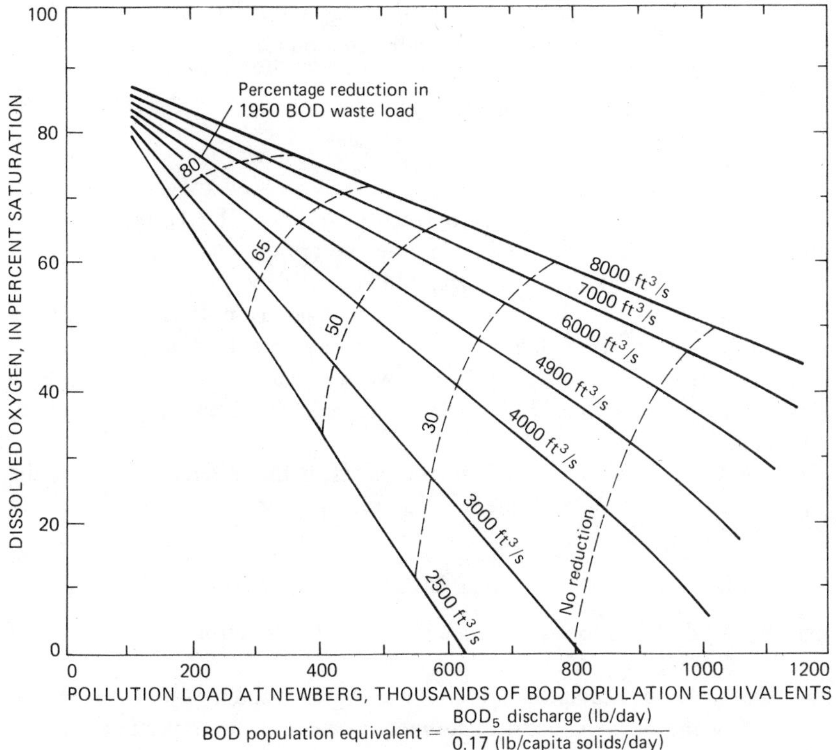

FIG. 13-3. Relation between DO and BOD waste loading for the Willamette River, Oreg. (1950–51), below river mile 50. Curves represent DO-BOD load relations at various river discharges as measured at Salem (river mile 85). (Adapted from Velz, 1951, p. 75a.)

stream. In a single-dimensional analysis the equation of the model can be expressed as follows:

$$\frac{\partial C}{\partial t} = -\frac{Q}{A}\frac{\partial C}{\partial x} - k_1 L(x) - k_N N(x) - A_1(x) - B(x) + k_2(C_s - C) + A_2(x, t)$$

where

C = concentration of dissolved oxygen
C_s = saturation of dissolved oxygen
k_1 = carbonaceous BOD reaction rate
k_N = nitrogenous oxidation reaction rate
k_2 = reaeration reaction rate
$L(x)$ = single longitudinal distance of carbonaceous BOD

$N(x)$ = single longitudinal distance of nitrogenous BOD
$A_2(x, t)$ = algal photosynthesis oxygen production rate
$A_1(x)$ = algal respiration rate
$B(x)$ = Benthic microbiological oxygen demand
Q/A = flow per unit area of the river

The quantities above containing k_1, k_N, and k_2 are well known and are computed by various methods, as shown in Chapters 4 and 5. The A_1 and A_2 as well as the B term are more difficult to compute and are the major reasons for verifications of true river models when algae and benthic deposits are involved.

Erdmann (1979) has proposed a method to assist us in determining reasonable values of A_1 and A_2 in diurnal analysis. He measures the diurnal variation in dissolved oxygen and flow data between two selected points on a river, Essentially, the method is a time derivative of changes in dissolved oxygen as caused by algal and bacterial action.

A_1, the respiration rate for all aquatic plants, is the difference between the total respiration rate and the bacterial respiration rate on both carbonaceous and nitrogenous matter, as follows:

$$A_1 = R - K_1 C \, \text{BOD} - K_N N \, \text{BOD}$$

where R = algal respiration rate, which can be measured during the darkness and equals $K_1(C_s - C) - (DC/Dt)$

A_2 is obtained by accounting for all the above rates and measuring the overall result during the daylight hours. The unaccounted for difference in oxygen is due to A_2.

13-3. MODELS FOR LAKES

Lakes are being modelled to predict the effect of nutrient loadings on eutrophication problems. These models are primarily of the statistical type because they are more dependable and easier to derive than ones based on detailed scientific inputs.

Essentially, statistical models are developed by plotting values of selected measures of lake effects against parameters causing eutrophication (see Chapters 1-6). Lake effects are normally some measure of algal growth while causative agents are nutrients usually phosphorous concentration. Since lakes differ in morphology as well as other chemical and biological factors, considerable dispersion of data can be expected when comparing lake responses. One must utilize the least-squares technique (described in Chapter 4, Section 1) to plot the line of best fit. Therefore, a number of typical, but varied lakes need to be analyzed and evaluated under both unaltered (high) and reduced nutrient loadings. The responses of these lakes (effects) can be used to derive a general model for use with any particular lake.

Most existing and current (1983) lake eutrophication models have been developed from the Cooperative Programme on Eutrophication Control of the Organization for Economic Cooperation and Development (OECD). Over 200 various types of lakes were used in developing the existing models. The most prevalent was by Lee, et al. (1978) and given in more detail by Rast and Lee (1978). Rast, Jones, and Lee now report (1983) from their examination of United States bodies of water, that statistical models have considerable validity and usefulness as quantitative tools for assessing the expected water quality responses of lakes and impoundments to changes in their phosphorous loads. The OECD models utilize as effects of eutrophication either: (1) chlorophyll A, (2) secchi depth, or (3) hypolimnetic oxygen depletion, while the causative nutrient is that of annual phosphorous loading. Certain constraints exist, however, when using these models. These constraints proposed by Rast, Jones, and Lee (1983) are summarized as follows:

1. Use only planktonic algae.
2. Discount lakes containing excessive nonalgal turbidity or color.
3. Hydraulic residence time during the growing season must be at least two weeks.
4. Lakes should act like completely mixed reactors.
5. Lakes should have constant volumes.
6. Phosphorous sedimentation should be proportional to in-lake concentration.

The first three of the above constraints are absolute requirements. Also, Rast, Jones, and Lee remind us that the response of a lake to a reduction in phosphorous load takes some time. This may be due to the periodic release of phosphorus from sediments.

The model loadings have been symbolized by the following relationship:

$$\frac{[L(P)/q_s]}{(1+\sqrt{\gamma_w})} = \text{mgs } P/m^3$$

where

$L(P)$ = areal annual phosphorous load, mgs $P/m^2/a$
q_s = mean depth/hydraulic residence time $\bar{z}/\gamma_w$
= meters/a
and γ_w = hydraulic residence time, years

The model effects have been presented by Rast, Jones, and Lee (1983) as chlorophyll A, secchi depth, and oxygen depletion rate. They found that most of the chlorophyll and secchi depth values were within a factor of 2 of the measured values, while nearly all predicted values were within a factor of 3. These findings indicate that these lake models, like stream models, leave much to be desired in improvement of correlation between cause and effect. Undoubtedly, refinement of input parameters in existing models is necessary. Data on a sufficient number of lakes is apparently lacking, and will be needed to refine existing model formations.

REFERENCES

Erdmann, J. B.: Systematic Diurnal Curve Analysis, *J. Water Poll. Control Fed.*, **51**:7, 1979.

Hines, W. G. et al.: Formulation and Use of Practical River Quality Assessment Models, *U.S. Geol. Survey Circular 715-B*, Second Printing, Washington, D.C., 1980.

Lee, G. F. et al.: Eutrophication of Waterbodies: Insights for an Age-Old Problem, *Environmental Science and Technology*, **12**:900, 1978.

O'Connor, D. J., and D. M. DiToro: Photosynthetic and Oxygen Balance in Streams, *J. San. Eng. Div., Proc.* ASCE, **96**:547, 1970.

Rast, W., and G. F. Lee: Summary Analysis of the North American OECD Eutrophication Project: Nutrient Loading-Lake Response Relationships and Trophic State Indices, U.S.E.P.A., 600/3-78-008, Corvallis, Oreg., 1978.

Rast, W., R. A. Jones, and G. F. Lee: Predictive Capability of U.S. OECD Phosphorous Loading-Eutrophication Response Models, *Federation Water Poll. Control Journal*, **55**(7):990, July 1983.

Velz, C. J.: Report of Natural Purification Capacities, Willamette River, *Nat. Council of Stream Improvement of the Pump, Paper and Paperboard Industries, Inc.*, School of Public Health, Univ. Michigan, Ann Arbor, Mich., 28 pp., 1951.

Weber, J. E., C. C. Kisiel, and L. Duckstein: On the Mismatch Between Data and Models of Hydrologic and Water Resource Systems, *Water Resources Bull.*, 9(6):1075–88, 1973.

SUMMARY

We have now reviewed the basic concepts that determine the quality of our watercourses: biology, chemistry, and hydrology. We have used our knowledge of mathematics to compute rates of depletion and renewal of oxygen, a vital natural stream resource. With these data it has been possible to predict resulting stream oxygen sag curves which define the overall condition of "the stream patient" as it meanders from its youth (headwaters) to its demise (oceans). In order to judge whether our "stream patient" is viable, useful, healthy, or sick, we studied generally accepted water quality criteria for the various major uses of water resources. A method was suggested for integrating many contaminant parameters into one index value to produce an overall water quality value. Since most waters play an important role just before they cease to exist as streams, we reviewed some basic concepts and methods for evaluating estuaries. It must have been apparent to the reader that proper evaluation of streams is not enough to ensure their survival and optimal use; they must be properly administered–and, therefore, we learned about some problems concerning ownership and control of streams. Your author made a plea for regional river basin management and proposed a system of marketing available resources. This, he feels, is an equitable, workable, and much needed resource allocation system. In fact, it may be the only method that over the long term will offer enough incentive to society to preserve our dwindling stream pollution carrying capacity. The reader was then subjected to an unadulterated illustration of a stream pollution evaluation problem. It is hoped that the reader learned that all theories do not hold true in actual practice, and that streams are complicated organisms requiring great knowledge, care, and patience in their proper evaluation.

With increasing demand for water, by reuse, many rivers pass either naturally or artificially through lakes. We reviewed some of the unique problems of adverse water quality arising from existence of these lakes. It will be up to us to devise means for protecting these valuable resources. One of the laws of nature is that water runs "downhill," and thus, the portion that is not consumed, evaporated, or leaked to underground aquifers finally reaches the coastal zone.

The contaminants in this water continue to exert residual effects on our seacoasts. Many of these impacts are adverse and remain as one of the great challenges to the environmental scientist of the 80s.

Accurate models which predict such adverse impacts must be devised so that we can apply bold and novel methods of abatement.

APPENDICES

APPENDIX A. Conversion Table. Second Feet to Million U.S. Gallons/Day (646,323 gallons per cfs)

cfs	.00	.01	.02	.03	.04	.05	.06	.07	.08	.09
0.0		.00646	.0129	.0194	.0259	.0323	.0388	.0452	.0517	.0582
0.1	.0646	.0711	.0776	.0840	.0905	.0969	.103	.110	.116	.123
0.2	.129	.136	.142	.149	.155	.162	.168	.175	.181	.187
0.3	.194	.200	.207	.213	.220	.226	.233	.239	.246	.252
0.4	.259	.265	.271	.278	.284	.291	.297	.304	.310	.317
0.5	.323	.330	.336	.343	.349	.355	.362	.368	.375	.381
0.6	.388	.394	.401	.407	.414	.420	.427	.433	.439	.446
0.7	.452	.459	.465	.472	.478	.485	.491	.498	.504	.511
0.8	.517	.524	.530	.536	.543	.549	.556	.562	.569	.575
0.9	.582	.588	.595	.601	.608	.614	.620	.627	.633	.640
1.0	.646	.653	.659	.666	.672	.679	.685	.692	.698	.704
1.1	.711	.717	.724	.730	.737	.743	.750	.756	.763	.769
1.2	.776	.782	.789	.795	.801	.808	.814	.821	.827	.834
1.3	.840	.847	.853	.860	.866	.873	.879	.885	.892	.898
1.4	.905	.911	.918	.924	.931	.937	.944	.950	.957	.963
1.5	.969	.976	.982	.989	.995	1.002	1.008	1.015	1.021	1.028
1.6	1.034	1.041	1.047	1.054	1.060	1.066	1.073	1.079	1.086	1.092
1.7	1.099	1.105	1.112	1.118	1.125	1.131	1.138	1.144	1.150	1.157
1.8	1.163	1.170	1.176	1.183	1.189	1.196	1.202	1.209	1.215	1.222
1.9	1.228	1.234	1.241	1.247	1.254	1.260	1.267	1.273	1.280	1.286
2.0	1.293	1.299	1.306	1.312	1.318	1.325	1.331	1.338	1.344	1.351
2.1	1.357	1.364	1.370	1.377	1.383	1.390	1.396	1.403	1.409	1.415
2.2	1.422	1.428	1.435	1.441	1.448	1.454	1.461	1.467	1.474	1.480
2.3	1.487	1.493	1.499	1.506	1.512	1.519	1.525	1.532	1.538	1.545
2.4	1.551	1.558	1.564	1.571	1.577	1.583	1.590	1.596	1.603	1.609

2.5	1.616	1.622	1.629	1.635	1.642	1.648	1.655	1.661	1.668	1.674
2.6	1.680	1.687	1.693	1.700	1.706	1.713	1.719	1.726	1.732	1.739
2.7	1.745	1.752	1.758	1.764	1.771	1.777	1.784	1.790	1.797	1.803
2.8	1.810	1.816	1.823	1.829	1.836	1.842	1.848	1.855	1.861	1.868
2.9	1.874	1.881	1.887	1.894	1.900	1.907	1.913	1.920	1.926	1.933
3.0	1.939	1.945	1.952	1.958	1.965	1.971	1.978	1.984	1.991	1.997
3.1	2.004	2.010	2.017	2.023	2.029	2.036	2.042	2.049	2.055	2.062
3.2	2.068	2.075	2.081	2.088	2.094	2.101	2.107	2.113	2.120	2.126
3.3	2.133	2.139	2.146	2.152	2.159	2.165	2.172	2.178	2.185	2.191
3.4	2.197	2.204	2.210	2.217	2.223	2.230	2.236	2.243	2.249	2.256
3.5	2.262	2.269	2.275	2.282	2.288	2.294	2.301	2.307	2.314	2.320
3.6	2.327	2.333	2.340	2.346	2.353	2.359	2.366	2.372	2.378	2.385
3.7	2.391	2.398	2.404	2.411	2.417	2.424	2.430	2.437	2.443	2.450
3.8	2.456	2.462	2.469	2.475	2.482	2.488	2.495	2.501	2.508	2.514
3.9	2.521	2.527	2.534	2.540	2.547	2.553	2.559	2.566	2.572	2.579
4.0	2.585	2.592	2.598	2.605	2.611	2.618	2.624	2.631	2.637	2.643
4.1	2.650	2.656	2.663	2.669	2.676	2.682	2.689	2.695	2.702	2.708
4.2	2.715	2.721	2.727	2.734	2.740	2.747	2.753	2.760	2.766	2.773
4.3	2.779	2.786	2.792	2.799	2.805	2.812	2.818	2.824	2.831	2.837
4.4	2.843	2.850	2.857	2.863	2.870	2.876	2.883	2.889	2.896	2.902
4.5	2.908	2.915	2.921	2.928	2.934	2.941	2.947	2.954	2.960	2.967
4.6	2.973	2.980	2.986	2.992	2.999	3.005	3.012	3.018	3.025	3.031
4.7	3.038	3.044	3.051	3.057	3.064	3.070	3.076	3.083	3.089	3.096
4.8	3.102	3.109	3.115	3.122	3.128	3.135	3.141	3.148	3.154	3.161
4.9	3.167	3.173	3.180	3.186	3.193	3.199	3.206	3.212	3.219	3.225
5.0	3.232	3.238	3.245	3.251	3.257	3.264	3.270	3.277	3.283	3.290
5.1	3.296	3.303	3.309	3.316	3.322	3.329	3.335	3.341	3.348	3.354

APPENDIX A. [*Continued*]

cfs	.00	.01	.02	.03	.04	.05	.06	.07	.08	.09
5.2	3.361	3.367	3.374	3.380	3.387	3.393	3.400	3.406	3.413	3.419
5.3	3.426	3.432	3.438	3.445	3.451	3.458	3.464	3.471	3.477	3.484
5.4	3.490	3.497	3.503	3.510	3.516	3.522	3.529	3.535	3.542	3.548
5.5	3.555	3.561	3.568	3.574	3.581	3.587	3.594	3.600	3.606	3.613
5.6	3.619	3.626	3.632	3.639	3.645	3.652	3.658	3.665	3.671	3.678
5.7	3.684	3.691	3.697	3.703	3.710	3.716	3.723	3.729	3.736	3.742
5.8	3.749	3.755	3.762	3.768	3.775	3.781	3.787	3.794	3.800	3.807
5.9	3.813	3.820	3.826	3.833	3.839	3.846	3.852	3.859	3.865	3.871
6.0	3.878	3.884	3.891	3.897	3.904	3.910	3.917	3.923	3.930	3.936
6.1	3.943	3.949	3.955	3.962	3.968	3.975	3.981	3.988	3.994	4.001
6.2	4.007	4.014	4.020	4.027	4.033	4.040	4.046	4.052	4.059	4.065
6.3	4.072	4.078	4.085	4.091	4.098	4.104	4.111	4.117	4.124	4.130
6.4	4.136	4.143	4.149	4.156	4.162	4.169	4.175	4.182	4.188	4.195
6.5	4.201	4.208	4.214	4.220	4.227	4.233	4.240	4.246	4.253	4.259
6.6	4.266	4.272	4.279	4.285	4.292	4.298	4.305	4.311	4.317	4.324
6.7	4.330	4.337	4.343	4.350	4.356	4.363	4.369	4.376	4.382	4.389
6.8	4.395	4.401	4.408	4.414	4.421	4.427	4.434	4.440	4.447	4.453
6.9	4.460	4.466	4.473	4.479	4.485	4.492	4.498	4.505	4.511	4.518
7.0	4.524	4.531	4.537	4.544	4.550	4.557	4.563	4.570	4.576	4.582
7.1	4.589	4.595	4.602	4.608	4.615	4.621	4.628	4.634	4.641	4.647
7.2	4.654	4.660	4.666	4.673	4.679	4.686	4.692	4.699	4.705	4.712
7.3	4.718	4.725	4.731	4.738	4.744	4.750	4.757	4.763	4.770	4.776
7.4	4.783	4.789	4.796	4.802	4.809	4.815	4.822	4.828	4.834	4.841
7.5	4.847	4.854	4.860	4.867	4.873	4.880	4.886	4.893	4.899	4.906
7.6	4.912	4.919	4.925	4.931	4.938	4.944	4.951	4.957	4.964	4.970
7.7	4.977	4.983	4.990	4.996	5.003	5.009	5.015	5.022	5.028	5.035
7.8	5.041	5.048	5.054	5.061	5.067	5.074	5.080	5.087	5.093	5.099
7.9	5.106	5.112	5.119	5.125	5.132	5.138	5.145	5.151	5.158	5.164
8.0	5.171	5.177	5.184	5.190	5.196	5.203	5.209	5.216	5.222	5.229

8.1	5.235	5.242	5.248	5.255	5.261	5.268	5.274	5.280	5.287	5.293
8.2	5.300	5.306	5.313	5.319	5.326	5.332	5.339	5.345	5.352	5.358
8.3	5.364	5.371	5.377	5.384	5.390	5.397	5.403	5.410	5.416	5.423
8.4	5.429	5.436	5.442	5.449	5.455	5.461	5.468	5.474	5.481	5.487
8.5	5.494	5.500	5.507	5.513	5.520	5.526	5.533	5.539	5.545	5.552
8.6	5.558	5.565	5.571	5.578	5.584	5.591	5.597	5.604	5.610	5.617
8.7	5.623	5.629	5.636	5.642	5.649	5.655	5.662	5.668	5.675	5.681
8.8	5.688	5.694	5.701	5.707	5.713	5.720	5.726	5.733	5.739	5.746
8.9	5.752	5.759	5.765	5.772	5.778	5.785	5.791	5.798	5.804	5.810
9.0	5.817	5.823	5.830	5.836	5.843	5.849	5.856	5.862	5.869	5.875
9.1	5.882	5.888	5.894	5.901	5.907	5.914	5.920	5.927	5.933	5.940
9.2	5.946	5.953	5.959	5.966	5.972	5.978	5.985	5.991	5.998	6.004
9.3	6.011	6.017	6.024	6.030	6.037	6.043	6.050	6.056	6.063	6.069
9.4	6.075	6.082	6.088	6.095	6.101	6.108	6.114	6.121	6.127	6.134
9	5.817	5.882	5.946	6.011	6.075	6.140	6.205	6.269	6.334	6.399
10	6.463	6.528	6.592	6.657	6.722	6.786	6.851	6.916	6.980	7.045
11	7.110	7.174	7.239	7.303	7.368	7.433	7.497	7.562	7.627	7.691
12	7.756	7.821	7.885	7.950	8.014	8.079	8.144	8.208	8.273	8.338
13	8.402	8.467	8.531	8.596	8.661	8.725	8.790	8.855	8.919	8.984
14	9.049	9.113	9.178	9.242	9.307	9.372	9.436	9.501	9.566	9.630
15	9.695	9.759	9.824	9.889	9.953	10.02	10.08	10.15	10.21	10.28
16	10.34	10.41	10.47	10.54	10.60	10.66	10.73	10.79	10.86	10.92
17	10.99	11.05	11.12	11.18	11.25	11.31	11.38	11.44	11.50	11.57
18	11.63	11.70	11.76	11.83	11.89	11.96	12.02	12.09	12.15	12.22
19	12.28	12.34	12.41	12.47	12.54	12.60	12.67	12.73	12.80	12.86
20	12.93	12.99	13.06	13.12	13.18	13.25	13.31	13.38	13.44	13.51

APPENDIX A. [*Continued*]

cfs	.00	.01	.02	.03	.04	.05	.06	.07	.08	.09
21	13.57	13.64	13.70	13.77	13.83	13.90	13.96	14.03	14.09	14.15
22	14.22	14.28	14.35	14.41	14.48	14.54	14.61	14.67	14.74	14.80
23	14.87	14.93	14.99	15.06	15.12	15.19	15.25	15.32	15.38	15.45
24	15.51	15.58	15.64	15.71	15.77	15.83	15.90	15.96	16.03	16.09
25	16.16	16.22	16.29	16.35	16.42	16.48	16.55	16.61	16.68	16.74
26	16.80	16.87	16.93	17.00	17.06	17.13	17.19	17.26	17.32	17.39
27	17.45	17.52	17.58	17.64	17.71	17.77	17.84	17.90	17.97	18.03
28	18.10	18.16	18.23	18.29	18.36	18.42	18.48	18.55	18.61	18.68
29	18.74	18.81	18.87	18.94	19.00	19.07	19.13	19.20	19.26	19.33
30	19.39	19.45	19.52	19.58	19.65	19.71	19.78	19.84	19.91	19.97
31	20.04	20.10	20.17	20.23	20.29	20.36	20.42	20.49	20.55	20.62
32	20.68	20.75	20.81	20.88	20.94	21.01	21.07	21.13	21.20	21.26
33	21.33	21.39	21.46	21.52	21.59	21.65	21.72	21.78	21.85	21.91
34	21.97	22.04	22.10	22.17	22.23	22.30	22.36	22.43	22.49	22.56
35	22.62	22.69	22.75	22.82	22.88	22.94	23.01	23.07	23.14	23.20
36	23.27	23.33	23.40	23.46	23.53	23.59	23.66	23.72	23.78	23.85
37	23.91	23.98	24.04	24.11	25.17	24.24	24.30	24.37	24.43	24.50
38	24.56	24.62	24.69	24.75	24.82	24.88	24.95	25.01	25.08	25.14

39	25.21	25.27	25.34	25.40	25.47	25.53	25.59	25.66	25.72	25.79
40	25.85	25.92	25.98	26.05	26.11	26.18	26.24	26.31	26.37	26.43
41	26.50	26.56	26.63	26.69	26.76	26.82	26.89	26.95	27.02	27.08
42	27.15	27.21	27.27	27.34	27.40	27.47	27.53	27.60	27.66	27.73
43	27.79	27.86	27.92	27.99	28.05	28.12	28.18	28.24	28.31	38.37
44	28.44	28.50	28.57	28.63	28.70	28.76	28.83	28.89	28.96	29.02
45	29.08	29.15	29.21	29.28	29.34	29.41	29.47	29.54	29.60	29.67
46	29.73	29.80	29.86	29.92	29.99	30.05	30.12	30.18	30.25	30.31
47	30.38	30.44	30.51	30.57	30.64	30.70	30.76	30.83	30.89	30.96
48	31.02	31.09	31.15	31.22	31.28	31.35	31.41	31.48	31.54	31.61
49	31.67	31.73	31.80	31.86	31.93	31.99	32.06	32.12	32.19	32.25
50	32.32	32.38	32.45	32.51	32.57	32.64	32.70	32.77	32.83	32.90
51	32.96	33.03	33.09	33.16	33.22	33.29	33.35	33.41	33.48	33.54
52	33.61	33.67	33.74	33.80	33.87	33.93	34.00	34.06	34.13	34.19
53	34.26	34.32	34.38	34.45	34.51	34.58	34.64	34.71	34.77	34.84
54	34.90	34.97	35.03	35.10	35.16	35.22	35.29	35.35	35.42	35.48
55	35.55	35.61	35.68	35.74	35.81	35.87	35.94	36.00	36.06	36.13
56	36.19	36.26	36.32	36.39	36.45	36.52	36.58	36.65	36.71	36.78
57	36.84	36.91	36.97	37.03	37.10	37.16	37.23	37.29	37.36	37.42
58	37.49	37.55	37.62	37.68	37.75	37.81	37.87	37.94	38.00	38.07

APPENDIX B-1. Dissolved Oxygen Saturation Values in Distilled Water. *[After Churchill, American Society of Civil Engineers Report 29, 1961. By permission of publishers.]*

Temp. °C	Dissolved oxygen as mg/l									
	0.0	0.1	0.2	0.3	0.4	0.5	0.6	0.7	0.8	0.9
0	14.65	14.61	14.57	14.53	14.49	14.45	14.41	14.37	14.33	14.29
1	14.25	14.21	14.17	14.13	14.09	14.05	14.02	13.98	13.94	13.90
2	13.86	13.82	13.79	13.75	13.71	13.68	13.64	13.60	13.56	13.53
3	13.49	13.46	13.42	13.38	13.35	13.31	13.28	13.24	13.20	13.17
4	13.13	13.10	13.06	13.03	13.00	12.96	12.93	12.89	12.86	12.82
5	12.79	12.76	12.72	12.69	12.66	12.62	12.59	12.56	12.53	12.49
6	12.46	12.43	12.40	12.36	12.33	12.30	12.27	12.24	12.21	12.18
7	12.14	12.11	12.08	12.05	12.02	11.99	11.96	11.93	11.90	11.87
8	11.84	11.81	11.78	11.75	11.72	11.70	11.67	11.64	11.61	11.58
9	11.55	11.52	11.49	11.47	11.44	11.41	11.38	11.35	11.33	11.30
10	11.27	11.24	11.22	11.19	11.16	11.14	11.11	11.08	11.06	11.03
11	11.00	10.98	10.95	10.93	10.90	10.87	10.85	10.82	10.80	10.77
12	10.75	10.72	10.70	10.67	10.65	10.62	10.60	10.57	10.55	10.52
13	10.50	10.48	10.45	10.43	10.40	10.38	10.36	10.33	10.31	10.28
14	10.26	10.24	10.22	10.19	10.17	10.15	10.12	10.10	10.08	10.06
15	10.03	10.01	9.99	9.97	9.95	9.92	9.90	9.88	9.86	9.84
16	9.82	9.79	9.77	9.75	9.73	9.71	9.69	9.67	9.65	9.63
17	9.61	9.58	9.56	9.54	9.52	9.50	9.48	9.46	9.44	9.42
18	9.40	9.38	9.36	9.34	9.32	9.30	9.29	9.27	9.25	9.23
19	9.21	9.19	9.17	9.15	9.13	9.12	9.10	9.08	9.06	9.04
20	9.02	9.00	8.98	8.97	8.95	8.93	8.91	8.90	8.88	8.86
21	8.84	8.82	8.81	8.79	8.77	8.75	8.74	8.72	8.70	8.68
22	8.67	8.65	8.63	8.62	8.60	8.58	8.56	8.55	8.53	8.52
23	8.50	8.48	8.46	8.45	8.43	8.42	8.40	8.38	8.37	8.35
24	8.33	8.32	8.30	8.29	8.27	8.25	8.24	8.22	8.21	8.19
25	8.18	8.16	8.14	8.13	8.11	8.10	8.08	8.07	8.05	8.04
26	8.02	8.01	7.99	7.98	7.96	7.95	7.93	7.92	7.90	7.89
27	7.87	7.86	7.84	7.83	7.81	7.80	7.78	7.77	7.75	7.74
28	7.72	7.71	7.69	7.68	7.66	7.65	7.64	7.62	7.61	7.59
29	7.58	7.56	7.55	7.54	7.52	7.51	7.49	7.48	7.47	7.45
30	7.44	7.42	7.41	7.40	7.38	7.37	7.35	7.34	7.32	7.31

APPENDIX B-2. Solubility of Oxygen in Water. (*From Water and Sewage Works,* **October 1957. By permission of publishers.**)

For the solubility of atmospheric oxygen in water at temperatures between 0° and 35°C, Hatfield (1941) gives the equation

$$S = \frac{0.678\,(P - \mu)}{t + 35}$$

where

S = solubility of oxygen in ppm
P = barometric pressure in mm of mercury
μ = vapor pressure of water at t°C.

A nomograph by **D. S. Davis,** Professor of Engineering
University of Alabama, Tuscaloosa, Ala.

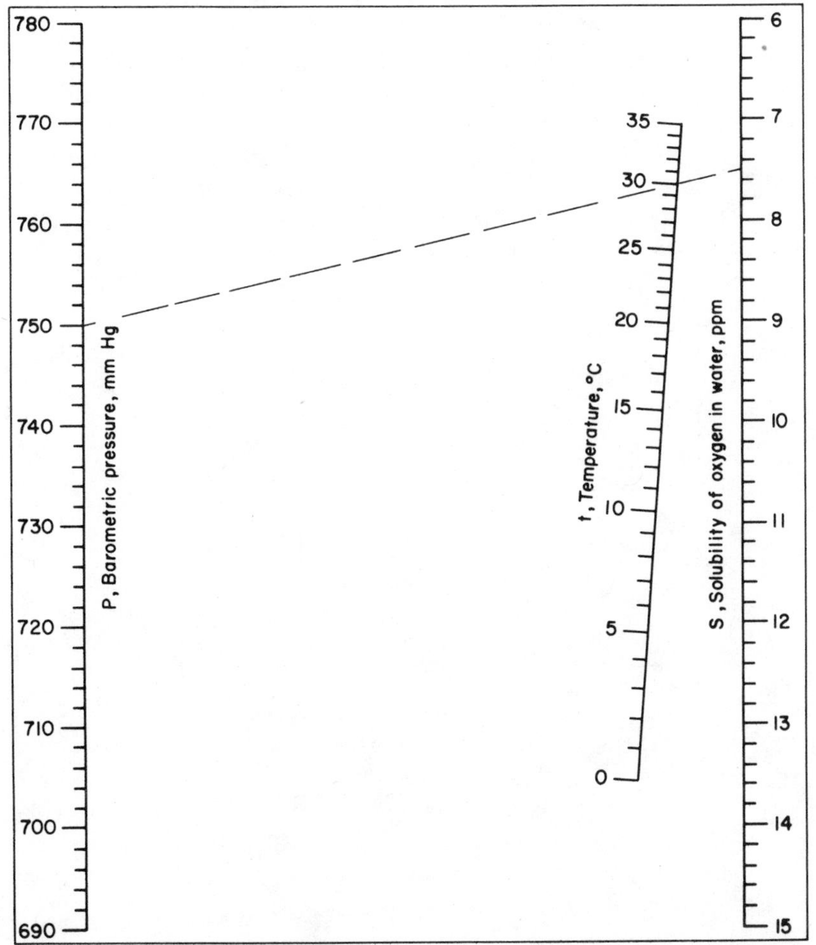

(*Text continued on p. 418*)

The nomograph, constructed through recourse to line coordinate methods, (Davis, 1955) enables rapid and accurate solution of the equation. The use of the chart is illustrated as follows: What is the solubility of oxygen in water at 30°C, in equilibrium with air, when the barometric pressure is 750 mm of mercury? Connect 750 on the *P*-scale and 30 on the *t*-scale with a straight line. Note the intersection with the *S*-scale at solubility of 7.5 ppm of oxygen.

APPENDIX C. Values of 10^{-kt} for Various Values of k and t.

k/t	1	2	3	4	5	6	7	8	9	10	11	12	13	14	15	k
.01	.9772	.9550	.9333	.9120	.8912	.8710	.8511	.8318	.8128	.7943	.7763	.7586	.7413	.7244	.7080	.01
.02	.9550	.9120	.8710	.8318	.7943	.7586	.7244	.6918	.6607	.6310	.6026	.5754	.5495	.5248	.5012	.02
.03	.9333	.8710	.8128	.7586	.7080	.6607	.6166	.5754	.5370	.5012	.4677	.4365	.4074	.3802	.3548	.03
.04	.9120	.8318	.7586	.6918	.6310	.5754	.5248	.4786	.4365	.3981	.3631	.3311	.3020	.2754	.2512	.04
.05	.8912	.7943	.7080	.6310	.5623	.5012	.4467	.3981	.3548	.3162	.2818	.2512	.2239	.1995	.1778	.05
.06	.8710	.7586	.6607	.5754	.5012	.4365	.3802	.3311	.2884	.2512	.2188	.1905	.1660	.1445	.1259	.06
.07	.8511	.7244	.6166	.5248	.4467	.3802	.3236	.2754	.2344	.1995	.1698	.1445	.1230	.1047	.08912	.07
.08	.8318	.6918	.5754	.4786	.3981	.3311	.2754	.2291	.1905	.1585	.1318	.1096	.09120	.07586	.06310	.08
.09	.8128	.6607	.5370	.4365	.3548	.2884	.2344	.1905	.1549	.1259	.1023	.08318	.06761	.05495	.04467	.09
.10	.7943	.6310	.5012	.3981	.3162	.2512	.1995	.1585	.1259	.1000	.07943	.06310	.05012	.03981	.03162	.10
.11	.7763	.6026	.4677	.3631	.2818	.2188	.1698	.1318	.1023	.07943	.06166	.04786	.03715	.02884	.02239	.11
.12	.7586	.5754	.4365	.3311	.2512	.1905	.1445	.1096	.08318	.06310	.04786	.03631	.02754	.02089	.01585	.12
.13	.7413	.5495	.4074	.3020	.2239	.1660	.1230	.09120	.06761	.05012	.03715	.02754	.02042	.01514	.01122	.13
.14	.7244	.5248	.3802	.2754	.1995	.1445	.1047	.07586	.05495	.03981	.02884	.02089	.01514	.01096	.007943	.14
.15	.7080	.5012	.3548	.2512	.1778	.1259	.08912	.06310	.04467	.03162	.02239	.01585	.01122	.007943	.005623	.15
.16	.6918	.4786	.3311	.2291	.1585	.1096	.07586	.05248	.03631	.02512	.01738	.01202	.008318	.005754	.003981	.16
.17	.6761	.4571	.3090	.2089	.1413	.09550	.06457	.04365	.02951	.01995	.01349	.009120	.006166	.004169	.002818	.17
.18	.6607	.4365	.2884	.1905	.1259	.08318	.05495	.03631	.02399	.01585	.01047	.006918	.004571	.003020	.001995	.18
.19	.6457	.4169	.2692	.1738	.1122	.07244	.04677	.03020	.01950	.01259	.008128	.005248	.003388	.002188	.001413	.19
.20	.6310	.3981	.2512	.1585	.1000	.06310	.03981	.02512	.01585	.01000	.006310	.003981	.002512	.001585	.001000	.20
.21	.6166	.3802	.2344	.1445	.08912	.05495	.03388	.02089	.01288	.007943	.004898	.003020	.001862	.001148	.000708	.21
.22	.6026	.3631	.2188	.1318	.07943	.04786	.02884	.01738	.01047	.006310	.003802	.002291	.001380	.000832	.000501	.22
.23	.5888	.3467	.2042	.1202	.07080	.04169	.02455	.01445	.008511	.005012	.002951	.001738	.001023	.000603	.000355	.23
.24	.5754	.3311	.1905	.1096	.06310	.03631	.02089	.01202	.006918	.003981	.002291	.001318	.000759	.000436	.000251	.24
.25	.5623	.3162	.1778	.1000	.05623	.03162	.01778	.01000	.005623	.003162	.001778	.001000	.000562	.000316	.000178	.25
.26	.5495	.3020	.1660	.09120	.05012	.02754	.01514	.008318	.004571	.002512	.001380	.000759	.000417	.000229	.000126	.26
.27	.5370	.2884	.1549	.08318	.04467	.02399	.01288	.006918	.003715	.001995	.001072	.000575	.000309	.000166	.000089	.27
.28	.5248	.2754	.1445	.07586	.03981	.02089	.01096	.005754	.003020	.001585	.000832	.000436	.000229	.000120	.000063	.28
.29	.5129	.2630	.1349	.06918	.03548	.01820	.009333	.004786	.002455	.001259	.000646	.000331	.000170	.000087	.000045	.29
.30	.5012	.2512	.1259	.06310	.03162	.01585	.007943	.003981	.001995	.001000	.000501	.000251	.000126	.000063	.000032	.30
.31	.4898	.2399	.1175	.05754	.02818	.01380	.006761	.003311	.001622	.000794	.000389	.000190	.000093	.000046	.000022	.31
.32	.4786	.2291	.1096	.05248	.02512	.01202	.005754	.002754	.001318	.000631	.000302	.000144	.000069	.000033	.000016	.32
.33	.4677	.2188	.1023	.04786	.02239	.01047	.004898	.002291	.001072	.000501	.000234	.000110	.000051	.000024	.000011	.33

APPENDIX C. [*Continued*]

k/t	1	2	3	4	5	6	7	8	9	10	11	12	13	14	15	k
.34	.4571	.2089	.09550	.04365	.01995	.009120	.004169	.001905	.000871	.000398	.000182	.000083	.000038	.000017	.000008	.34
.35	.4467	.1995	.08912	.03981	.01778	.007943	.003548	.001585	.000708	.000316	.000141	.000063	.000028	.000013		.35
.36	.4365	.1905	.08318	.03631	.01585	.006918	.003020	.001318	.000575	.000251	.000110	.000048	.000021	.000009		.36
.37	.4266	.1820	.07763	.03311	.01413	.006026	.002570	.001096	.000468	.000200	.000085	.000036	.000015			.37
.38	.4169	.1738	.07244	.03020	.01259	.005248	.002188	.000912	.000380	.000158	.000066	.000028	.000011			.38
.39	.4074	.1660	.06761	.02754	.01122	.004571	.001862	.000759	.000309	.000126	.000051	.000021	.000008			.39
.40	.3981	.1585	.06310	.02512	.01000	.003981	.001585	.000631	.000251	.000100	.000040	.000016				.40
.41	.3890	.1514	.05888	.02291	.008912	.003467	.001349	.000525	.000204	.000079	.000031	.000012				.41
.42	.3802	.1445	.05495	.02089	.007943	.003020	.001148	.000437	.000166	.000063	.000024	.000009				.42
.43	.3715	.1380	.05129	.01905	.007080	.002630	.0009772	.000363	.000135	.000050	.000019					.43
.44	.3631	.1318	.04786	.01738	.006310	.002291	.000832	.000302	.000110	.000040	.000014					.44
.45	.3548	.1259	.04467	.01585	.005623	.001995	.000708	.000251	.000089	.000032	.000011					.45
.46	.3467	.1202	.04169	.01445	.005012	.001738	.000603	.000209	.000072	.000025	.0000087					.46
.47	.3388	.1148	.03890	.01318	.004467	.001514	.000513	.000174	.000059	.000020						.47
.48	.3311	.1096	.03631	.01202	.003981	.001318	.000436	.000145	.000048	.000016						.48
.49	.3236	.1047	.03388	.01096	.003548	.001148	.000372	.000120	.000039	.000013						.49
.50	.3162	.1000	.03162	.01000	.003162	.001000	.000316	.000100	.000037	.000010						.50
.51	.3090	.09550	.02951	.009120	.002818	.000871	.000269	.000083	.000026							.51
.52	.3020	.09120	.02754	.008318	.002512	.000759	.000229	.000069	.000021							.52
.53	.2951	.08710	.02570	.007586	.002239	.000661	.000195	.000058	.000017							.53
.54	.2884	.08318	.02399	.006918	.001995	.000575	.000166	.000049	.000014							.54
.55	.2818	.07943	.02239	.006310	.001778	.000501	.000141	.000040	.000011							.55
.56	.2754	.07586	.02089	.005754	.001585	.000436	.000120	.000033	.000009							.56
.57	.2692	.07244	.01950	.005248	.001413	.000380	.000102	.000028								.57
.58	.2630	.06918	.01820	.004786	.001259	.000331	.000087	.000023								.58
.59	.2570	.06607	.01698	.004365	.001122	.000288	.000074	.000019								.59
.60	.2512	.06310	.01585	.003981	.001000	.000251	.000063	.000016								.60
.61	.2455	.06026	.01479	.003631	.000891	.000219	.000054	.000013								.61
.62	.2399	.05754	.01380	.003311	.000794	.000190	.000045	.000011								.62
.63	.2344	.05495	.01288	.003020	.000708	.000166	.000039	.000009								.63
.64	.2291	.05248	.01202	.002754	.000631	.000144	.000033									.64
.65	.2239	.05012	.01122	.002512	.000562	.000126	.000028									.65

.66	.2188	.04786	.01047	.02291	.000501	.000110	.000024		.66
.67	.2138	.04571	.009772	.002089	.000447	.000096	.000020		.67
.68	.2089	.04365	.009120	.001905	.000398	.000083	.000017		.68
.69	.2042	.04169	.008511	.001738	.000355	.000072	.000015		.69
.70	.1995	.03981	.007943	.001585	.000316	.000063	.000013		.70
.71	.1950	.03802	.007413	.001445	.000282	.000055	.000011		.71
.72	.1905	.03631	.006918	.001318	.000251	.000048	.000009		.72
.73	.1862	.03467	.006457	.001202	.000224	.000042			.73
.74	.1820	.03311	.006026	.001096	.000120	.000036			.74
.75	.1778	.03162	.005623	.001000	.000178	.000032			.75
.76	.1738	.03020	.005248	.000912	.000158	.000028			.76
.77	.1698	.02884	.004898	.000832	.000141	.000024			.77
.78	.1660	.02754	.004571	.000759	.000126	.000021			.78
.79	.1622	.02630	.004266	.000692	.000112	.000018			.79
.80	.1585	.02512	.003981	.000631	.000100	.000016			.80
.81	.1549	.02399	.003715	.000575	.000089	.000014			.81
.82	.1514	.02291	.003467	.000525	.000079	.000012			.82
.83	.1479	.02188	.003236	.000479	.000071	.000010			.83
.84	.1445	.02089	.003020	.000436	.000063	.000009			.84
.85	.1413	.01995	.002818	.000398	.000056				.85
.86	.1380	.01905	.002630	.000363	.000050				.86
.87	.1349	.01820	.002455	.000331	.000045				.87
.88	.1318	.01738	.002291	.000302	.000040				.88
.89	.1288	.01660	.002138	.000275	.000035				.89
.90	.1259	.01585	.001995	.000251	.000032				.90
.91	.1230	.01514	.001862	.000229	.000028				.91
.92	.1202	.01445	.001738	.000209	.000025				.92
.93	.1175	.01380	.001622	.000190	.000022				.93
.94	.1148	.01318	.001514	.000174	.000020				.94
.95	.1122	.01259	.001413	.000158	.000018				.95
.96	.1096	.01202	.001318	.000144	.000016				.96
.97	.1072	.01148	.001230	.000132	.000014				.97
.98	.1047	.01096	.001148	.000120	.000013				.98
.99	.1023	.01047	.001072	.000110	.000011				.99
1.00	.1000	.01000	.001000	.000100	.000010				1.00

APPENDIX D. Values of $1 - 10^{-kt}$ for Various Values of k and t.

k/t	1	2	3	4	5	6	7	8	9	10	11	12	13	14	15	k
.01	.0228	.0450	.0667	.0880	.1088	.1290	.1489	.1682	.1872	.2057	.2237	.2414	.2587	.2756	.2920	.01
.02	.0450	.0880	.1290	.1682	.2057	.2414	.2756	.3082	.3393	.3690	.3974	.4246	.4505	.4752	.4988	.02
.03	.0667	.1290	.1872	.2414	.2920	.3393	.3834	.4246	.4630	.4988	.5323	.5635	.5926	.6198	.6452	.03
.04	.0880	.1682	.2414	.3082	.3690	.4246	.4752	.5214	.5635	.6019	.6369	.6689	.6980	.7246	.7488	.04
.05	.1088	.2057	.2920	.3690	.4377	.4988	.5533	.6019	.6442	.6838	.7182	.7488	.7761	.8005	.8222	.05
.06	.1290	.2414	.3393	.4246	.4988	.5635	.6198	.6689	.7116	.7488	.7812	.8095	.8340	.8555	.8741	.06
.07	.1489	.2756	.3834	.4752	.5533	.6198	.6764	.7246	.7656	.8005	.8302	.8555	.8770	.8953	.91088	.07
.08	.1682	.3082	.4246	.4214	.6019	.6689	.7246	.7709	.8095	.8415	.8682	.8904	.90880	.92414	.93690	.08
.09	.1872	.3393	.4630	.5635	.6452	.7116	.7656	.8095	.8451	.8741	.8977	.91682	.93239	.94505	.95533	.09
.10	.2057	.3690	.4988	.6019	.6838	.7488	.8005	.8451	.8741	.9000	.92057	.93690	.94988	.96019	.96838	.10
.11	.2237	.3974	.5323	.6369	.7182	.7812	.8302	.8682	.8977	.92057	.93834	.95214	.96285	.97116	.97761	.11
.12	.2414	.4246	.5635	.6689	.7488	.8095	.8555	.8904	.91692	.93690	.95214	.96369	.97246	.97911	.98415	.12
.13	.2587	.4505	.5926	.6980	.7761	.8340	.8770	.90880	.93239	.94988	.96285	.97246	.97958	.98486	.98878	.13
.14	.2756	.4752	.6198	.7246	.8005	.8555	.8953	.92414	.94505	.96019	.97116	.97911	.98486	.98904	.992057	.14
.15	.2920	.4988	.6452	.7488	.8222	.8741	.91088	.93690	.95533	.96838	.97761	.98415	.98878	.992057	.994377	.15
.16	.3082	.5214	.6689	.7709	.8415	.8904	.92414	.94752	.96369	.97488	.98262	.98798	.991682	.994246	.996019	.16
.17	.3239	.5429	.6910	.7911	.8587	.90450	.93543	.95635	.97049	.98005	.98651	.99088	.993834	.995831	.997182	.17
.18	.3393	.5635	.7116	.8095	.8741	.91682	.94505	.96369	.97601	.98415	.98953	.993082	.995429	.996980	.998005	.18
.19	.3543	.5831	.7308	.8262	.8878	.92756	.95323	.96980	.98050	.98741	.991872	.994752	.996612	.997812	.998587	.19
.20	.3690	.6019	.7488	.8415	.9000	.93690	.96019	.97488	.98415	.99000	.993690	.996019	.997488	.998415	.999000	.20
.21	.3834	.6198	.7656	.8555	.91088	.94505	.96612	.97911	.98712	.992057	.995192	.996980	.998138	.998852	.999292	.21
.22	.3974	.6369	.7812	.8682	.92057	.95214	.97116	.98262	.98953	.993690	.996198	.997709	.998620	.999168	.999499	.22
.23	.4112	.6533	.7958	.8798	.92920	.95831	.97545	.98555	.991489	.994988	.997049	.998262	.998977	.999397	.999645	.23
.24	.4246	.6689	.8095	.8904	.93690	.96369	.97911	.98798	.993082	.996019	.997709	.998682	.999241	.999564	.999749	.24
.25	.4377	.6838	.8222	.9000	.94377	.96838	.98222	.99000	.994377	.996838	.998222	.999000	.999438	.999684	.999822	.25
.26	.4505	.6980	.8340	.90880	.94988	.97246	.98486	.991682	.995429	.997488	.998620	.9992414	.999583	.999771	.999874	.26
.27	.4630	.7116	.8451	.91682	.95533	.97601	.98712	.993082	.996285	.998005	.998928	.999425	.999691	.999834	.999911	.27
.28	.4752	.7246	.8555	.92414	.96019	.97911	.98904	.994246	.996980	.998415	.999168	.999563	.999771	.999880		.28
.29	.4871	.7370	.8651	.93082	.96452	.98180	.990667	.995214	.997545	.998741	.999354	.999669	.999830	.999913		.29
.30	.4988	.7488	.8741	.93690	.96838	.98415	.992057	.996019	.998005	.999000	.999499	.999749	.999874			.30
.31	.5102	.7601	.8825	.94246	.97182	.98620	.993239	.996689	.998378	.999206	.999611	.999810	.999907			.31
.32	.5214	.7709	.8904	.94752	.97488	.98798	.994246	.997246	.998682	.999369	.999698	.999856				.32

.33	.5323	.7812	.8977	.95214	.97761	.98953	.995102	.997709	.998928	.999499	.999766	.999890	.33
.34	.5429	.7911	.90450	.95635	.98005	.990880	.995831	.998095	.999129	.999602	.999818	.999917	.34
.35	.5533	.8005	.91088	.96019	.98222	.992057	.996452	.998415	.999292	.999684	.999859		.35
.36	.5635	.8095	.91682	.96369	.98415	.993082	.996980	.998682	.999425	.999749	.999890		.36
.37	.5734	.8180	.92237	.96689	.98587	.993974	.997430	.998904	.999532	.999800	.999915		.37
.38	.5831	.8262	.92756	.96980	.98741	.994752	.997812	.999088	.999620	.999842			.38
.39	.5926	.8340	.93239	.97246	.98878	.995429	.998138	.999241	.999681	.999874			.39
.40	.6019	.8415	.93690	.97488	.99000	.996019	.998415	.999369	.999749	.999900			.40
.41	.6110	.8486	.94112	.97709	.991088	.996533	.998651	.999475	.999796				.41
.42	.6198	.8555	.94505	.97911	.992057	.996980	.998852	.999564	.999834				.42
.43	.6285	.8620	.94871	.98095	.992920	.997370	.999023	.999637	.999865				.43
.44	.6369	.8682	.95214	.98262	.993690	.997709	.999168	.999698	.999890				.44
.45	.6452	.8741	.95533	.98415	.994377	.998005	.999292	.999749	.999919				.45
.46	.6533	.8798	.95831	.98555	.994988	.998262	.999397	.999791					.46
.47	.6612	.8852	.96110	.98682	.995533	.998486	.999487	.999826					.47
.48	.6689	.8904	.96369	.98798	.996019	.998682	.999564	.999856					.48
.49	.6764	.8953	.96612	.98904	.996452	.998852	.999628	.999880					.49
.50	.6838	.9000	.96838	.99000	.996838	.999000	.999684	.999900					.50
.51	.6910	.90450	.97049	.990880	.997182	.999129	.999731						.51
.52	.6980	.90880	.97246	.991682	.997488	.999241	.999771						.52
.53	.7049	.91290	.94730	.992414	.997761	.999339	.999805						.53
.54	.7116	.91682	.97601	.993082	.998005	.999425	.999834						.54
.55	.7182	.92057	.97761	.993690	.998222	.999499	.999859						.55
.56	.7246	.92414	.97911	.994246	.998415	.999563	.999880						.56
.57	.7308	.92756	.98050	.994752	.998587	.999620	.999898						.57
.58	.7370	.93082	.98180	.995214	.998741	.999669	.999913						.58
.59	.7430	.93393	.98302	.995635	.998878	.999712							.59
.60	.7488	.93690	.98415	.996019	.999000	.999749							.60
.61	.7545	.93974	.98521	.996369	.999109	.999781							.61
.62	.7601	.94246	.98620	.996689	.999206	.999810							.62
.63	.7656	.94505	.98712	.996980	.999292	.999834							.63
.64	.7709	.94752	.98798	.997246	.999369	.999856							.64
.65	.7761	.94988	.98878	.997488	.999438	.999874							.65

APPENDIX D. [*Continued*]

k/t	1	2	3	4	5	6	7	8	9	10	11	12	13	14	15	k
.66	.7812	.95214	.98953	.997709	.999499	.999890										.66
.67	.7862	.95429	.990228	.997911	.999553	.999905										.67
.68	.7911	.95635	.990880	.998095	.999602											.68
.69	.7958	.95831	.991489	.998262	.999645											.69
.70	.8005	.96019	.992057	.998415	.999684											.70
.71	.8050	.96198	.992587	.998555	.999718											.71
.72	.8095	.96369	.993082	.998682	.999749											.72
.73	.8138	.96533	.993543	.998798	.999776											.73
.74	.8180	.96689	.993974	.998904	.999801											.74
.75	.8222	.96838	.994377	.999000	.999822											.75
.76	.8262	.96980	.994752	.999088	.999842											.76
.77	.8302	.97116	.995102	.999168	.999859											.77
.78	.8340	.97246	.995429	.999241	.999874											.78
.79	.8378	.97370	.995734	.999308	.999888											.79
.80	.8415	.97488	.996019	.999369	.999900											.80
.81	.8451	.97601	.996285	.999425												.81
.82	.8486	.97709	.996533	.999475												.82
.83	.8521	.97812	.996764	.999521												.83
.84	.8555	.97911	.996980	.999564												.84
.85	.8587	.98005	.997182	.999602												.85
.86	.8620	.98095	.997370	.999637												.86
.87	.8651	.98180	.997545	.999669												.87
.88	.8682	.98262	.997709	.999698												.88
.89	.8712	.98340	.997862	.999725												.89
.90	.8741	.98415	.998005	.999749												.90
.91	.8770	.98486	.998138	.999771												.91
.92	.8798	.98555	.998262	.999791												.92
.93	.8825	.98620	.998378	.999810												.93
.94	.8852	.98682	.998486	.999826												.94
.95	.8878	.98741	.998587	.999842												.95
.96	.8904	.98798	.998682	.999856												.96
.97	.8928	.98852	.998770	.999868												.97
.98	.8953	.98904	.998852	.999880												.98
.99	.8977	.98953	.998928	.999890												.99
1.00	.9000	.99000	.999000	.999900												1.00

APPENDIX E. Ratios of BOD's for Various Values of *k*.

k	$\frac{y1}{y2}$	$\frac{y1}{y3}$	$\frac{y1}{y4}$	$\frac{y1}{y5}$	$\frac{y1}{y6}$	$\frac{y2}{y3}$	$\frac{y2}{y4}$	$\frac{y2}{y5}$	$\frac{y2}{y6}$	$\frac{y3}{y4}$	$\frac{y3}{y5}$	$\frac{y3}{y6}$	$\frac{y4}{y5}$	$\frac{y4}{y6}$	$\frac{y5}{y6}$	k
.01	.5067	.3418	.2591	.2096	.1767	.6747	.5114	.4136	.3488	.7580	.6130	.5170	.8088	.6822	.8434	.01
.02	.5114	.3488	.2675	.2188	.1864	.6822	.5232	.4278	.3645	.7669	.6271	.5344	.8177	.6968	.8521	.02
.03	.5170	.3563	.2763	.2284	.1966	.6891	.5344	.4418	.3802	.7755	.6411	.5517	.8267	.7115	.8606	.03
.04	.5232	.3645	.2855	.2385	.2072	.6968	.5457	.4558	.3961	.7832	.6542	.5685	.8352	.7258	.8690	.04
.05	.5289	.3726	.2948	.2486	.2181	.7044	.5574	.4700	.4124	.7913	.6671	.5854	.8430	.7398	.8775	.05
.06	.5344	.3802	.3038	.2586	.2289	.7115	.5685	.4840	.4284	.7991	.6802	.6021	.8512	.7535	.8852	.06
.07	.5403	.3884	.3133	.2691	.2402	.7188	.5800	.4981	.4446	.8068	.6929	.6186	.8588	.7667	.8927	.07
.08	.5457	.3961	.3226	.2794	.2514	.7258	.5911	.5120	.4608	.8143	.7054	.6348	.8662	.7795	.8998	.08
.09	.5517	.4043	.3322	.2901	.2631	.7328	.6021	.5259	.4768	.8216	.7176	.6506	.8734	.7919	.9067	.09
.10	.5574	.4124	.3418	.3008	.2747	.7398	.6130	.5396	.4929	.8287	.7294	.6661	.8802	.8038	.9132	.10
.11	.5629	.4202	.3512	.3115	.2864	.7466	.6240	.5533	.5087	.8358	.7412	.6814	.8868	.8153	.9194	.11
.12	.5685	.4284	.3609	.3224	.2982	.7535	.6348	.5670	.5245	.8424	.7525	.6961	.8933	.8263	.9250	.12
.13	.5742	.4366	.3706	.3333	.3102	.7602	.6454	.5805	.5402	.8490	.7636	.7106	.8994	.8369	.9306	.13
.14	.5800	.4446	.3803	.3443	.3222	.7667	.6558	.5936	.5555	.8554	.7743	.7245	.9052	.8470	.9357	.14
.15	.5854	.4526	.3900	.3551	.3341	.7731	.6661	.6067	.5706	.8616	.7847	.7281	.9107	.8566	.9406	.15
.16	.5911	.4608	.3998	.3662	.3461	.7795	.6764	.6196	.5856	.8677	.7949	.7512	.9161	.8658	.9451	.16
.17	.5966	.4687	.4094	.3772	.3581	.7857	.6862	.6322	.6002	.8735	.8047	.7640	.9213	.8746	.9494	.17
.18	.6021	.4768	.4191	.3882	.3701	.7919	.6961	.6447	.6146	.8791	.8141	.7762	.9261	.8829	.9534	.18
.19	.6076	.4848	.4288	.3991	.3820	.7979	.7058	.6568	.6286	.8845	.8232	.7879	.9306	.8907	.9571	.19
.20	.6130	.4928	.4385	.4100	.3938	.8038	.7153	.6688	.6424	.8898	.8320	.7992	.9350	.8982	.9606	.20
.21	.6186	.5008	.4482	.4209	.4057	.8096	.7245	.6804	.6558	.8949	.8410	.8101	.9392	.9052	.9638	.21
.22	.6240	.5087	.4577	.4317	.4174	.8153	.7336	.6918	.6689	.8998	.8486	.8205	.9431	.9118	.9668	.22
.23	.6294	.5167	.4674	.4425	.4291	.8209	.7426	.7031	.6817	.9045	.8564	.8304	.9468	.9181	.9696	.23
.24	.6348	.5245	.4769	.4532	.4406	.8263	.7512	.7140	.6941	.9091	.8640	.8400	.9504	.9239	.9722	.24
.25	.6401	.5324	.3863	.4638	.4520	.8317	.7598	.7245	.7061	.9136	.8712	.8490	.9536	.9294	.9746	.25

APPENDIX E. *(Continued)*

k	$\frac{y1}{y2}$	$\frac{y1}{y3}$	$\frac{y1}{y4}$	$\frac{y1}{y5}$	$\frac{y1}{y6}$	$\frac{y2}{y3}$	$\frac{y2}{y4}$	$\frac{y2}{y5}$	$\frac{y2}{y6}$	$\frac{y3}{y4}$	$\frac{y3}{y5}$	$\frac{y3}{y6}$	$\frac{y4}{y5}$	$\frac{y4}{y6}$	$\frac{y5}{y6}$	k
.26	.6454	.5402	.4957	.4743	.4632	.8369	.7680	.7348	.7178	.9177	.8780	.8576	.9568	.9345	.9768	.26
.27	.6506	.5479	.5050	.4846	.4744	.8420	.7762	.7449	.7291	.9218	.8846	.8659	.9597	.9388	.9788	.27
.28	.6558	.5555	.5142	.4949	.4853	.8470	.7841	.7546	.7400	.9257	.8910	.8738	.9624	.9438	.9807	.28
.29	.6609	.5630	.5233	.5050	.4961	.8519	.7918	.7641	.7507	.9294	.8969	.8811	.9651	.9481	.9824	.29
.30	.6661	.5706	.5324	.5151	.5068	.8566	.7992	.7732	.7608	.9330	.9026	.8882	.9675	.9520	.9840	.30
.31	.6712	.5781	.5413	.5250	.5173	.8613	.8065	.7821	.7707	.9364	.9081	.8948	.9698	.9556	.9854	.31
.32	.6764	.5856	.5503	.5348	.5278	.8658	.8136	.7908	.7803	.9397	.9133	.9012	.9719	.9590	.9867	.32
.33	.6814	.5930	.5590	.5445	.5379	.8702	.8205	.7991	.7895	.9428	.9182	.9072	.9739	.9622	.9880	.33
.34	.6862	.6002	.5677	.5540	.5479	.8746	.8272	.8072	.7984	.9458	.9229	.9128	.9758	.9652	.9891	.34
.35	.6912	.6074	.5762	.5633	.5577	.8788	.8337	.8150	.8069	.9486	.9274	.9182	.9776	.9679	.9901	.35
.36	.6961	.6146	.5847	.5726	.5674	.8829	.8400	.8225	.8151	.9513	.9316	.9232	.9792	.9704	.9910	.36
.37	.7010	.6216	.5930	.5816	.5769	.8868	.8460	.8297	.8230	.9540	.9356	.9280	.9807	.9728	.9918	.37
.38	.7058	.6286	.6012	.5905	.5862	.8907	.8519	.8367	.8306	.9564	.9394	.9324	.9822	.9749	.9926	.38
.39	.7106	.6356	.6094	.5993	.5953	.8945	.8576	.8435	.8378	.9588	.9430	.9367	.9835	.9769	.9933	.39
.40	.7153	.6424	.6174	.6080	.6043	.8982	.8632	.8500	.8449	.9610	.9464	.9406	.9846	.9788	.9940	.40
.41	.7200	.6592	.6253	.6165	.6131	.9017	.8685	.8562	.8516	.9632	.9496	.9444	.9859	.9805	.9945	.41
.42	.7245	.6558	.6330	.6248	.6217	.9052	.8738	.8623	.8581	.9652	.9526	.9479	.9869	.9821	.9951	.42
.43	.7291	.6625	.6407	.6330	.6302	.9086	.8787	.8681	.8643	.9671	.9555	.9512	.9879	.9835	.9955	.43
.44	.7336	.6689	.6482	.6409	.6384	.9118	.8836	.8737	.8702	.9690	.9582	.9543	.9889	.9849	.9960	.44
.45	.7381	.6754	.6556	.6488	.6465	.9150	.8882	.8790	.8758	.9707	.9607	.9572	.9897	.9862	.9964	.45
.46	.7426	.6817	.6629	.6566	.6544	.9181	.8927	.8842	.8813	.9724	.9631	.9600	.9905	.9873	.9967	.46
.47	.7469	.6880	.6700	.6642	.6622	.9210	.8970	.8892	.8865	.9740	.9654	.9626	.9912	.9883	.9970	.47
.48	.7512	.6941	.6770	.6716	.6698	.9239	.9012	.8940	.8916	.9754	.9675	.9650	.9919	.9894	.9973	.48
.49	.7555	.7001	.6839	.6788	.6772	.9267	.9052	.8985	.8963	.9768	.9696	.9672	.9926	.9902	.9976	.49
.50	.7598	.7061	.6907	.6860	.6845	.9294	.9091	.9028	.9009	.9782	.9714	.9693	.9931	.9910	.9978	.50

APPENDIX F. Values of $1 - 10^{-kt}$ for Various Values of kt.

k/t	0	1	2	3	4	5	6	7	8	9
.00	–	.0023	.0046	.0069	.0092	.0114	.0137	.0160	.0183	.0205
.01	.0228	.0250	.0273	.0295	.0317	.0339	.0362	.0384	.0406	.0428
.02	.0450	.0472	.0494	.0516	.0538	.0559	.0581	.0603	.0624	.0646
.03	.0667	.0689	.0710	.0732	.0753	.0774	.0796	.0818	.0838	.0859
.04	.0880	.0901	.0922	.0943	.0964	.0984	.1005	.1026	.1046	.1067
.05	.1087	.1108	.1128	.1149	.1169	.1190	.1210	.1230	.1250	.1270
.06	.1290	.1310	.1330	.1350	.1370	.1390	.1410	.1430	.1449	.1469
.07	.1489	.1508	.1528	.1547	.1567	.1586	.1605	.1625	.1644	.1663
.08	.1682	.1701	.1721	.1740	.1759	.1778	.1796	.1815	.1834	.1853
.09	.1872	.1890	.1909	.1928	.1946	.1965	.1983	.2002	.2020	.2038
.10	.2057	.2075	.2093	.2111	.2130	.2148	.2166	.2184	.2202	.2220
.11	.2237	.2255	.2273	.2291	.2309	.2326	.2344	.2362	.2379	.2397
.12	.2414	.2432	.2449	.2466	.2484	.2501	.2518	.2536	.2553	.2570
.13	.2587	.2604	.2621	.2638	.2655	.2672	.2689	.2704	.2722	.2739
.14	.2756	.2772	.2789	.2806	.2822	.2839	.2855	.2871	.2888	.2904
.15	.2920	.2937	.2953	.2969	.2985	.3002	.3018	.3034	.3050	.3066
.16	.3082	.3098	.3113	.3129	.3145	.3161	.3177	.3192	.3208	.3224
.17	.3239	.3255	.3270	.3286	.3301	.3317	.3332	.3347	.3363	.3378
.18	.3393	.3408	.3423	.3439	.3454	.3469	.3484	.3499	.3514	.3529
.19	.3543	.3558	.3573	.3588	.3603	.3617	.3632	.3647	.3661	.3676
.20	.3690	.3705	.3719	.3734	.3748	.3763	.3777	.3791	.3806	.3820
.21	.3834	.3848	.3862	.3876	.3891	.3905	.3919	.3933	.3947	.3961
.22	.3974	.3988	.4002	.4016	.4030	.4043	.4057	.4071	.4084	.4098
.23	.4112	.4125	.4139	.4152	.4166	.4179	.4192	.4206	.4219	.4232
.24	.4246	.4259	.4272	.4285	.4298	.4311	.4325	.4338	.4351	.4364
.25	.4377	.4389	.4402	.4415	.4428	.4441	.4454	.4466	.4479	.4492
.26	.4505	.4517	.4530	.4542	.4555	.4567	.4580	.4592	.4605	.4617
.27	.4630	.4642	.4654	.4667	.4679	.4691	.4703	.4716	.4728	.4740
.28	.4752	.4764	.4776	.4788	.4800	.4812	.4824	.4836	.4848	.4860
.29	.4871	.4883	.4895	.4907	.4918	.4930	.4942	.4953	.4965	.4977
.30	.4988	.5000	5.011	.5023	.5034	.5045	.5057	.5068	.5080	.5091
.31	.5102	.5113	.5125	.5136	.5147	.5158	.5169	.5181	.5192	.5203
.32	.5214	.5225	.5236	.5247	.5258	.5268	.5279	.5290	.5301	.5312
.33	.5323	.5333	.5344	.5355	.5366	.5376	.5387	.5397	.5404	.5419
.34	.5429	.5440	.5450	.5461	.5471	.5481	.5492	.5502	.5113	.5523
.35	.5533	.5543	.5554	.5564	.5574	.5584	.5594	.5605	.5615	.5625
.36	.5635	.5645	.5655	.5665	.5675	.5685	.5695	.5705	.5715	.5724
.37	.5734	.5744	.5754	.5764	.5773	.5783	.5793	.5802	.5812	.5822
.38	.5831	.5841	.5850	.5860	.5870	.5879	.5889	.5898	.5907	.5917
.39	.5926	.5936	.5945	.5954	.5964	.5973	.5982	.5991	.6001	.6010
.40	.6019	.6028	.6037	.6046	.6055	.6064	.6074	.6083	.6092	.6101
.41	.6110	.6118	.6127	.6136	.6145	.6154	.6163	.6172	.6181	.6189
.42	.6198	.6207	.6216	.6224	.6233	.6242	.6250	.6259	.6267	.6276
.43	.6285	.6293	.6302	.6310	.6319	.6327	.6336	.6344	.6352	.6361

APPENDIX F. [*Continued*]

k/*t*	0	1	2	3	4	5	6	7	8	9
.44	.6369	.6378	.6386	.6394	.6403	.6411	.6419	.6427	.6435	.6444
.45	.6452	.6460	.6468	.6476	.6484	.6492	.6501	.6509	.6517	.6525
.46	.6533	.6541	.6549	.6557	.6564	.6572	.6580	.6588	.6596	.6604
.47	.6612	.6619	.6627	.6635	.6643	.6650	.6658	.6666	.6673	.6684
.48	.6689	.6696	.6704	.6711	.6719	.6727	.6734	.6742	.6749	.6757
.49	.6764	.6772	.6779	.6786	.6794	.6801	.6808	.6816	.6823	.6830
.50	.6838	.6845	.6852	.6859	.6867	.6874	.6881	.6888	.6895	.6903
.51	.6910	.6917	.6924	.6931	.6938	.6945	.6952	.6959	.6966	.6973
.52	.6980	.6987	.6994	.7001	.7008	.7015	.7021	.7028	.7035	.7042
.53	.7049	.7056	.7062	.7069	.7076	.7083	.7089	.7096	.7103	.7110
.54	.7116	.7123	.7129	.7136	.7138	.7145	.7152	.7158	.7165	.7171
.55	.7182	.7188	.7195	.7201	.7207	.7214	.7220	.7227	.7233	.7239
.56	.7246	.7252	.7258	.7265	.7271	.7277	.7284	.7290	.7296	.7302
.57	.7308	.7315	.7321	.7327	.7333	.7339	.7345	.7351	.7358	.7364
.58	.7370	.7376	.7382	.7388	.7394	.7400	.7406	.7412	.7418	.7424
.59	.7430	.7436	.7441	.7447	.7453	.7459	.7465	.7471	.7477	.7482
.60	.7488	.7494	.7500	.7505	.7511	.7517	.7523	.7528	.7534	.7540
.61	.7545	.7551	.7557	.7562	.7568	.7573	.7579	.7585	.7590	.7596
.62	.7601	.7607	.7612	.7618	.7623	.7629	.7634	.7640	.7645	.7650
.63	.7656	.7661	.7667	.7672	.7677	.7683	.7688	.7693	.7699	.7704
.64	.7709	.7714	.7720	.7725	.7730	.7735	.7741	.7746	.7751	.7756
.65	.7761	.7766	.7772	.7777	.7782	.7787	.7792	.7797	.7802	.7807
.66	.7812	.7817	.7822	.7827	.7832	.7837	.7842	.7847	.7852	.7857
.67	.7862	.7867	.7872	.7877	.7882	.7887	.7891	.7896	.7901	.7905
.68	.7911	.7916	.7920	.7925	.7930	.7935	.7939	.7944	.7949	.7954
.69	.7958	.7963	.7968	.7972	.7977	.7982	.7986	.7991	.7996	.8000
.70	.8005	.8009	.8014	.8018	.8023	.8028	.8032	.8037	.8041	.8046
.71	.8050	.8055	.8059	.8064	.8068	.8072	.8077	.8081	.8086	.8090
.72	.8095	.8099	.8103	.8108	.8112	.8116	.8121	.8125	.8129	.8134
.73	.8138	.8142	.8146	.8151	.8155	.8159	.8163	.8168	.8172	.8176
.74	.8180	.8184	.8189	.8193	.8197	.8201	.8205	.8209	.8214	.8218
.75	.8222	.8226	.8230	.8234	.8238	.8242	.8246	.8250	.8254	.8258
.76	.8262	.8266	.8270	.8274	.8278	.8282	.8286	.8290	.8294	.8298
.77	.8302	.8306	.8313	.8318	.8317	.8321	.8325	.8329	.8333	.8337
.78	.8340	.8344	.8348	.8352	.8356	.8359	.8363	.8367	.8371	.8374
.79	.8378	.8382	.8386	.8389	.8393	.8397	.8400	.8404	.8408	.8411
.80	.8415	.8419	.8422	.8426	.8430	.8433	.8437	.8440	.8444	.8448
.81	.8451	.8455	.8458	.8462	.8465	.8469	.8472	.8475	.8479	.8483
.82	.8486	.8490	.8493	.8497	.8500	.8504	.8507	.8511	.8514	.8517
.83	.8521	.8524	.8528	.8531	.8534	.8538	.8541	.8545	.8548	.8551
.84	.8555	.8558	.8561	.8565	.8568	.8571	.8574	.8578	.8581	.8584
.85	.8587	.8591	.8594	.8597	.8600	.8604	.8607	.8610	.8613	.8616
.86	.8620	.8623	.8626	.8629	.8632	.8635	.8639	.8642	.8645	.8648
.87	.8651	.8654	.8657	.8660	.8663	.8666	.8669	.8672	.8675	.8678

APPENDIX F. [*Continued*]

k/t	0	1	2	3	4	5	6	7	8	9
.88	.8681	.8685	.8688	.8691	.8694	.8697	.8700	.8703	.8706	.8709
.89	.8712	.8715	.8718	.8721	.8724	.8726	.8729	.8732	.8735	.8738
.90	.8741	.8744	.8747	.8750	.8753	.8755	.8758	.8761	.8764	.8767
.91	.8770	.8773	.8775	.8778	.8781	.8784	.8787	.8789	.8792	.8795
.92	.8798	.8801	.8803	.8806	.8809	.8811	.8814	.8817	.8820	.8822
.93	.8825	.8828	.8831	.8833	.8836	.8839	.8841	.8844	.8847	.8849
.94	.8852	.8854	.8857	.8860	.8862	.8865	.8868	.8870	.8873	.8875
.95	.8878	.8881	.8883	.8886	.8890	.8891	.8893	.8896	.8898	.8901
.96	.8904	.8906	.8909	.8911	.8914	.8916	.8919	.8921	.8924	.8926
.97	.8928	.8931	.8933	.8936	.8938	.8941	.8943	.8946	.8948	.8950
.98	.8953	.8955	.8958	.8960	.8962	.8965	.8967	.8970	.8972	.8974
.99	.8977	.8979	.8981	.8984	.8986	.8988	.8991	.8993	.8995	.8998
1.00	.9000	.9002	.9005	.9007	.9009	.9011	.9014	.9016	.9018	.9020
1.01	.9023		.9027		.9032		.9036		.9041	.9043
1.02	.9045		.9049		.9054		.9058		.9062	.9065
1.03	.9067		.9070		.9075		.9080		.9086	.9088
1.04	.9090		.9092		.9096		.9100		.9105	.9127
1.05	.9109		.9113		.9117		.9120		.9125	.9127
1.06	.9129		.9133		.9137		.9141		.9145	.9147
1.07	.9149		.9153		.9157		.9161		.9164	.9166
1.08	.9168		.9174		.9176		.9179		.9183	.9185
1.09	.9187		.9190		.9195		.9198		.9202	.9204
1.10	.9206		.9209		.9213		.9217		.9220	.9222
1.11	.9224		.9227		.9231		.9234		.9238	.9240
1.12	.9241		.9245		.9248		.9252		.9255	.9257
1.13	.9259		.9262		.9266		.9269		.9272	.9274
1.14	.9276		.9279		.9282		.9286		.9289	.9290
1.15	.9292		.9295		.9298		.9302		.9305	.9307
1.16	.9308		.9311		.9315		.9318		.9321	.9322
1.17	.9324		.9327		.9330		.9333		.9336	.9337
1.18	.9339		.9342		.9345		.9348		.9351	.9353
1.19	.9354		.9357		.9360		.9363		.9366	.9368
1.20	.9369		.9372		.9375		.9378		.9381	.9382

APPENDIX G. Nomograms to Calculate *C*, Dissolved-Oxygen Content. (From Hart, I. C., "Nomograms to Calculate Dissolved-Oxygen Contents and Exchange Coefficients," *Water Research*, vol. I, Pergamon Press, New York, 1969, p. 393. By permission of publishers.)

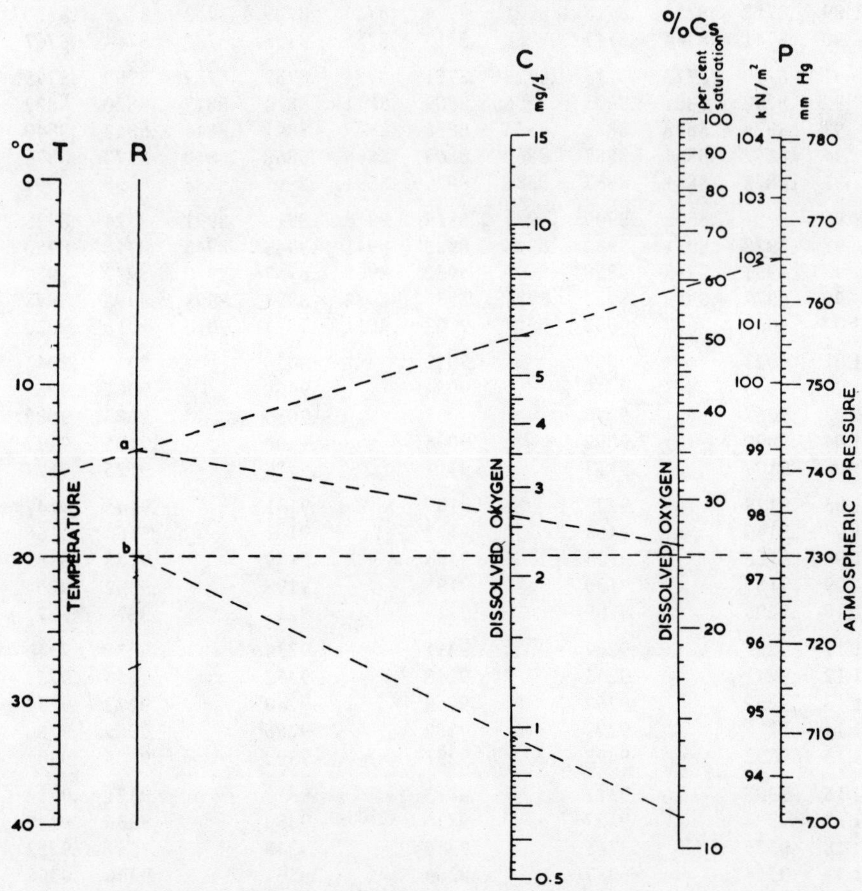

FIG. 1. Nomogram to calculate *C*, dissolved-oxygen content (mg/l) from % *Cs*, dissolved oxygen (percent saturation) at *T*, temperature (°C) and *P*, atmospheric pressure (mm Hg or kN/m^2).

Connect *T* and *P* to give intersection on *R*. Connect point of intersection and % *Cs*. Read *C*.

APPENDIX H. Calculating Expected Exchange Coefficients in Streams from Mean Velocities and Depths. (From Hart, I. C., "Nomograms to Calculate Dissolved-Oxygen Contents and Exchange Coefficients," *Water Research*, vol. 1, Pergamon Press, New York, 1969, p. 394. By permission of publishers.)

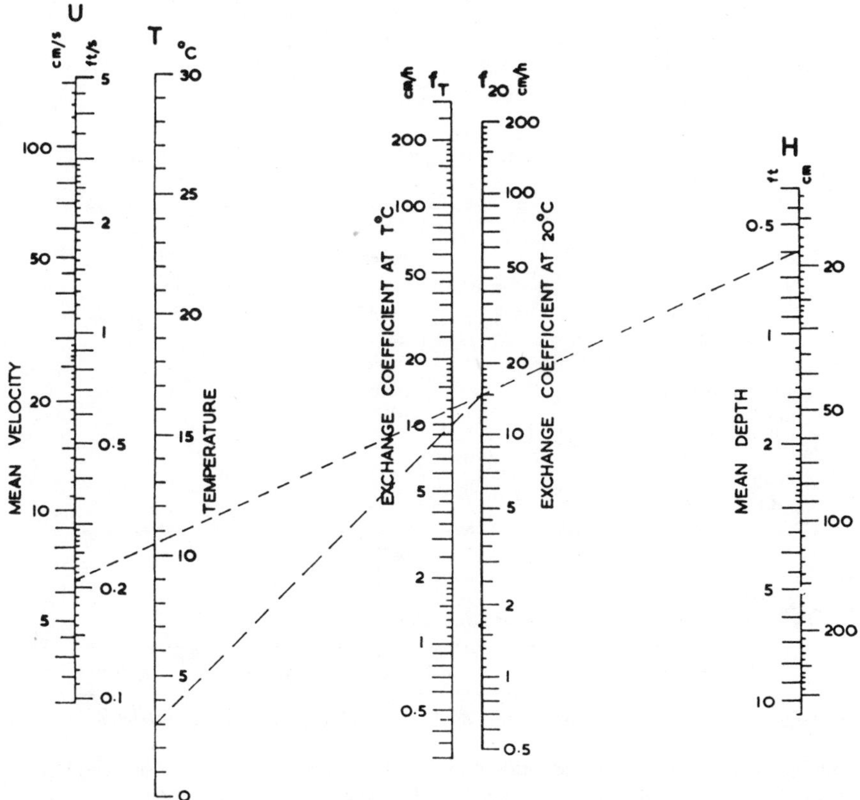

FIG. 2. Nomogram to calculate expected exchange coefficients in streams for mean velocities and depths using equations $f_{20} = KU^{0.67}H^{-0.85}$ (f_{20}, exchange coefficient (cm/h) at 20°C, K, constant = 27·5 if U, mean velocity in ft/sec, and H, mean depth in ft, or = 50·8 if U in cm/sec and H in cm) and $f_T = f_{20}\ 1\cdot 024^{T-20}$ (f_T, exchange coefficient (cm/hr) at temperature T°C).

To calculate f_{20}: Connect U and H; read f_{20}. To calculate f_T: Connect U and H to give intersection on f_{20}; connect point of intersection and T; read f_T.

APPENDIX I

The following six tables were originally prepared for the Universities Council on Water Resources by Peter C. Klingeman of Oregon State University in June 1976. They should be helpful to all readers engaged in many areas of water resources research and education.

TABLE I-1.

Quantity	Unit	SI Symbol
Base Units:		
length	meter	m
mass	kilogram	kg
time	second	s
electric current	ampere	A
thermodynamic temperature	kelvin	K
amount of substance	mole	mol
luminous intensity	candela	cd
Supplementary Units:		
plane angle	radian	rad
solid angle	steradian	sr

TABLE I-2. Derived Units in SI Measurement

Quantity	Unit	SI Symbol	Formula
acceleration	meter per second squared	...	m/s^2
activity (of a radioactive source)	disintegration per second	...	(disintegration)/s
angular acceleration	radian per second squared	...	rad/s^2
angular velocity	radian per second	...	rad/s
area	square meter	...	m^2
density	kilogram per cubic meter	...	kg/m^3
electric capacitance	farad	F	$A \cdot s/V$
electrical conductance	siemens	S	A/V
electric field strength	volt per meter	...	V/m
electric inductance	henry	H	$V \cdot s/A$
electric potential difference	volt	V	W/A
electric resistance	ohm	Ω	v/A
electromotive force	volt	V	W/A
energy	joule	J	$N \cdot m$
entropy	joule per kelvin	...	J/K
force	newton	N	$kg \cdot m/s^2$

TABLE I-2. *(Continued)*

Quantity	Unit	SI Symbol	Formula
frequency	hertz	Hz	s^{-1}
illuminance	lux	lx	lm/m^2
luminance	candela per square meter	. . .	cd/m^2
luminous flux	lumen	lm	$cd \cdot sr$
magnetic field strength	ampere per meter	. . .	A/m
magnetic flux	weber	Wb	$V \cdot s$
magnetic flux density	tesla	T	Wb/m^2
magnetomotive force	ampere	A	. . .
power	watt	W	J/s
pressure	pascal	Pa	N/m^2
quantity of electricity	coulomb	C	$A \cdot s$
quantity of heat	joule	J	$N \cdot m$
radiant intensity	watt per steradian	. . .	W/sr
specific heat	joule per kilogram-kelvin	. . .	$J/kg \cdot K$
stress	pascal	Pa	N/m^2
thermal conductivity	watt per meter-kelvin	. . .	$W/m \cdot K$
velocity	meter per second	. . .	m/s
viscosity, dynamic	pascal-second	. . .	$Pa \cdot s$
viscosity, kinematic	square meter per second	. . .	m^2/s
voltage	volt	V	W/A
volume	cubic meter	. . .	m^3
wavenumber	reciprocal meter	. . .	(wave)/m
work	joule	J	$N \cdot m$

TABLE I-3. SI Prefixes

Prefix	SI Symbol	Multiplication Factor
tera	T	$1\,000\,000\,000\,000 = 10^{12} = E + 12$
giga	G	$1\,000\,000\,000 = 10^{9} = E + 9$
mega	M	$1\,000\,000 = 10^{6} = E + 6$
kilo	k	$1\,000 = 10^{3} = E + 3$
hecto**	h	$100 = 10^{2} = E + 2$
deka**	da	$10 = 10^{1} = E + 1$
deci**	d	$0.1 = 10^{-1} = E - 1$
centi**	c	$0.01 = 10^{-2} = E - 2$
milli	m	$0.001 = 10^{-3} = E - 3$
micro	μ	$0.000\,001 = 10^{-6} = E - 6$
nano	n	$0.000\,000\,001 = 10^{-9} = E - 9$
pico	p	$0.000\,000\,000\,001 = 10^{-12} = E - 12$
femto	f	$0.000\,000\,000\,000\,001 = 10^{-15} = E - 15$
atto	a	$0.000\,000\,000\,000\,000\,001 = 10^{-18} = E - 18$

**Avoid use of this prefix where possible.

TABLE I-4. Conversion to SI Units for Physical Quantities Frequently Used in Water Resources Practice

To convert from:		Multiply by:	To obtain:	Units
Acceleration				
foot/second^2	×	0.304 8	= meter/second^2	m/s^2
Area				
acre	×	4 047	= meter^2	m^2
foot^2	×	0.092 90	= meter^2	m^2
hectare	×	1.000 E + 4	= meter^2	m^2
inch^2	×	6.452 E − 4	= meter^2	m^2
mile^2	×	2.590 E + 6	= meter^2	m^2
yard^2	×	0.836 1	= meter^2	m^2
Bending Moment or Torque or Moment of Force				
foot-pound$_{force}$	×	1.356	= newton-meter	N·m
Capacity (see Volume)				
Density (see Mass/Volume)				
Discharge (see Volume/Time)				
Energy or Work				
BTU	×	1 055	= joule	J
calorie	×	4.187	= joule	J
foot-pound$_{force}$	×	1.356	= joule	J
kilowatt-hour	×	3.600 E + 6	= joule	J
Energy/Area Time				
BTU/foot^2-second	×	1.135 E + 4	= watt/meter^2	W/m^2
BTU/foot^2-hour	×	3.152	= watt/meter^2	W/m^2
calorie/cm^2-minute	×	697.3	= watt/meter^2	W/m^2
Flow Rate (see Mass/Time or Volume/Time)				
Force				
dyne	×	1.000 E − 5	= newton	N
kilogram$_{force}$	×	9.807	= newton	N
kip	×	4 448	= newton	N
pound$_{force}$	×	4.448	= newton	N
Force/Area (see Pressure)				
Heat				
BTU/foot^2	×	1.136 E + 4	= joule/meter^2	J/m^2
BTU/hour-foot^2–°F	×	5.678	= watt/meter^2–kelvin	W/m^{2}K
BTU/pound$_{mass}$	×	2 326	= joule/kilogram	J/kg
BTU/pound$_{mass}$–°F	×	4 187	= joule/kilogram–kelvin	J/kg·K
BTU/second-foot^2–°F	×	2.044	= Watt/meter^2–kelvin	W/m^2·K
calorie/cm^2	×	4.184 E + 4	= joule/meter^2	J/m^2
calorie/gram	×	4 187	= joule/kilogram	J/kg
calorie/gram–°C	×	4 187	= joule/kilogram–kelvin	J/kg·K
Length				
foot	×	0.304 8	= meter	m
inch	×	0.025 40	= meter	m
mile, nautical	×	1 852	= meter	m
mile	×	1 609	= meter	m
yard	×	0.914 4	= meter	m

TABLE I-4. *(Continued)*

To convert from:	Multiply by:	To obtain:	Units
Mass			
pound_{mass}(avoird.)	X 0.453 6	= kilogram	kg
slug	X 14.59	= kilogram	kg
ton (long, 2240 lb_m)	X 1 016	= kilogram	kg
ton (short, 2000 lb_m)	X 907.2	= kilogram	kg
ton (metric)	X 1 000	= kilogram	kg
Mass/Time			
pound_{mass}/second	X 0.453 6	= kilogram/second	kg/s
Mass/Volume or Density			
gram/cm^3	X 1 000	= kilogram/meter^3	kg/m^3
pound_{mass}/foot^3	X 16.02	= kilogram/meter^3	kg/m^3
pound_{mass}/gallon	X 119.8	= kilogram/meter^3	kg/m^3
slug/foot^3	X 515.4	= kilogram/meter^3	kg/m^3
Power			
BTU/second	X 1 054	= watt	W
BTU/minute	X 17.57	= watt	W
BTU/hour	X 0.293 1	= watt	W
calorie/second	X 4.184	= watt	W
calorie/minute	X 0.069 73	= watt	W
foot-pound_{force}/second	X 1.356	= watt	W
horsepower (550 ft·lb_f/s)	X 745.7	= watt	W
horsepower (electric)	X 746.0	= watt	W
horsepower (water)	X 746.0	= watt	W
Pressure or Stress or Force/Area			
atmosphere	X 1.013 E+5	= pascal	Pa
bar	X 1.000 E+5	= pascal	Pa
dyne/cm^2	X 0.100 0	= pascal	Pa
foot of water (39.4°F)	X 2 989	= pascal	Pa
gram_{force}/cm^2	X 98.07	= pascal	Pa
inches of mercury	X 3 386	= pascal	Pa
inches of water (39.4°F)	X 249.1	= pascal	Pa
kilogram_{force}/meter^2	X 9.807	= pascal	Pa
kip/inch^2 (ksi)	X 6.895 E+6	= pascal	Pa
millibar	X 100	= pascal	Pa
millimeters of mercury	X 133.3	= pascal	Pa
pound_{force}/foot^2	X 47.88	= pascal	Pa
pound_{force}/inch^2 (psi)	X 6 895	= pascal	Pa
Stress (see Pressure)			
Temperature			
degree Celsius	+ 273.15	= kelvin	K
degree Fahrenheit	+ 459.67 ÷ 1.8	= kelvin	K
degree Rankine	÷ 1.8	= kelvin	K
degree Fahrenheit	− 32 ÷ 1.8	= degree Celsius	

TABLE I-4. (*Continued*)

To convert from:		Multiply by:	To obtain:	Units
Time				
day	X	8.640 E + 4	= second	s
hour	X	3 600	= second	s
minute	X	60.00	= second	s
month (mean calendar)	X	2.628 E + 6	= second	s
year (calendar)	X	3.154 E + 7	= second	s
Torque (see Bending Moment)				
Velocity or Speed				
foot/second	X	0.304 8	= meter/second	m/s
foot/minute	X	0.005 080	= meter/second	m/s
foot/hour	X	8.467 E – 5	= meter/second	m/s
kilometer/hour	X	0.277 8	= meter/second	m/s
mile/hour	X	0.447 0	= meter/second	m/s
Viscosity				
centipose	X	0.001 000	= pascal–second	Pa · s
centistokes	X	1.000 E – 6	= meter^2/second	m^2/s
foot^2/second	X	0.092 90	= meter^2/second	m^2/s
poise	X	0.100 0	= pascal–second	Pa · s
pound$_{force}$–second/foot^2	X	47.88	= pascal–second	Pa · s
slug/foot-second	X	47.88	= pascal–second	Pa · s
stokes	X	1.000 E – 4	= meter^2/second	m^2/s
Volume or Capacity				
acre-foot	X	1 233	= meter^3	m^3
barrel (oil, 42 gal)	X	0.159 0	= meter^3	m^3
board foot	X	0.002 360	= meter^3	m^3
bushel (U.S.)	X	0.035 24	= meter^3	m^3
foot^3	X	0.028 32	= meter^3	m^3
gallon (U.S. liquid)	X	0.003 785	= meter^3	m^3
inch^3	X	1.639 E – 5	= meter^3	m^3
liter	X	0.001 000	= meter^3	m^3
yard^3	X	0.764 6	= meter^3	m^3
Volume/Time or Discharge or Flow Rate				
foot^3/second	X	0.028 32	= meter^3/second	m^3/s
foot^3/minute	X	4.719 E – 4	= meter^3/second	m^3/s
gallon/minute	X	6.309 E – 5	= meter^3/second	m^3/s
gallon/day	X	4.381 E – 8	= meter^3/second	m^3/s
Work (see Energy)				

TABLE I-5. Commonly Used Equivalent Units in Water Resources[1]

A. LENGTH:

Unit	Equivalent(a)(b)					
	Millimeter	Inch	Foot	Meter(c)	Kilometer	Mile
millimeter	1	0.039 37	0.003 281	0.001 000	1 E−6	0.621 4 E−6
inch	25.40	1	0.083 3	0.025 40	25.40 E−6	15.78 E−6
foot	304.8	12	1	0.304 8	304.8 E−6	189.4 E−6
meter(c)	1 000	39.37	3.281	1	0.001	621.4 E−6
kilometer	1 000 000	39 370	3 281	1 000	1	0.621 4
mile	1 609 000	63 360	5 280	1 609	1.609	1

B. AREA

Unit	Equivalent(a)(b)						
	Sq. inch	Sq. foot	Sq. meter(c)	Acre	Hectare	Sq. kilometer	Sq. mile
sq. inch	1	0.006 944	645.2 E−6	0.159 4 E−6	64.52 E−9	645.2 E−12	249.1 E−12
sq. foot	144	1	0.092 90	22.96 E−6	9.290 E−9	92.90 E−9	35.87 E−9
sq. meter(c)	1 550	10.76	1	247.1 E−6	1 E−4	1 E−6	386.1 E−9
acre	6 273 000	43 560	4 047	1	0.404 7	0.004 047	0.001 563
hectare	15 500 000	107 600	10 000	2.471	1	0.01	0.003 861
sq. kilometer	1.550 E+9	10 764 000	1 000 000	247.1	100	1	0.386 1
sq. mile	4.014 E+9	27 880 000	2 590 000	640	259	2.590	1

C. VOLUME:

Unit	Equivalent(a)(b)							
	Cu. inch	Liter	U.S. gallon	Cu. foot	Cu. yard	Cu. meter(c)	Acre-foot	Sec-foot-day
cubic inch	1	0.016 39	0.004 329	578.7 E−6	21.43 E−6	16.39 E−6	13.29 E−9	6.698 E−9
liter	61.02	1	0.264 2	0.035 31	0.001 308	0.001	810.6 E−9	408.7 E−9
U.S. gallon	231.0	3.785	1	0.133 7	0.004 951	0.003 785	3.068 E−6	1.547 E−6
cubic foot	1 728	28.32	7.481	1	0.037 04	0.028 32	22.96 E−6	11.57 E−6
cubic yard	46 660	764.6	202.0	27	1	0.764 6	619.8 E−6	312.5 E−6
cubic meter(c)	61 020	1 000	264.2	35.31	1.308	1	810.6 E−6	408.7 E−6
acre-foot	75.27 E+6	1 233 000	325 900	43 560	1 613	1 233	1	0.504 2
second-foot-day	149.3 E+6	2 447 000	646 400	86 400	3 200	2 447	1.983	1

D. DISCHARGE (FLOW RATE, VOLUME/TIME):

Unit	Equivalent(a)(b)					
	Gallon/min	Liter/sec	Acre-foot/day	Foot^3/sec	Million gal/day	Meter^3 sec(c)
gallon/minute	1	0.063 09	0.004 419	0.002 228	0.001 440	63.09 E−6
liter/second	15.85	1	0.070 05	0.035 31	0.022 82	0.001
acre-foot/day	226.3	14.28	1	0.504 2	0.325 9	0.014 28
foot^3/second	448.8	28.32	1.983	1	0.646 3	0.028 32
million gallons/day	694.4	43.81	3.069	1.547	1	0.043 81
meter^3/second(c)	15 850	1 000	70.04	35.31	22.82	1

TABLE I-5. (*Continued*)

E. VELOCITY:

Unit	Equivalent[(a)(b)]				
	Foot/day	Kilometer/hour	Foot/sec	Mile/hour	Meter/sec[(c)]
foot/day	1	12.70 E−6	11.57 E−6	7.891 E−6	3.528 E−6
kilometer/hour	78 740	1	0.911 3	0.621 4	0.277 8
foot/second	86 400	1.097	1	0.681 8	0.304 8
mile/hour	126 700	1.609	1.467	1	0.447 0
meter/second[(c)]	283 500	3.600	3.281	2.237	1

F. MASS:

Unit	Equivalent[(a)(b)]					
	$\text{Pound}_{\text{mass}}$	Kilogram[(c)]	Metric slug	Slug	Metric ton	Long ton
$\text{pound}_{\text{mass}}$(avoird.)	1	0.453 6	0.046 25	0.031 08	453.6 E−6	446.4 E−6
kilogram[(c)]	2.205	1	0.102 0	0.068 52	0.001	984.2 E−6
metric slug	21.62	9.807	1	0.672 1	0.009 807	0.009 651
slug	32.17	15.49	1.490	1	0.014 59	0.014 36
metric ton	2 205	1 000	102.0	68.52	1	0.984 2
long ton	2 240	1 016	103.7	69.63	1.016	1

G. FORCE:

Unit	Equivalent[(a)(b)]			
	Dyne	Newton[(c)]	$\text{Pound}_{\text{force}}$	$\text{Kilogram}_{\text{force}}$
dyne	1	1 E−5	2.248 E−6	1.020 E−6
newton[(c)]	100 000	1	0.224 8	0.102 0
$\text{pound}_{\text{force}}$	444 800	4.448	1	0.453 6
$\text{kilogram}_{\text{force}}$	980 700	9.807	2.205	1

[1] Footnotes for all parts of Table 5.

(a) Equivalent values are shown to 4 significant figures.

(b) Multiply the numerical amount of the given unit by the equivalent value shown (per single amount of given unit) to obtain the numerical amount of the equivalent unit (e.g.: 5 inches × 0.025 40 m/inch = 0.127 0 m).

(c) This is the SI expression, in base units or derived units, for the physical quantiy.

TABLE I-6. Commonly Used Numerical Values for Physical Properties

Quantity	U.S. FLT System[1]	SI System
Gravitational acceleration g (std., free fall)	32.2 ft/s^2	9.807 m/s^2
Density of water, ρ_{H_2O} (a 50°F/10°C	1.94 $slugs/ft^3$	1.000 kg/m^3
Specific weight of water, γ_{H_2O} (a 50°F/10°C	62.4 lb/ft^3	9.807 kN/m^3
Dynamic viscosity of water, μ_{H_2O} (a 50°F/10°C	2.73×10^{-5} $lb \cdot s/ft^2$	1.30×10^{-3} Pa · s
(a 70°F/20°C	2.05×10^{-5} $lb \cdot s/ft^2$	1.00×10^{-3} Pa · s
Kinematic viscosity of water, ν_{H_2O} (a 50°F/10°C	1.41×10^{-5} ft^2/s	1.30×10^{-6} m^2/s
(a 70°F/20°C	1.06×10^{-5} ft^2/s	1.00×10^{-6} m^2/s
Atmospheric pressure, p (std.)	14.70 psia 2,116 psfa	101.3 kP/a^2

[1] This is presented for illustrative purposes only, as the C.E. Textbooks in fluid mechanics have generally used the FLT System alone, prior to introduction of SI units in new textbooks beginning in 1975.

REFERENCES

Davis, D. S., *Nomography and Empirical Equations*, Chap. 10, Reinhold Publishing Corporation, 1955.

Hatfield, W. D., *Sewage Works J.*, p. 557, May 1941.

INDEX

INDEX